Algorithmic Modernity

Algorithmic Modernity

Mechanizing Thought and Action, 1500–2000

Edited by

Morgan G. Ames and Massimo Mazzotti

OXFORD
UNIVERSITY PRESS

OXFORD
UNIVERSITY PRESS

Oxford University Press is a department of the University of Oxford. It furthers
the University's objective of excellence in research, scholarship, and education
by publishing worldwide. Oxford is a registered trade mark of Oxford University
Press in the UK and certain other countries.

Published in the United States of America by Oxford University Press
198 Madison Avenue, New York, NY 10016, United States of America.

Library of Congress Cataloging-in-Publication Data
Names: Ames, Morgan G., editor. | Mazzotti, Massimo, editor.
Title: Algorithmic modernity : mechanizing thought and action, 1500–2000 /
edited by Morgan G. Ames and Massimo Mazzotti.
Description: New York, NY : Oxford University Press, [2023] |
Includes bibliographical references and index.
Identifiers: LCCN 2022018922 (print) | LCCN 2022018923 (ebook) |
ISBN 9780197502426 (hardback) | ISBN 9780197502440 (epub)
Subjects: LCSH: Algorithms—History. | Mathematics—History.
Classification: LCC QA9.58 .A2535 2022 (print) | LCC QA9.58 (ebook) |
DDC 518/.1—dc23/eng20220718
LC record available at https://lccn.loc.gov/2022018922
LC ebook record available at https://lccn.loc.gov/2022018923

DOI: 10.1093/oso/9780197502426.001.0001

Printed by Integrated Books International, United States of America

Contents

vi Content

10. Decision Trees, Random Forests, and the Genealogy of
the Black Box
Jonathan F. Kahn

Notes
Bibliography
Index

List of Contributors

Amir Alexander teaches the history of science at UCLA. His latest book is *Proof! How the World Became Geometrical* (2019), which recounts how the ancient science of geometry came to shape the world we know today. The interdependence of mathematics and the culture of modernity was also the subject of his previous books: *Geometrical Landscapes* (2002), *Duel at Dawn: Heroes, Martyrs, and the Rise of Modern Mathematics* (2010), and *Infinitesimal: The Dangerous Mathematical Theory That Shaped the Modern World* (2014). Alexander's latest project focuses on how the Cartesian mathematical grid codified a vision of the New World and was inscribed onto the landscape of North America.

Morgan G. Ames is Assistant Professor of Practice in the School of Information and Associate Director of Research for the Center for Science, Technology, Medicine and Society (CSTMS) at the University of California, Berkeley. Ames researches the ideological origins of inequality in the technology world, with a focus on utopianism, childhood, and learning. Her book *The Charisma Machine: The Life, Death, and Legacy of One Laptop per Child* (MIT Press, 2019), winner of the 2020 Best Information Science Book Award, the 2020 Sally Hacker Prize, and the 2021 Computer History Museum Prize, draws on archival research and ethnographic fieldwork in Paraguay to explore the cultural history, results, and legacy of the OLPC project—and what it tells us about the many other technology projects that draw on similar utopian ideals.

Michael J. Barany is Lecturer in the History of Science at the University of Edinburgh. Barany researches and teaches the history, sociology, and culture of modern mathematics and science. His main current project examines the institutional, infrastructural, intellectual, political, and cultural transformations associated with the globalization of mathematical research. The essay shared here derives from a long-running investigation into the social and material foundations and uses of elementary mathematical concepts, from the Euclidean Point to the basic counting numbers. With Kirsti Niskanen, he coedited the volume *Gender, Embodiment, and the History of the Scholarly Persona* (Palgrave, 2021). His historical research has figured in recent discussions of politics and diversity in contemporary mathematics and education, and his essays have appeared five times in the Best Writing on Mathematics anthology series.

Theodora Dryer PhD is a writer, historian, and critical policy analyst. Her research centers on data and technology in the climate crisis and the political functions of algorithms and digital data systems in water and natural resource management. She teaches on technology and environmental justice at New York University.

Andrew Fiss received graduate degrees in History and Philosophy of Science and postdoctoral fellowships in Science, Technology, and Society (STS) and writing studies, which have positioned him to work at the intersections of STS and Scientific and Technical Communication. His research articles have appeared in the journals *Science & Education*,

History of Education Quarterly, New York History, Peitho, Configurations, and *Technical Communication Quarterly.* Fiss's book *Performing Math: A History of Communication and Anxiety in the American Mathematics Classroom* (Rutgers University Press, 2020) argues that we must understand mathematics as communication-based, particularly in order to deal with Americans' widespread math hatred and high rates of math anxiety. With evidence from a dozen educational archives, it recovers the performative dimensions of communicating about mathematics during the emergence of many assumptions, techniques, and frameworks of American higher education.

Matthew L. Jones is James R. Barker Professor of Contemporary Civilization at Columbia University, where he focuses on the history of science and technology in early modern Europe and on recent information technologies. With his collaborator Chris Wiggins, he is completing *How Data Happened: A History of the Science, Politics, and Power of Data, Statistics, and Machine Learning from the 1800s to the Present* (Norton, forthcoming). His publications include *Reckoning with Matter: Calculating Machines, Improvement, and Thinking about Thinking from Pascal to Babbage* (University of Chicago Press, 2016) and *The Good Life in the Scientific Revolution* (University of Chicago Press, 2006).

Abram Kaplan is a junior fellow at the Harvard Society of Fellows. His research focuses on humanism and the exact sciences in early modern Europe. Themes discussed in the present essay, notably the relationship between demonstration and means of persuasion, are also discussed in his 2018 article for *Notes and Records.* He is working on a book manuscript, based on his dissertation, about the Renaissance roots of the idea of progress in mathematics.

Kevin Lambert is Professor in the Department of Liberal Studies at California State University Fullerton and a historian of modern science and mathematics. His book *Symbols and Things: Material Mathematics in the Eighteenth and Nineteenth Centuries* (University of Pittsburgh Press, 2021) investigates British mathematics as a way of thinking outside the body.

Massimo Mazzotti is the Thomas M. Seibel Professor of the History of Science and Director of the Center for Science, Technology, Medicine, and Society (CSTMS) at the University of California, Berkeley. He studies the history of science from the early modern period to the present, with a focus on the mathematical sciences and technology. He is interested in the social dimension of technical knowledge—and particularly in the way this knowledge is made, and made credible and authoritative. He has published on the gendering of mathematics, mathematics and religion, Enlightenment science, and the politics of various processes of quantification, standardization, and mechanization. His current projects explore the political dimension of mathematical reasoning in revolutionary Europe; the intersection of technology, design, and social planning in postwar Italy; and the social life of algorithms.

Christopher J. Phillips is Associate Professor of History at Carnegie Mellon University, where he teaches the history of science. He is the author of *The New Math: A Political History* (University of Chicago Press, 2014) and *Scouting and Scoring: How We Know What We Know about Baseball* (Princeton University Press, 2019). He is currently finishing a book on the history of biostatistics and the way in which modern statistical methods transformed medicine in the twentieth century. He previously taught at NYU and holds a PhD in History of Science from Harvard University.

Caitlin C. Rosenthal is Associate Professor of History at the University of California, Berkeley, and a historian of eighteenth- and nineteenth-century US history. Rosenthal's research focuses on the development of management practices, especially those based on data analysis. Rosenthal's first book, *Accounting for Slavery: Masters and Management* (Harvard University Press, 2018), explored the development of business practices on slave plantations and used this history to understand the relationship between violence and innovation. The book won the Simpkins Award of the Southern Historical Association as well as the first book prize of the Economic History Society. It was featured as a "Five Books" best book in economics for 2018 and was honored by the San Francisco Public Library Laureates. Her current research explores the history of human resources departments.

J. B. Shank is Distinguished University Teaching Professor of History and the former Director of the Center for Early Modern History and the Andrew W. Mellon Foundation Consortium for the Study of the Premodern World at the University of Minnesota. His books include *The Newton Wars and the Beginning of the French Enlightenment* (University of Chicago Press, 2008) and *Before Voltaire: The French Origins of "Newtonian" Mechanics, 1680–1715* (University of Chicago Press, 2018). His current research focuses on Enlightenment cosmological science from the fourteenth to the nineteenth centuries.

Introduction

Morgan G. Ames and Massimo Mazzotti

"Algorithm" is a word whose time has come.

In the last few years, the term "algorithm" has been widely used in public discourse, evoking mixed images of liberation and alienation. When we now imagine our future, we think of algorithms. To some commentators, algorithms are sublime problem-solvers that will help us end famine and cheat death, and may even usher in the emergence of artificial superintelligence—a high-tech version of the theological "union with God." For others, the algorithms' promise of power is fraught with ambiguity. Algorithms, we have realized, can be carriers of shady interests and vehicles of corporate guile; they can amplify forms of discrimination and reinforce social stratification. The dream of unbiased algorithmic decision-making seems to have remained just that, a dream. Even so, the rhetoric of algorithmic neutrality is more alive than ever—why? The answer, we believe, needs to include an analysis of long-term historical processes. This book explores key moments in the historical emergence of algorithmic practices and in the constitution of their credibility and authority through the modern period. If algorithms are historical objects and their associated meanings and values are situated and contingent—and if we are to push back against rhetorical claims of otherwise—then a genealogical investigation is essential to understand the power of the algorithm.

But what does "algorithm" even mean in the present conversations? Both in public and in specialized discourses, the meaning of the term has expanded prodigiously in just a few years. This transformation is reminiscent of what happened to the term "technology" in the first part of the twentieth century, only more rapid. Langdon Winner noted how the meaning of "technology" morphed from something relatively precise, limited, and unimportant to something vague, expansive, and highly significant, laden with both utopic and dystopic import. The word had become "amorphous in the extreme," a site of semantic confusion—surely a sign, he concluded, that the language of ordinary life as well as those of the sciences had "failed to keep pace with the reality that needs to be discussed." The political implications included

Morgan G. Ames and Massimo Mazzotti, *Introduction* In: *Algorithmic Modernity*. Edited by: Morgan G. Ames and Massimo Mazzotti, Oxford University Press. © Oxford University Press 2023. DOI: 10.1093/oso/9780197502426.003.0001 radicalization of the debate around technology, a proliferation of useless dichotomies, and the disappearance of public spaces for discussing technological change in an informed manner. At the root of this situation, and of the "technics-out-of-control" theme, Winner saw the dangerous concealing of technology's political dimension (Winner 1978). The recent expansion of the meaning of "algorithm" signals that we are experiencing a comparable moment of political and semantic inadequacy.

Historians tracing the development of algorithms have traditionally assumed the object of their study was relatively well defined, and certainly neither glamorous nor threatening. "Algorithms are simply a set of step-by-step instructions, to be carried out quite mechanically, so as to achieve some desired result" is a standard opening for such presentations. Algorithms, in this sense, are routine problem-solving methods, recipes, and as such are certainly not limited to mathematics and have existed well before a special word had been coined to describe them. Their history, when restricted to mathematics, is that of a variety of methods for arithmetic operations across different times and cultures, with some crucial pivots such as the algebra of the eponymous al-Khwarizmi, from which the word algorithm derives, or d'Alembert's encyclopedic definition of the word, which expands its meaning from certain routine arithmetic procedures to "the method and notation of all types of calculation." By the age of Enlightenment, the term had come to mean any process of systematic calculation—a process that could be carried out *automatically*. In the nineteenth and twentieth centuries this meaning of algorithm was sharpened by adding a set of specific conditions, most importantly the condition of finiteness: the procedure must reach its result in a finite number of steps. While this concern emerged in mathematical logic, finiteness acquired new and consequential significance in the age of computing (Chabert 1999, 1–6).

Of course, algorithms as mechanizable instructions have been around long before the dawn of electronic computers. Not only were they implemented in mechanical calculating devices, they were used by humans who behaved in machine-like fashion. Indeed, in the pre-digital world, the very term "computer" referred to a human who performed calculations according to precise instructions—as shown in Jennifer Light's classic article on the women who performed ballistic calculations at the University of Pennsylvania during World War II (Light 1999). It is clear, however, that we now rarely use the word algorithm to refer solely to a set of instructions. Terse definitions do not seem to capture current usages. Rather, the word now usually signifies a program running on a physical machine—*as well as its effects on other systems*. Algorithms have thus become agents, which is partly why they give rise to so many suggestive metaphors. Algorithms now *do* things. They determine important aspects of our social reality. They generate new forms of subjectivity and new social relationships. They are how a billion-plus people get to where they're going. They free us from sorting through multitudes of irrelevant results. They drive cars. They manufacture goods. They decide whether a client is creditworthy. They buy and sell stocks, thus shaping all-powerful financial markets. They can even be creative; indeed, apparently, they have already composed symphonies "as moving as those composed by Beethoven" (Steiner 2012).

Algorithms have captured the scholarly imagination every bit as much as the popular one. Academics variously describe them as a new technology, a particular form of decision-making, the incarnation of a new epistemology, the carriers of a new ideology, and even a veritable modern myth—a way of saying something, a type of speech that naturalizes beliefs and worldviews (Barthes 1972). A few years ago, sociologist David Beer, in describing algorithms as expressions of a new rationality and form of social organization, was on to a fundamental point. Machines are never just neutral instruments. They embody, express, and naturalize specific cultures—and shape how we live according to the assumptions and priorities of those cultures (Beer 2016).

How we talk about algorithms can be vague and contradictory, but also evocative and revealing. Semantic confusion may in fact signal a threshold moment when it behooves us to revise entrenched assumptions about people and machines. Hence we shift our attention from possible definitions of the term "algorithm" to the omnipresent *figure* of the algorithm as an object that refracts collective expectations and anxieties. Historians of science are familiar with technological artifacts that capture and transform people's imagination, becoming emblematic of certain eras. They are, at once, tools for doing and tools for thinking (Bertoloni Meli 2006). The mechanical clock in Newton's time is a prime example. Consider the heuristic power and momentous implications—scientific, philosophical, and cultural—of seeing the universe as an immense piece of machinery, whose parts relate to one another like those of a sophisticated precision clock. A clockwork universe means, for example, that one can expect to discover regular and immutable laws governing phenomena. But a clock is not simply an image. A clock that can measure fractions of a second inevitably changes our perception of time. It turns time into something that can be broken down into small units, scientifically measured—and accurately priced. The precision clock helped spawn new temporalities, oceanic navigation, the Industrial Revolution. At the same time, it was taken to be the best representation of the world it was shaping: a mechanistic, quantifiable, and predictable world, made up of simple elementary components and mechanical forces.

In the twentieth century, seeing the workings of the human mind as analogous to the operations of a hefty Cold War electronic computer signals a similarly momentous cognitive and social shift. Lorraine Daston describes it as the transition from Enlightenment reason to Cold War rationality, a form of cognition literally black-boxed in shiny-cased machines evoked by the inhumanly perfect monolith in the film *2001: A Space Odyssey* (Daston 2011). Many sharp minds of the postwar period believed that the machine's algorithmic procedures, free of emotions and bias, could solve all kinds of problems, including the most urgent ones arising from the confrontation between the two superpowers. It did not work out that way—the world was too complicated to be reduced to game theory or cybernetics. Like clocks, Cold War computers were emblematic artifacts. They shaped how people understood and acted within the world. Clocks and computers also shaped how people understood themselves, and how they imagined their future. It is in this sense that we now live in the age of the algorithm (Erickson et al. 2013).

Preeminent among the figures of the algorithm is the algorithm-as-doer. This image captures a salient feature of our experience: algorithms can alter reality, changing us and the world around us. The Quantified Self movement, which promotes "self-knowledge through numbers," is an example of how subjectivities can be reshaped algorithmically—in this case by monitoring vital functions and processing data relative to lifestyles to enable self-disciplining practices. But algorithms do not just shape subjectivities. The world of our experience is pervaded by code that runs on machines. In the current expanded sense of the word, algorithms generate infrastructures—like social media—that shape our social interactions. They don't just select information for us, they also define its degree of relevance, how it can be acquired, and how we can participate in a public discussion about it. In other words, they have the power to enable and assign a "level of meaningfulness," thus setting the conditions for our participation in social and political life (Langlois 2014).

The fact that algorithms create the conditions for many of our encounters with social reality contrasts starkly with their relative invisibility. Once we become habituated to infrastructures, we are likely to take them for granted. They become transparent and ready-to-hand (Bowker and Star 1999). But there is something distinctive about the invisibility of algorithms, about the way they are embedded in the world we inhabit. This has to do with their liminal, elusive materiality (Dourish 2017), and with the way they can be deliberately obfuscated or epistemologically opaque (Pasquale 2015; Burrell 2016).

There might well be "something impenetrable about algorithms," writes Tarleton Gillespie, while Malte Ziewitz has detected a strange tension in the current debate on algorithms: they judge, regulate, choose, classify, and discipline but they are described as strange, elusive, inscrutable entities; it has even been argued that they are virtually unstudiable (Gillespie 2014; Ziewitz 2016).

More than other artifacts, algorithms are easily black-boxed and thus shielded from the scrutiny of users and analysts, who cease seeing them as contingent and modifiable, accepting them instead as obvious and unproblematic. Historically, black-boxing has been particularly effective when the technology in question depends on a high degree of logical and mathematical knowledge. This is not because mathematics is obvious to the general public, but because of the widespread assumption that mathematics consists of deductive knowledge that—as such—is merely instrumental. What could possibly be social or biased about manipulating strings of numbers or code according to formal rules? Aren't these operations purely technical and therefore neutral?

The deterministic view of the algorithm—the figure of the algorithm that does things—certainly helps understand how, as a technological artifact, it can change the world we live in. In this type of speech, the word "algorithm" functions as a synecdoche for software and larger sociotechnical systems. The algorithm-as-doer, however, is also misleading because it hides its larger ecological context; it represents the algorithm as a self-contained entity, a tiny portable machine whose inner workings are fixed and whose outcomes are determined. The figure of the algorithm-as-doer reinforces an image of the algorithm as a tiny machine crunching data and mixing them up to produce the desired result—as in the tongue-in-cheek visual rendering of EdgeRank as a nineteenth-century set of three cast-iron grinders, one of each of its main "ingredients": affinity, weight, and time (Widman n.d). Such images suggest that algorithms exist independently of their ecology, invariably producing the same effects wherever and whenever deployed. Ludwig Wittgenstein commented on the potency of such idealized images of machines; when we look at one of them, he wrote, "everything else, that is its movements, seems to be already completely determined." All the possible movements of the "machine-as-symbol" are already there "in some mysterious way," much as the correct result of 2 + 2 is already there, as a shadow, when one writes those three symbols (Wittgenstein 1953). It's a rigid and therefore absolutely reliable and predictable object—the stuff of technological dreams, or nightmares. Experience tells that machines do not behave like that. Rather, the idealized machine is a projection of the alleged rigidity of logical rules and mathematical reasoning. In particular, that's not what happens with sets of instructions running on physical machines interacting with other systems. The empirical study of algorithms suggests that we can understand their functioning—and their meaning—only by considering the sociotechnical ecologies within which they are embedded.

There is another reason the algorithm-as-doer is misleading: it conceals the design process of the algorithm, and therefore the human intentions and material conditions that shaped it. Thus far, we have argued for the significance of the ecology of algorithms, which is primarily a spatial and synchronic notion. It emphasizes the algorithm's relational properties—how it interacts with machines and human collectives. But we need something more. Consider, for example, the algorithms that produce and certify information. In exploring their ecology we can address important questions about authority, trust, and reliability. But what about the logic that shaped their design in the first place? Who decided the criteria to be adopted and their relative weight in the decision-making process? Why were the algorithms designed in one particular way and not another? To answer these questions, we need to see the technical features of an algorithm as the outcome of a process. In other words, we need a historical—indeed *genealogical*—understanding of the algorithm. The notion of genealogy is rooted in temporality and diachronicity; it calls attention to the historical emergence of the algorithm's properties, their contingency and precarious stability. It invites us to question technical features that would otherwise seem obvious and self-evident.

A historical sensibility allows us to situate algorithms within a longer quest for mechanization and automation. Like clocks and the programs of early electronic computers before them, digital algorithms embody an aspiration to mechanize human thought and action in order to make them more efficient and reliable. This is a familiar and yet also unsettling story, constitutive of our modernity. In the most iconic frame of *Modern Times*, Charlie Chaplin is literally consumed by the cogs of a mechanized factory. As machines become more efficient, the image suggests, they become more deadly. But what does "efficiency" even mean in this context? Theorizing the antagonistic relationship between labor and machines in nineteenth-century factories, Marx argued that the new machines not only increased production and reduced costs, but were "the means of enslaving, exploiting, and impoverishing the laborer." Mechanization, he argued, was a weapon in class warfare (Marx 1867).

Scholars in the Marxist tradition have continued to pay attention to mechanized production. Harry Braverman, for example, argued that the kind of automation prevailing in postwar American factories was far from being an obvious imperative. Rather, numerical control machinery had been designed to reshape the relations between management and labor in order to wrest control over the production process away from the workshop floor. Other kinds of automation were possible, he argued, but were not pursued because they would not have the same *social* effects (Braverman 1974). Processes of automation and mechanization, in short, are not simply about productivity and profit. Design features are shaped by social relations. In the 1980s a wave of social studies expanded this intuition to argue that technological artifacts are inherently social—in other words, any technology bears the mark of the specific culture and interests that shaped it (MacKenzie and Wajcman 1985). The technical sphere is never truly separated from the social sphere: the technical *is* the social. Technological change is thus never neutral, let alone natural. To say a mechanized procedure is more efficient than another is not an adequate historical explanation for its success. The notion of efficiency itself is always relative to a set of assumptions and goals; making them visible is a prerequisite for any informed discussion about technological change and its implications.

How exactly do these insights apply to the study of algorithms? Consider the work of Donald MacKenzie on the assumptions and negotiations that shaped certain algorithms now embedded into software used by financial traders worldwide. Their design could have been different; there is never just one way to automate a given financial transaction and, more generally, to regulate a market. Choices are made and these choices do not follow a neutral universal logic; they are the outcomes of contingent interests and negotiations (MacKenzie 2006). In a similar vein, Elizabeth van Couvering and Astrid Mager have shown how algorithms behind for-profit search engines are shaped by specific business models, based mainly on targeted advertising. They have also shown that these search algorithms stabilize and reinforce the socioeconomic practices they embody (van Couvering 2008; Mager 2014). Precisely because of their unique combination of pervasiveness and invisibility, algorithms can effectively embed and amplify existing social stratifications. The neutrality of algorithms is therefore an illusion. Or, in some cases, a powerful rhetorical tool.

The past few years have seen a flourishing of research that explores algorithms and their ecologies through the lenses of the humanities and the interpretive social sciences. Powerful arguments for the non-neutrality of algorithms have been articulated in exemplary work on the algorithmic management of information (Gillespie 2014), the racist biases embedded in search engines (Noble 2018), the economic and racial stratification reinforced by automated policies (Benjamin 2019; Eubanks 2018; Gangadharan and Niklas 2019; O'Neil 2016). Scholars have emphasized the limits of automation and criticized facile forms of techno-optimism (Broussard 2018; Roberts 2019;

Stark 2018), and in special issues have explored questions of algorithmic governance (Ziewitz 2016), of the social power of algorithms (Beer 2016), and of algorithms in (and as) culture (Ames 2018).

This volume aims to contribute to this growing research field with a distinctive and multifaceted exploration of the genealogies of algorithms, algorithmic thinking, and the way we currently talk about algorithms and their power. This genealogical approach results from taking a historical perspective that constructs contemporary discussions and interpretations of algorithms as the outcome of long-term social processes. This means mapping the historical emergence, stabilization, crisis, and reinterpretation of priorities, concepts, and methods that are often taken for granted in contemporary conversations. The arc of this volume's contributions thus traces key moments, ruptures, and shifts within the intellectual and cultural history of algorithms, taking the reader from sixteenth-century algebra as "The Great Art" to the world of Enlightenment, the rise of cybernetics, and the twenty-first century's breakthroughs in big data and machine learning.

Throughout its essays, this volume keeps a clear focus on the emergence and continuous reconstitution of algorithmic practices alongside the ascendance of modernity. In this way, the essays contained in this book identify and throw into sharp relief moments of the trajectory of an algorithmic modernity, one characterized by attitudes and practices that are best emblematized by the modernist aesthetic and inhuman efficacy of the algorithm. This connection between algorithms and modernity is one of our central concerns; the essays address it through detailed historical reconstructions of specific moments, thinkers, and cultural phenomena over the last 500 years. They collectively lead us to the definitions of algorithm most legible today and to the pervasiveness of both algorithmic procedures and rhetoric. Taken together, these essays offer a genealogy of the distinctly modernist faith in algorithms as neutral tools that merely illuminate the natural and social world. They account for the rise in credibility of the belief that algorithms are able to regularize, systematize, and generalize the natural and social world, providing insights that would remain otherwise hidden and inaccessible.

The initial four chapters chart the development and spread of algorithmic thinking first in Renaissance and then Enlightenment Europe, where it not only reformed mathematics pedagogy but modes of knowledge production more broadly. In the process, "algorithm" went from a vague and multifaceted signifier to a stable and recognizable concept. The protagonists of the first chapter are Renaissance mathematicians in Italy and France, who practiced the "The Great Art," modern algebra's major predecessor, and saw the generality and persuasiveness of demonstrations in the repeatability of procedures.

Reforming mathematics pedagogy was a central plank of humanist education at the turn of the sixteenth century, and humanists framed algebra as a
perfected form of dialectic, Abram Kaplan explains, replacing the *trivium* of
logic, rhetoric, and grammar with a new kind of learning: *mathesis*. In their
quest for new modes of persuasion well adapted to the algorithm, he argues,
these mathematicians assimilated algorithms to the authoritative model of
the geometrical demonstration, so that worked examples could take the ar-
gumentative place of diagrams. A historical epistemology of the Renaissance
algorithm thus reveals the key role of the humanist-rhetorical tradition in
shaping a range of persuasive techniques that supported the emergence of a
fully symbolic algebra.

Chapter 2 delves into key aspects of early modern algorithmic practices
through a close reading of polymath Robert Recorde's presentation of long
division and discussion of the terminology of "arithmetic" and "algorism."
Michael Barany shows that interest in algorithms grew significantly
throughout the sixteenth century, particularly among those interested in
early modern mathematical pedagogy, through vernacular arithmetics like
Recorde's 1543 *The Ground of Artes*. Here, Recorde presented rule-based algorithmic methods as powerful but error-prone, requiring practice and
expertise to apply effectively. In contrast to later and more mechanistic
understandings of algorithms, sixteenth-century algorithmic thinking before
the development of calculus thus reflected expert judgment and discernment,
involving both the manipulation of numbers as well as an appreciation of
the context and limitations of mathematical knowledge. "If algorithms seem
now to evacuate judgement rather than demand it," Barany concludes, "that
is only because the early modern embrace of algorism has been joined by a
later modern move to write misprision out of calculation's essential operation.
Even where context and contingency have been embraced as essential to the
practice of reasoning, formalist fantasies have sterilized such practices of their
error-bound roots."

Chapter 3 charts the distinct shift away from Renaissance expert discernment and toward more mechanistic understandings of the algorithm in the
Enlightenment. Amir Alexander focuses on Isaac Newton's development of
calculus to illustrate this transformation of algorithmic practice and its distance from previous mathematical approaches. This chapter provides a close
analysis of the algorithmic nature of Newtonian calculus—where the algorithm dictates a series of predetermined steps that could be systematically
applied—and shows that previous related techniques like the "method of
indivisibles" were broader and more flexible: they often relied on geometric
properties that had to be tailored to each particular question and were not

generalizable across classes of questions. Rather than relying on the mathematical ingenuity that these previous techniques demanded, calculus, Alexander argues, "strove to be a series of fixed rules that could be applied uniformly and unthinkingly to any question of a given type, and guaranteed to produce the correct results." To demonstrate how this shift took place, this chapter traces Newton's development of increasingly systematic and abstract methods from the binomial theorem to *De Quadratura Curvarum*—a path that parallels Newton's quest for divine order in the messy and seemingly contradictory physical world.

In chapter 4 we move from this close reading of the antecedents and developments of calculus to a reconstruction of the ways in which algorithmic thinking reshaped methods of knowledge production across mathematics, the physical sciences, and much of Enlightenment thought. J. B. Shank calls this ascendance and stabilization of an algorithmic paradigm the Algorithmic Enlightenment. This chapter connects seemingly disparate cultural shifts, including a focus on quantification and precision in the sciences, and a widespread instrumentalization of mathematics, showing that calculus was exemplary of a shift away from demonstrative geometric mathematics toward utilitarian mathematical tools capable of rendering complex problems solvable. From there, quantification, algorithmic methods, and a focus on seemingly universal tools for empirical analysis moved through mathematics to astronomy, mechanics, and into Enlightenment philosophy more generally, transforming eighteenth-century understandings of the ways knowledge works. This primacy of algorithmic thinking, understood as a new appreciation for quantitative, rule-based conceptualizations and problem-solving methods, sits underneath many of the classic scholarly answers to the perennial modern question: What is Enlightenment?

From here, chapters 5, 6, and 7 chart the growing cultural significance of algorithms through the Industrial Revolution and into the beginning of the Machine Age. While the concept of algorithm was still largely confined to intellectual elites, popular algorithms nevertheless began to shape ways of thinking and knowing across Europe and beyond, as Caitlin Rosenthal explores in chapter 5. Tracing the use and diffusion of the "rule of three" algorithm—a method of cross-multiplication—on both sides of the Atlantic Ocean in the eighteenth and nineteenth centuries, Rosenthal reveals that this algorithm gave both merchants and wage laborers newly engaged in commercial enterprises an important tool that bootstrapped them toward numeracy. Considered the pinnacle of colonial math education, the rule of three enabled new forms of negotiation and shifts in power: on the one hand, it democratized the kind of leverage numeracy enables in a capitalist system, while on the other, it privileged the kinds of exchanges that this algorithm made legible. While simple compared to today's algorithms, the rule of three—along with other arithmetical rules and table books of the day—was a black box in a world of innumeracy and semi-numeracy, which offered a diverse range of people the ability to negotiate more effectively while obscuring and stabilizing the power relations it enabled. Understanding how the rule of three was used and discussed thus not only demonstrates the pervasiveness of algorithmic processes in everyday life and popular culture (as well as their connection to colonialist expansion), but reveals how this simple algorithm helped legitimate particular types of exchange and commensuration, and with them the emerging capitalist economy.

Chapter 6 rounds out our examination of how the algorithmic thinking of algebra seeped into popular understandings of modernity in the nineteenth century, and connects it to the coming computational turn. In this chapter, Kevin Lambert uses the work of Victorian mathematicians George Boole and Augustus De Morgan to frame Victorian mathematics as essentially algorithmic. The interest these two mathematicians had in empiricism and inductive reasoning brought about what Lambert terms an "algorithmic mathematics." Its explicit emphasis on the material dimension and mechanical procedures contrasted starkly with contemporary continental mathematics, characterized instead by priorities such as rigor and purity—which is why they have been often described as marginal and eccentric to nineteenth-century mathematics. Focusing on Boole's pioneering research in mathematical logic, as well as De Morgan's view that divergent infinite series can be described entirely in terms of ideas, this chapter invites us to take seriously the work of these mathematicians—their ideas, intentions, but also their material practices and library procedures. Lambert concludes that this mathematics, as well as British algebra more broadly, was ultimately practiced as an experimental science that foreshadows the use of computers in mathematics research.

In chapter 7, Andrew Fiss further elaborates the connection between manual algorithmic practices with early computational cultures through the labor of human computers needed to enact them. In particular, this chapter examines the Harvard Observatory—an organization that subsequently set the standard for hiring this human computer workforce—which in the 1870s shifted away from male servants and toward women. It explores how this shift was understood by the staff at the time and in the decades after this transition through the ways it was encoded and interpreted in a Harvard Observatory adaptation of a comic opera by Gilbert and Sullivan, *H.M.S. Pinafore*, written in 1879 but first performed in 1929. Considering the *Observatory Pinafore* as a dramatization of workplace stratification, Fiss uses the main character of Joseph/ine—jointly female and male, computer and observer, domestic and professional—as a humorous lens into these negotiations. He then examines the ways in which the history of human computers was rewritten to be almost entirely feminized in subsequent decades. Throughout, the chapter makes the case that histories of computers and algorithms should consider how computing work has been differently professionalized through time, particularly the different paths these computers took into the profession, the ways in which this work was negotiated, and ways in which it blurred boundaries between professional-domestic spaces and servant-familial roles.

The last three chapters of this volume bring us into the late modern era, during which computers supplement and then supplant algebra and calculus as the paradigmatic locus of algorithmic thinking. Chapters 1–7 articulate how algorithmic thinking pervaded and shaped notions of modernity in the Western world, spreading first among Enlightenment elites and then to merchants, laborers, and beyond. Algorithms promised to render the world regulatable, controllable, and therefore colonizable (Edwards 2010; Law 1986, 1994; Scott 1999), even as the concept of "algorithm" was still little known outside the world of mathematics. The subsequent chapters reveal how algorithms overflowed their mathematical frame to become a powerful symbolic referent for a machinated, computational future, a sociotechnical imaginary (Jasanoff 2015) on which the hopes and fears of computer-assisted social control have been hung. As Dryer explains in this volume, "At its roots, the history of algorithmic computing is a history of how information and data are conceptualized, ordered, and put to use in economic decision making" and how it impacts human lives (see also Dryer 2018, 2019).

Alongside the heady transdisciplinary rubric of cybernetics and the transcendence of statistics-backed behaviorism in the mid-twentieth century, the first artificial intelligence algorithms sought not just to define rules for human-machine systems, but to govern them through far-reaching feedback mechanisms (Dupuy 2009; Dick 2011, 2015), conjuring *sublime* worlds of automatically regulated harmony (Ames 2018; Mosco 2004; Nye 1996). This utopian dream was accompanied by a dystopian nightmare of a mechanized society in which individual agency was subsumed by coercive algorithmic control via institutions and governments. The 1960s Berkeley Free Speech Movement protested the latter, riffing on the text printed on computer punch cards to make signs that read, "I am a UC student. Please don't bend, fold, spindle, or mutilate me" (Lubar 1992). Both aspects of this algorithmic sublime largely turned to disillusionment as these algorithms failed to deliver on their utopian promises—even as some of the dystopian fears of algorithmic control were quietly implemented by corporations and governments in the decades since. At the same time, algorithmic developments and practices influenced popular understandings of the contemporary moment, which then in turn influenced the ways in which these algorithms were developed, deployed, interpreted, and trusted. Algorithmic thinking in the twentieth century has increasingly regulated labor, supply chains, political processes, institutions, decision-making, science, and our social worlds and imaginations. The final chapters take us from the gendered labor of human "computers" described in chapter 7 up to the antecedents of the machine-learning revolution of the 2010s, rooted throughout in the cybernetic and behaviorist assumptions of twentieth-century algorithmic thought.

Chapter 8 demonstrates how algorithmic management strategies became an apparatus of state-run control systems in the twentieth century. There, Theodora Dryer presents an origin story of data-driven and algorithmically controlled industrial agriculture in the interwar and New Deal United States of the 1920s and 1930s, during which the Statistical Quality Control Chart transformed sugar beet fields into laboratories for mathematical testing. The development of these control systems was symbolically important in this era—a moment of radical instability and restructuring of state policy over agricultural resources, labor, and land. It moreover set the stage for more computationally intensive state control systems in the coming decades, both within and beyond agriculture. Dryer focuses on the role that instability and statistical probability—statistical randomization, the law of large numbers, inferential management, and industrial data—played in the development of these control systems. She also traces the allure of control systems as effective modes of governance in the popular imagination. In the process, this chapter illustrates how the New Deal moment entrenched not only the computational oversight of farming, but enduring sociotechnical imaginaries about the power of algorithmic control of data. Dryer shows how new visions of a data-driven agricultural economy link the mathematical mechanization of control to much larger political and economic processes, in the name of a control society.

In chapter 9, Christopher Phillips shows how the reliance on statistically defined and algorithmically regulated control systems described by Dryer soon expanded throughout the social, behavioral, and managerial sciences in the United States and beyond starting in the 1930s, fundamentally shaping these fields with consequences that some are only starting to grapple with (as evinced by psychology's replication crisis). This chapter focuses in particular on the heavy reliance on statistical inference and significance testing, illustrating how these methods have been used ritualistically to move researchers from experimental design to interpretation to conclusion with little critical thought about the statistical methods themselves. As hypothesis-testing replaced descriptive statistics and significance-testing became *de rigueur*, the devotion to these forms of algorithmic thinking and their results became quasi-religious in nature, algorithmically transforming "profane and mundane experimental findings into sacred theories and hypotheses." Even as these methods continued to be examined and critiqued within the field of statistics itself, this chapter shows that their mechanical application across the sciences more broadly—much to statisticians' chagrin—ultimately reflects of Cold War models of decision-making as well as broader transitions in how practitioners of the human sciences dealt with the "problem" of causality. These ideas have, in turn, shaped causal assumptions in the era of big data, where some acolytes claim that conclusions can arise from enough data theory-free.

"*Both* the number crunching *and* the philosophical foundations of statistical inference were black-boxed," Phillips argues in this volume: "given a set of data, researchers can just run the software to quickly figure out whether they have 'significant' (and therefore publishable) results." Matthew Jones further explores the contemporary notion of black-boxing and automation in our final chapter, focusing on the iterative genealogy, from the 1960s to the 2020s, of a popular class of supervised learning algorithms—decision trees—which continue to form the basis of many automated classification systems. Like many machine-learning algorithms, decision trees were developed with and for computers—the computations they require are simply not feasible for the human computers discussed by Fiss or the statistical tables discussed by Dryer. As such, Jones's examination brings us fully into the computational age, a fitting endpoint to our exploration of the role of algorithmic thinking in shaping our notions of modernity. At the same time, several themes developed throughout this section come together in this final chapter: a faith that algorithmic management can adequately reflect and regulate the world, an erasure of inconvenient histories and labor from algorithmic processes, and above all a ritualized application of algorithms—a practice that has led some to claim that big data and machine learning will soon lead to the "end of theory" across the sciences more broadly. Along with other chapters in this volume that explore the complex interrelations between the social and technical, this final chapter pushes back against the technologically deterministic account that the recent boom in machine learning is entirely due to the availability of massive benchmark data sets, increased computing power, and new techniques that take advantage of both. Indeed, without the long-term social, structural, and ideological shifts described throughout this volume, the Western world would not have simultaneously taken certain forms and assumptions of algorithmic thinking so deeply for granted as integral aspects of modernity, and elevated others—such as machine learning—to the sublime.

1
Algorithm and Demonstration in the Sixteenth-Century *Ars Magna*

Abram Kaplan

Renaissance mathematicians in sixteenth-century Italy and France sought to humanize algorithmic mathematics. By then, algorithm had been well enough known in a nascent Europe for some time.[1] Manuals like Fibonacci's *Liber abaci* disseminated to merchants and others the practice of doing arithmetical calculations using Arabic numerals.[2] In the sixteenth century, new works were written to teach to students the rules of numeral calculation, which was sometimes called "algorism" in English.[3] In these contexts, algorithm was a calculation practice, and so it remained in the textbooks on algebra and the *ars magna*, or great art, that I discuss here. But in these works it was not just a practice. Incorporating algorithm into established educational institutions meant assimilating it to mathematical learning as already understood. This entailed theories—of education, of demonstration, of metaphysics.

When mathematical authors sought to naturalize algorithmic practice within humanist learning, they did so by associating it with the seven liberal arts that together constituted preliminary university education in Renaissance Europe. At times, they leaned heavily on the identification of algebra with arithmetic. But there were other options. As has been recognized, university pedagogy played a central role in the diffusion of humanism in France and the Low Countries. Reforming mathematics pedagogy was a central plank of humanist education at the turn of the sixteenth century.[4] A series of studies by Giovanna Cifoletti forms the background to my argument here; she shows that French humanists took advantage of this institution by framing algebra as a perfected form of dialectic and thus as a replacement for the *trivium* of logic, rhetoric, and grammar no less than as a replacement for the *quadrivium*. This new kind of learning was called *mathesis*.[5] And in yet another interpretation of *mathesis*, namely a metaphysical one consistent with Aristotelian theories of science, algebra seemed to some writers to unify arithmetic and geometry by identifying a subject common to both.[6]

Abram Kaplan, *Algorithm and Demonstration in the Sixteenth-Century* Ars Magna In: *Algorithmic Modernity*. Edited by: Morgan G. Ames and Massimo Mazzotti, Oxford University Press. © Oxford University Press 2023. DOI: 10.1093/oso/9780197502426.003.0002

All these approaches involved efforts to change algorithmic practice no less than new justifications for it.

I focus here on a narrow slice of these efforts, namely the integration of algorithms into Euclidean geometric demonstration. Algebra was one subject in which this integration took place.[7] In book II of the *Elements*, Euclid used geometric techniques to explore the relationship between sides and squares; in later books he broached the subjects of proportions between numbers, in book V, and between figures, in book VII. The study of proportions was central to medieval and early modern mathematics, and Greek mathematics provided ample resources for addressing it.[8] Meanwhile, just this material was addressed efficaciously by the algorithms of the great art. How this complementarity in disciplinary subject matter was recognized and accommodated theoretically has been much discussed.[9] It suffices here to note that the integration of algorithms into demonstrations implicated elements from both interpretations of algebra as a form of *mathesis*. Not the only discipline of university mathematics in the sixteenth century, Euclidean geometry was nevertheless then of high status both cognitively and metaphysically.[10] Accordingly, the use of arithmetical and algorithmic demonstrations had potential ramifications both for cognitive dimensions of persuasion (hence for Renaissance dialectic) and for the relationship between arithmetic and geometry (hence for the ontology of mathematics).

My subject here is how mathematical authors combined, within particular demonstrations, the genre of demonstration and the algorithmic rule. My interest is in the relationship between algorithmic procedure and persuasion: can following an algorithm lead to some kind of knowledge about the truth of a mathematical statement or belief about the reliability of a rule? Or must this knowledge or persuasion come from elsewhere? Algebra books from Italy, the Netherlands, and France—by Girolamo Cardano, Simon Stevin, and Guillaume Gosselin—illustrate instances of this relationship during the half-century during when algorithms took center stage in the study of proportions.

I begin by looking at Cardano's mid-century *Book of the Great Art or of Algebraic Rules*, which uses diagram-based geometry to demonstrate rules for solving cubic equations. In Cardano's text the reliability of rules is a consequence of traditional geometric demonstrations, and each rule reflects its demonstration and so its justification. In contrast to this approach, algorithms play a central role in the demonstrations developed several decades later by both Stevin and Gosselin. Both authors integrate algorithms into Euclidean demonstration by accommodating the standard form of the demonstration to the cognitive, and paper, practices made available by the algorithm.

My goal is to identify some possibilities for mathematical cognition that were implicit in the incorporation of algorithms into Euclidean demonstration. This investigation ramifies because in our contemporary world the use of algorithms is often presented as somehow anti-cognitive, as requiring humans to behave like machines. The relationship between putatively automatic algorithms and eminently intellectual demonstration is one place to explore the historical truth of this claim.

Form as a Guide to Procedure: Cardano's Diagrams

In his *Great Art*, Cardano founded an algorithmic approach to cubic equations on a geometric basis. Consistent with the kind of claim typically found in algorithmic texts, Cardano formulated "rules" for the solution of cubic equations.[11] He justified these rules using typical geometric demonstrations that he carried out on a figure. Recollections from high school algebra might suggest to some readers that Cardano used equations that could be manipulated. On the contrary, the stable forms of Cardano's equations served a different purpose. To wit, where geometrical demonstrations justified rules that could be applied independently of the diagram, the forms of the equations served to demarcate the generality of their applicability.

The form of each equation both anchors the generality of Cardano's rule and marks out its limits. A modern reader might expect a general form of the equation and a single, closed-form procedure for its solution, such as $ax^2 + bx + c = 0$ and the well-known quadratic formula taught in high school, $x = -b \pm \dfrac{\sqrt{b^2 - 4ac}}{2a}$. Instead, Cardano proposed multiple rules following differences in the equations' forms, differences such as that between $x^3 + bx = c$ and $x^3 = bx + c$ (note the shifted equals sign).[12] Each case—or in Cardano's words, "head"—had its own "mode" of the demonstration and its own rule.[13] For Cardano, the form facilitated procedure in two interrelated ways. First, it identified which rule to carry out. Second, it shaped the question so as to provide material for those procedures, much as the general equation $ax^2 + bx + c$ fixes values for a, b, and c that are used in the quadratic formula.

Beyond its use in marking out generality, the form of each particular case also participated in the justification that Cardano gave for each rule. Cardano connected each form to a diagrammatic representation of a cubic equation. Cardano used each diagram in order to demonstrate the corresponding rule. In the demonstrations, the diagrams kept track of the various parts of the form, namely the cubes, squares, sides, and number. By keeping track, they also facilitated calculations. By anchoring his rules in diagrams, Cardano claimed to give demonstrations that could produce "belief" [*fides*].[14] Let's look at an example.

The sixth chapter introduced a diagram that would figure centrally in the rest of the book as well as a key concept: "substitution."[15] The diagram (Figure 1.1) expressed "three greatly useful substitutions [*supposita*]" that enabled his comprehensive treatment of all cases.

Cardano elevated a cube on the square AC and collected its parts: "the cube *AE* will consist of eight bodies, four of which are made from line *AB* times the surfaces *DA*, *DC*, *DE*, and *DF*, and the remaining four from the line *BC* times the same four surfaces." Of these bodies, two are cubes (*AB* × *DF* and *BC* × *DC*) and six are parallelepipeds, of two different kinds (*AB* × *DA*, *AB* × *DE*, and *BC* × *DF* are equal; *AB* × *DC*, *BC* × *DA*, and *BC* × *DE* are equal). The cube imagined to rise upon the diagram arranged all the parts of the cube of a general binomial: if the side AC of a square is cut at point B, then $AC^3 = (AB + BC)^3 = AB^3 + 3AB^2 \times BC + 3AB \times BC^2 + BC^3$. This substitution of eight parts (two cubes and six parallelepipeds) for AC³ furnished Cardano with a fixed relationship of volumes expressed in terms of the cubes, squares, and sides of two values, AB and BC.

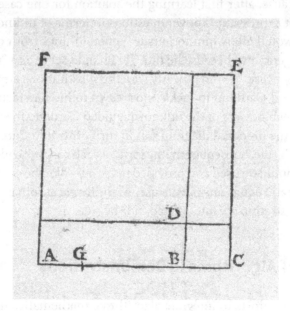

Figure 1.1 Erect a cube on the side AC.

Source: Girolamo Cardano. *Artis magnae, sive, De regulis algebraicis, lib. unus* (Nuremberg: Johannes Petreius, 1545), fol. 16r. Call number *IC5 C1782 545a, Houghton Library, Harvard University.

Cardano used the substitution to organize the argument for each case. In order to do so, he had to assign appropriate values to the sides of the diagram. Take the case where "the cube is equal to things and a number" ($x^3 = bx + c$). In typical Euclidean style, Cardano constructed the solution (see again Figure 1.1) before he demonstrated it: "let DC and DF be two cubes whose sides AB and BC, when they are extended, produce by multiplication the third part of the number of things," namely the coefficient b of bx. Further, the sum of the cubes is equal to the number c. Then "I say that AC is the value of the sought number," x. Cardano's construction was fitting: by making AB × BC = $b/3$, Cardano obtained the equation 3 AB × BC = a and so the useful equation AC × (3 AB × BC) = ax. Since AC = AB + BC, the product AC × (3 AB × BC) makes all six parallelepipeds in the original cube. (Cardano advanced a subsidiary proof of this claim.) Recall that the sum of the cubes CD and DF is c. By appeal to the basic substitution, since the cube AC is composed by the six parallelepipeds and the two cubes CD and DF, it is equal to AC × (3 AB × BC) plus CD + DF, thus to $bx + c$, and thus to the cube of x.[16] By assigning appropriate values, such as AB × BC = $b/3$, Cardano could embed certain relationships in the diagram and then exploit them through substitutions. Substitutions in Cardano's great art were not algebraic procedures to be used in solving a particular problem, but ways of relating rules to their geometric justifications.

Cardano recalled that he turned to the great art and its use of unknown quantities because, after first learning the solution for one case of the cubic, he judged that expressing "known questions in terms of unknown positions [*positiones*]" would allow him to pursue "general things from one question." He turned to geometry because he thought it could serve as a "royal road for coming to every case."[17] And it nearly was: the substitutions embedded in the diagram allowed Cardano to tackle most cases to his satisfaction. The relationship between powers of the unknown guided Cardano in setting up each diagram, and the prepared diagram was in turn central to Cardano's demonstration of each rule. Subsequently, the form of each case both identified when a given rule should be used and provided material—like the coefficients b and c in the quadratic equation—for its use. Algorithmic automaticity and intellectual demonstration are interrelated, side by side.

The Role of Algorithms in Demonstrations

Reflecting their efforts to integrate algorithmic mathematics into the liberal arts, sixteenth-century writers about algorithms drew on two different organizational styles for presenting their mathematics: a style of Greek origin composed of theorems and demonstrations, typically used for university geometry, and a vernacular style composed of rules and examples, typically used for arithmetic.[18] As we saw, Cardano used the geometric style to justify arithmetical rules that could be applied independently of the diagram. Other writers integrated the two styles differently, and in ways that pushed against the division of intellectual labor that Cardano's organization suggested: a division between justification as the preserve of geometry and practice as the preserve of algorithms. This section looks at how mathematical writers following Cardano integrated algorithms into the Euclidean demonstrative scheme and thus used algorithms to produce knowledge or belief.

In mid-century arithmetical works, geometry often played a justificatory role. Both Jacques Peletier and Pierre Forcadel, who wrote French-language texts on arithmetic and algebra in the 1540s and 1550s, appealed to propositions from Euclid in order to justify (Peletier) or explain (Forcadel) the direct and inverse rules of three.[19] As Cifoletti has argued, French humanists took recourse to the language of theory and practice in order to explain their appeals to geometry. In terms that Guillaume Gosselin used in his 1583 *Renewed lecture on the way of teaching and learning mathematics*, the knowing part of arithmetic makes laws and puts together rules, while the active part uses those rules for action.[20] Arithmetical writers appealed to geometry for its theorems and put them to use in solving arithmetical or algebraic problems.[21]

Today's reader might suppose that geometry's justificatory role derived from the generality of its theorems, as distinguished from the particularity of any application of a rule. But in arithmetical writings, geometrical demonstrations were frequently particular. Cardano demonstrated several of his rules by assigning definite values to the coefficients of the unknown: his demonstration for "the cube and square equal to the number" began "Let $AB^3 + 6AB^2 = 100$" rather than with the indefinite coefficients that we might expect and that would make his demonstration general.[22] When Forcadel appealed to geometry to supply "the understanding, perfection, and true intelligence of the rules of three," he instructed the reader to "impress in yourself the idea of the forty-fourth proposition of the first book of Euclid, and propose yourself these three numbers, 2, 3, and 6" through which the fourth proportional might be found.[23] Forcadel apparently believed that working through the particular example would bring understanding more easily than recurring to the generality of Euclid's demonstration.

But what if Euclid's discussion wasn't general either? In the mid-seventeenth century the English mathematician John Wallis advanced just this interpretation of Greek geometric demonstration. A traditional scheme perhaps originating with the ancient philosopher Proclus divided Euclid's theorems into six parts: where the *protasis* (enunciation) and *superasma* (conclusion) stated a theorem in general terms, the four intervening steps stated and proved the same theorem by referring to a particular diagram using lettered objects.[24] In Wallis's view, the relationship between general result and particular demonstration was established by induction, by extending the results of one or several cases to all of them. Wallis summarized Euclid's construction of an equilateral triangle, the same example that Proclus had used to discuss the scheme: "He sets forth one right line AB, and demonstrates the triangle constructed on it to be equilateral ... . By the same method, on any other right line, a triangle can be constructed in this way and also demonstrated, therefore, in any one you like." Although Euclid performed the construction on just one line, it could be repeated on "any other." According to Wallis, the force of the demonstration rested in just this repeatability: "The force of this or of almost any other demonstration is not considered to lie anywhere else than in the supposition, that no case would be able to be pressed to the contrary to which the demonstration set forth could not be applied."[25] Wallis's argument was an interested one: he was hoping to legitimate his own use of mathematical induction in the 1655 *Arithmetica Infinitorum* by portraying Euclidean method as also based on induction. Nevertheless, his discussion of Euclid's very first proposition sheds light on the sixteenth-century practice of making demonstrations using single examples.

Sixteenth-century algebraic writers adapted the Euclidean scheme to arithmetic by placing worked algorithms in the place of geometric constructions. A rediscovered ancient work, Diophantus of Alexandria's *Arithmetica*, encouraged these authors in conceiving of a demonstrative arithmetic on the model of Euclidean geometry.[26] For Diophantus himself seemed to employ a version of the distinction between general enunciation and particular demonstration, setting out general algebraic results (like those demonstrated by Cardano) and then showing them by means of a particular example. The 1575 Latin translation by the German humanist Wilhelm Holtzman (usually called Xylander) underscored the parallel between geometric constructions and algorithms by transforming each problem into a "canon" or rule. "Since the arts master [*artifex*] is able to construct infinite canons even concerning things of great moment from algebraic operations," he wrote, "we also make a general canon from what has been demonstrated." Like a Euclidean conclusion, Xylander expressed the canon generally and at the end of the demonstration (even after his own scholium).[27] Apparently sixteenth-century scholars saw little to distinguish geometric constructions made out of particular lines from algorithms using particular numbers, since they fulfilled similar roles in the demonstrative scheme.[28]

The assimilation of algorithms to geometric constructions is even clearer in Simon Stevin's 1585 *L'arithmetique* and accompanying edition of the first four books of Diophantus' *Arithmetic*.[29] Well known as an engineer and pioneer in mechanics, Stevin was a polymath whose reflections on number and language reflected his classical learning.[30] His edition of Diophantus distinguished between general theorems and particular problems or, here, "questions."[31] In framing his theorems, Stevin employed a version of the traditional Euclidean scheme, whose parts he marked out using headings: a general statement and conclusion bracketed the "explanation of the given" and "of the sought" and the particular demonstration based on them. In his theorems, Stevin left out the "construction" of each demonstration. But Stevin's questions made explicit his understanding that a sequence of arithmetical operations was a sort of "construction," akin to Euclid's creation of the diagram (see Figure 1.2).

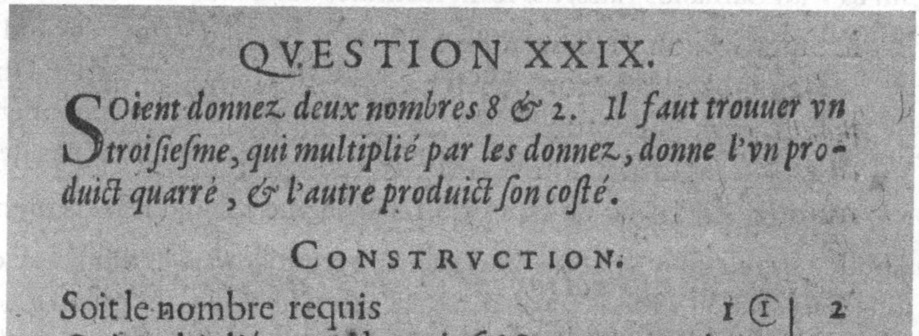

QVESTION XXIX.

Soient donnez deux nombres 8 & 2. Il faut trouuer vn troisiesme, qui multiplié par les donnez, donne l'vn produict quarré, & l'autre produict son costé.

CONSTRVCTION.

Soit le nombre requis	1 ①	2
Qui multiplié par 8 donné, faict pour premier produict	8 ①	16
Et multiplié ledict nombre requis par 2 donné, faict pour second produict	2 ①	4
Qui doibt estre la racine du premier produict, ergo son quarré	4 ②	16
Est egal au second en l'ordre	8 ①	16

Lesquels reduicts, 4 ① seront egales à 8, & par le 67 probleme 1 ① vaudra 2.

Ie di que 2 est le nombre requis. *Demonstration.* Multiplié 2 par 8, faict 16, & le mesme 2 multiplié par le 2 donné faict 4, qui est racine quarrée, du quarré 16 selon le requis; ce qu'il falloit demonstrer.

Figure 1.2 Stevin constructs a number.
Source: Simon Stevin. L'Arithmetique (Leiden: Christophe Plantin, 1585), 409. Call number *NC6 St485 585a, Houghton Library, Harvard University.

The right margin of the construction makes explicit the running total produced by the series of procedures. Stevin used print formatting to show the progressive construction of the sought value whose veracity could then be demonstrated by plugging it back into the statement of the question.[32]

The French humanist Guillaume Gosselin, a well-connected courtier whose interests ranged from scholarship to pedagogy, gave "arithmetical demonstrations" for the rules he offered in his 1577 *Books on the great art or on the hidden region of numbers, which is vulgarly called algebra and almucabala*.[33] The term would catch on: in his 1657 *Mathesis Universalis* Wallis would use it in a theoretical discussion of arithmetical, geometrical, and algebraic ways to demonstrate books II and V of Euclid's *Elements*.[34] Using a term employed in Petrus Ramus' *Algebra* that (as we saw) Xylander also used to generalize Diophantus' results, Gosselin called his rules "canons."[35] Like Cardano's rules, Gosselin's canons are rules applicable to particular forms of the equation. Again as with Cardano's rules, the form identifies values that are used in calculating the root. In the second canon, a square is equal to some sides and a number ($x^2 = bx + c$). Take half the number of sides and square them. Add the number. To the side for which this new number is the square (or in modern terminology, to the square root of this number), add half the number of sides: now we have $\frac{b}{2} + \sqrt{\frac{b^2}{4} + c}$, an expression that should remind readers of the quadratic formula for $a = 1$ (the difference in sign follows Gosselin's way of writing the equation). This is equal to x, the value of the side. The canon is general; the form it depends on, here "1 Q is equal to sides and numbers," marks out the limits of the procedure's generality.[36]

But the arithmetical demonstrations are particular, and through them, canons carried out on particular numbers serve to show their own generality. Gosselin adduced the example "6 L P. 16 equals 1 Q": six lengths plus sixteen is a square.[37] In order to solve the example, Gosselin employed his canon, justifying each step of its application.

We divide 6 plus 2 A into A and 6 plus A, half of 6 plus 2 A is 3 plus A, therefore following the fifth theorem of the second [book] of Euclid, the product made out of A and 6 plus A with the square of the excess of 6 plus A over 3 plus A, is equal to the square of half of the whole thing—that is, 3 plus A, and the product made out of A and 6 Plus A is 16, as is having to be demonstrated.[38]

Gosselin's arithmetical demonstrations could appeal to explicit axioms no less than to Euclid.[39] But these were not demonstrations whose validity rested on formal generality, for the particularity of the worked-out canon did not impede the demonstration from being general.[40] Arithmetical writers evidently believed that demonstrations based on particular cases were sufficient to produce belief.[41]

Visual and Noetic Means of Understanding

During the sixteenth century, algorithmic and geometric constructions resembled each other. On one hand, interest in both practical mathematics and mathematical pedagogy supported an explicit materialization of geometric construction. The materiality of constructions was reflected in the rising importance of mathematical instruments.[42] On the other hand, formalizing within arithmetic the use of Indo-Arabic numerals and other signs, as well as written procedures for their manipulation, involved making algorithmic constructions as visible as geometric ones. This section identifies two kinds of rhetorically inflected algorithmic demonstrations and distinguishes the different techniques of persuasion that they use.

Early modern education reformers in general sought new ways visually to organize and present classroom material in order to make it easier to navigate and learn, and mathematics was understood as a model in this regard.[43] Mathematics teachers emphasized the necessity of visual organization of various kinds for promoting understanding. Pierre Hérigone, author of a multivolume *Cursus Mathematicus* or "course in mathematics" printed double-column in both French and Latin, claimed to introduce a system of "notes" or signs into Euclid's *Elements* that would "without using any language" present each further statement "explained and allowed from the premises" that have come before.[44] For the English author William Oughtred, one of the great advantages of symbolic algebra was its ability to make arithmetical operations visible: symbols neither rack the memory "with multiplicity of words, nor chargeth the phantasie with comparing and laying things together; but plainly presenteth to the eye the whole course and processe of every operation and argumentation."[45] Making these processes transparent would continue to occupy certain mathematical writers later in the seventeenth century.[46]

Attempts to visualize mathematics drew on a tradition of Ciceronian rhetoric that figured centrally in sixteenth-century algebra. As late as 1640, Hérigone repeated a commonplace argument of the earlier rhetorical-mathematical tradition: he claimed that his new notes managed the difficult but necessary goal of joining "brevity and clarity" in a single pedagogical method.[47] In the decades around 1550, the algebraist and poet Jacques Peletier drew from Ciceronian rhetoric key concepts that served to organize

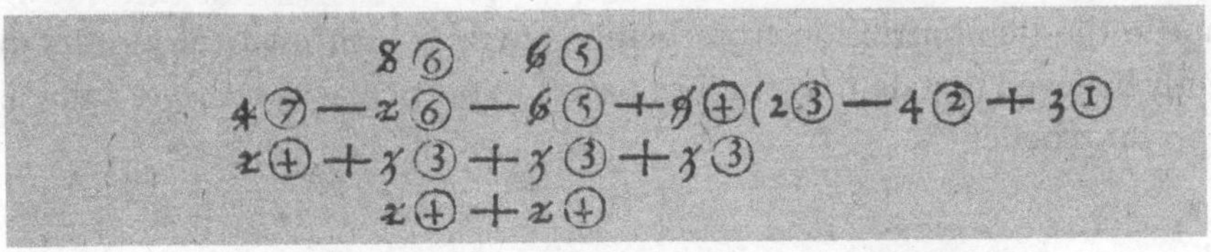

$$8\,\textcircled{6} \quad 6\,\textcircled{5}$$
$$4\,\textcircled{7} - z\,\textcircled{6} - 6\,\textcircled{5} + 5\,\textcircled{+}\,(2\,\textcircled{3} - 4\,\textcircled{2} + 3\,\textcircled{1}$$
$$z\,\textcircled{+} + 5\,\textcircled{3} + 5\,\textcircled{3} + 5\,\textcircled{3}$$
$$z\,\textcircled{+} + z\,\textcircled{+}$$

Figure 1.3 Long-division algorithm for polynomials.
Source: Simon Stevin. L'Arithmetique (Leiden: Christophe Plantin, 1585), 234. Call number *NC6 St485 585a, Houghton Library, Harvard University.

algebra as a new kind of discourse, notably "invention" and "disposition."[48] For Peletier, algebra served as an alternative to traditional modes of geometric presentation, one that involved only the "points necessary to resolve a difficulty."[49] A related attempt to use algebra to streamline geometrical argument is evident in the written constructions through which Stevin incorporated algorithms into the Euclidean demonstration scheme. To wit, he described as a "disposition" of numbers a long-division algorithm that could be used to divide polynomials (Figure 1.3).

First given numbers were disposed; then the parts of the quotient could be disposed, following theorems given earlier in the text, until the final quotient was reached.[50] For Stevin, disposition was the algorithmic practice that, in parallel to the construction of a geometric figure in a traditional geometric demonstration, manifested the sought quantity.

Because they are not visualized, meanwhile, Gosselin's canons instance another way in which algorithmic practice gave rise to understanding. Disposition was not unknown to Gosselin: elsewhere in his *Great Art* he showed readers how to set up an array in order to keep track of the many interim quantities produced by the rule of double false hypothesis.[51] But the arithmetical demonstrations of Gosselin's canons were supposed to produce understanding—or at least belief—without explicit reliance on visuality or other rhetorical resources. And on its face this seems like a problem for arithmetical demonstration in general, since particular numbers evidently lack the indeterminacy of magnitude that makes induction from particular geometric constructions to general geometrical theorems seem plausible to us.[52] At least some sixteenth-century algebraists understood their demonstrations to depend on rhetorical *loci communes*, commonplaces that they associated with internalized common notions.[53] This observation may help explain why the purpose of demonstrations was to produce belief, a classic goal of classical rhetoric. But it leaves unspecified how particular instantiations of more general axioms served to evidence general canons. Looking at Gosselin's particulars sheds light on how, within the framework of algebraic demonstrations' dependence on common notions, algorithms could produce belief. A posteriori, Gosselin's canons also shed light on how appeal to common notions worked.

Gosselin sandwiched the "demonstration" of each canon between two other parts. The first part introduced the rule, while the third provided the "use" of the rule for the resolution of a "problem."[54] All three parts involve particular, always different examples. The example in the demonstration is consistently of medium difficulty between the extremely simple example in the statement and the difficult example in the problem. For Gosselin, "uses" of the rule frequently involve fractions and three-digit numbers.[55] "Demonstrations" never do: here Gosselin prefers relatively small whole numbers, as in the example discussed above: "six lengths plus sixteen is a square." Meanwhile, the examples in the statement of the rule are even smaller: "A square equals four times its side plus twelve."[56]

Since the example of median difficulty is neither more nor less true than any other example, its suitability for the demonstration must come from elsewhere. Given Gosselin's consistency in choosing the difficulty of his examples, it seems that Gosselin tailored the difficulty of each to its purpose within his presentation. The very simple example given in the statement allows the rule to be immediately grasped as plausible. The most difficult example given in the form of a problem allowed Gosselin to show off the application of the rule in a case that exceeds unaided human powers of numerical reckoning, at least for most people. Meanwhile, I suggest, the example of median difficulty renders the demonstration more persuasive than either alternative would.

Using the middle example, the right reader can verify the procedure by following its steps in the mind. The level of difficulty of the median examples is low enough that the reader can verify each step rather quickly. Meanwhile, the difficulty is not so low that an unpracticed reader can see the answer immediately without working through the procedure. Such examples could be verified in the course of the demonstration, since the reader could compare his or her results against the intermediate sums and products Gosselin provided. The examples are thus persuasive in the sense that their truth is made available to the reader in an inward way, through the intellectual use of the procedure.

The notion of inward persuasion is consistent with a broad Platonist trend in the epistemology of mathematics adduced in sixteenth-century French arithmetic and algebra books. A vague but vividly illustrated Platonist epistemology often coexisted, in the same textbook, with efforts to ameliorate mathematical pedagogy through new formatting and rhetorical formalism and with correlated appeals to rhetorical epistemology. This vivid Platonism emphasized the interiority of understanding rather than the visualization of knowledge. Textbook authors drew on standard Platonic *topoi*, such as the image of ascent from the *Phaedrus* and the image of conversion from the *Republic*, to frame mathematical learning as the best way that students could first experience intellectual transformation. Some curricular reformers thus assigned mathematics a central role in liberal education.[57] For certain mathematicians, meanwhile, mathematical learning cultivated souls and gave direct, experiential knowledge of metaphysical principles.[58] In a somewhat hyperbolic adaptation of Plato's allegory of the cave, Peletier claimed that "there is no speculation that can serve man with a more spacious country to frolic, to explore his thoughts, to draw him outside of himself and then assume himself again, than the university of numbers."[59] Experience with mathematical things led, he believed, to self-knowledge and understanding.

The organization of Gosselin's canons, clearly, does not advance so comprehensive a view of the relationship between mathematics and philosophy. But Gosselin's use of examples reflected his own understanding that mathematical education aimed to cultivate inward understanding. The persuasiveness of Gosselin's arithmetical demonstrations depended on this previously cultivated understanding. A contrary emphasis on the generality of the mathematical object in Gosselin's *Renewed lecture* reflects the absence of such an understanding: that text is pitched at beginning arts students without meaningful familiarity with mathematics.[60]

Gosselin's commitment to the mind contrasted with the emphasis on visualization evident elsewhere in the algebraic tradition. Hérigone's method of notes, for instance, served not just to visualize the steps of a demonstration but to visualize *all* the steps. "There is no doubt that understanding is easier than with the common way of demonstrating," he wrote, "since in my method nothing is asserted unless it is corroborated by some citation. Other authors do not exactly observe this, but, measuring the necessity of citations from those things which seem open or obscure to them, they use many consequences without any citations, which nevertheless everyone knows are of great help to uneducated and less exercised [readers]."[61] Whereas Peletier and others turned to rhetorical algebra in order to cut away the fat from geometric demonstration, Hérigone wanted to bulk up the demonstrations that he expressed using symbols. Only such an exhaustive approach would make it easy "to convert" or "to resolve into syllogisms" each line of a given demonstration.[62] For Hérigone and other early modern textbook writers, making it explicit—through notes or other formalism—served pedagogy, no less than new modes of writing like Stevin's dispositions served effective practice. Both could serve as aids to those who lacked understanding, since canons and rules can be used without understanding their reasons, as they still are today by students of high school algebra. Perhaps these sorts of writing furnished a kind of knowledge that did not implicate the self. Meanwhile, Gosselin's canons depended on a firm grasp not just of common notions but also of how to use them to calculate. In the context of the *Great Art* this included experience with, and knowledge of, the behavior of numbers as well.

Understanding Algebra

By way of conclusion it will profit us to look briefly at the mathematics of François Viète. Like his contemporary Gosselin, Viète was inspired by the recovery of Diophantus' *Arithmetic*; unlike Gosselin, he took Diophantus' study of higher powers of the unknown in a mostly new direction.[63] Viète set out to understand indefinite equations, which he understood as proportions between "species" of quantities (the same sides, squares, and cubes that we have already seen) and which he represented using a combination of abbreviations, short words, and symbols reminiscent of Diophantus' own text. Viète's interest in and success at studying equations has led many scholars to see Viète's mathematics as a major step in the direction of contemporary mathematics. And this is true regardless of how they interpret Viète's algebra, which often means what ontology they attribute to his species: whether they think that Viète's algebra identifies some new object common to arithmetic and geometry or that it, rather, is an object-independent mathematical tool along the lines of the *trivium*.[64] Regardless of the interpretation, however, Viète's symbolic mathematics can hardly be depicted as purely formal and automatic. For it involves all three cognitive practices we have seen: form as an indicator of generality and its limits, algorithm as a means of demonstration, and a dependence on inward understanding for justification.

Viète drew on Diophantus to distinguish multiple "zetetics," finding aids, based on different givens.[65] Like Cardano's rules and Gosselin's canons, Viète's zetetics had their own domains of applicability. Once the right zetetic was identified, it could be used automatically to find the solution to a given problem. But Viète's justifications were not automatic. Seventeenth-century algebraists would resolve their equations through piecemeal manipulation, raising and lowering by degree until they arrived at a solution.[66] Viète's *Introduction to the Analytic Art* set out rules for doing just this.[67] But Viète did not always use such a piecemeal approach in his demonstrations. Instead, the demonstrations could employ relationships between different problems by referring the reader back to earlier zetetics. This reference practice instanced Viète's reliance on inward understanding for justification. Unlike Hérigone, he didn't spell out in great detail how to get a new demonstration from an earlier one; instead, he referred generally to operations that he expected his reader to know how to use.[68] Treatises on more advanced topics were pitched at a higher level, and Viète's use of earlier finding aids in new demonstrations was even less explicit.

Finally, with Viète we get a near-total fusion of algorithms and demonstrations, since the finding aids are at once part of the discourse and the practice. Like the relationship between a Euclidean figure and its text, Stevin's dispositions were printed in parallel to the language that explained them. If one were actually to divide an equation using Stevin's method, the justificatory language would be unnecessary; one could simply take recourse to the operative disposition. Viète exempted himself from using such a language. Although the parallel to Diophantus' *Arithmetic* might suggest it, Viète's freedom from the double-discourse was not simply a matter of symbols and abbreviations. It also derived from Viète's willingness to rely on the reader's powers of recognition even when he silently used earlier finding aids to transform one formal expression into another. The persuasive aspect of Viète's symbolic discourse depended on the recognition of equations, on the interaction between forms as markers and inward understanding.

To return to our orienting question: does the use of algorithms require humans to behave like machines? Of course, answering this depends on how (and whether) machines behave. But what's clear is that sixteenth-century algorithms offered many possibilities for use and understanding, from the automatic to the apprehending. I can make apposite comparison to two other chapters in this volume. Michael Barany next argues for the importance of experience and discernment to the correct use of early modern algorithms. In the examples we've seen, discernment is key to recognizing which algorithm to use, and experience plays a central role in understanding demonstrations and so reasons. But the kind of algorithmic facility with the rule of three that Caitlin Rosenthal finds in the Atlantic business life c. 1800 (see chapter 5) was also one possible outcome of an encounter with sixteenth century algebras.

Nor is this diversity limited to algorithms. In their foundational work on humanist education, Anthony Grafton and Lisa Jardine identified the emergence of the "humanities" with Ramus' decision to shift the emphasis of his pedagogy from cultivating the soul to inculcating practices,[69] as though different outcomes and different kinds of understanding were not always possible even within a single classroom, let alone from a whole pedagogical movement. The history of how algorithms have been understood should remind us that when it comes to understanding, there are degrees.

2

"Some Call It Arsmetrike, and Some Awgryme"

Misprision and Precision in Algorithmic Thinking and Learning in 1543 and Beyond

Michael J. Barany

From Arses to Algorisms

Robert Recorde's 1543 arithmetic primer *The Ground of Artes* taught anglo-phone readers the art and science of numbers and their innumerable uses, promising in its subtitle an "easier & exacter" exposition "moch necessary for all states of men."[1] While not the first English vernacular arithmetic, the text and subsequent editions revised and expanded by Recorde himself and by other notable authors stood out at a time of burgeoning vernacular printed science and remained in print for more than a century and a half.[2] Historians have long regarded the work as a significant milestone in the histories of education, numeracy, and arithmetic.[3] Often erroneously credited with inventing the "=" sign for equivalence,[4] Recorde incontrovertibly left a lasting mark on pedagogy, terminology, and symbolism through his coinages, explanations, and popularizations. Recent scholarship has situated *The Ground of Artes* (along with Recorde's other works) in the histories of reasoning and compu-tation, as well, with a 2011 Recorde biography even appearing in a History of Computing book series.[5]

The book's Pembrokeshire-born and Oxbridge-educated pun-plying poly-math author, one of British mathematics and science's most important early vernacularizers and expositors, chose his words with care and wit.[6] The book's opening dialogue begins with the Master impressing upon the Scholar (with a clearly intended pun) numbers' "unnumerable" "co[m]modities" as "the ground of all me[n]ns affayres" (1ᵛ). Summarizing in verse, "yf nombre be lackynge, it maketh men dumme, so that to most questions, they must answere mum" (2ᵛ). After an extended exchange on the value of numbers, the Master asks the Scholar, "what call you the science, that you desyre so greatly"

Michael J. Barany, *"Some Call It Arsmetrike, and Some Awgryme"* In: *Algorithmic Modernity*. Edited by: Morgan G. Ames and Massimo Mazzotti, Oxford University Press. © Oxford University Press 2023. DOI: 10.1093/oso/9780197502426.003.0003

(6^r). Despite having named it unproblematically moments earlier in the dia-logue (3^r), and having encountered the term no fewer than eleven times from the Master in the interim (3^r–5^r), the Scholar muffs the quiz: "Some call it Arsmetrike, and some Awgryme" (6^r).

Word choices like these carried significant implications in Recorde's writing, and the author's attention to etymology and flair for coinage allow one to im-pute meaning even in errant phrases offered so as to be corrected. Variants of both of the Scholar's terms can be found in English manuscripts dating to the fifteenth century and earlier, derived through multiple renderings from Greek (by way of Latin) and Arabic, respectively, and one scholar has suggested a Welsh influence in Recorde's latter term, as well.[7] If they were not neces-sarily in the air, neither were they pulled from nowhere. To those who may have encountered these names in the past, the Master's correction set them right. "Bothe names are corruptly writen," asserts the Master, "Arsmetrike for Arithmetyke (as the Grekes call it) and Awgrym for Algorisme (as Arabyans sounde it) whiche bothe betoken the science of nombrynge" (6^r). Indeed, the Oxford word oracle places Recorde's 1543 text at the culmination of a centuries-long chain that "gradually corrected" *arsmetrike* to the cur-rent standard of *arithmetic*, and notes a fifteenth-century arithmetic text that contrasted *algorym* to the "lewder use" of *Augrym*.[8]

Those seeing the Scholar's corruptions for the first time could still rec-ognize something in the words. The first presented a pun: science of num-bers, meet the mensuration of arses.[9] This pun had a distinguished pedigree in English letters, figuring pivotally in the *Canterbury Tales*, where Chaucer invokes "ars-metrike" in the Summoner's Tale in face of the problem of di-viding a fart equally.[10] The second misprision looked strange, and the Master's correction in terms of how it was "written" draws attention to this alien or-thography. But to appreciate the full measure of its strangeness one should try to say it. As "the Arabians" say it, algorism is new and foreign. As the Scholar misconstrues it, the term is something of a mumblesome mouthful, a kind of garbled speech well matched in the period's rhetorical traditions to the inar-ticulate expressions associated with the arse.[11]

At bottom, Recorde propounded arses and mumbles to make a point about pedagogy, a point this chapter shall interpret through Recorde's algorithmic-arithmetic practice in *The Ground of Artes*. Recorde's preface jus-tifies using the "fourme of a dyaloge, bycause I judge that to be the easyest waye of enstructio[n], when the scholer may aske every doubte orderly, and [the] mayster may answere to his questio[n] playnly."[12] This "reason of ryght teachynge" constitutes a claim about how to learn and think. *The Ground of Artes* proceeds dialogically through mistakes and corrections, training readers to figure correctly while also insisting that right thinking comes from practice, discernment, and mastery, not strict uncomprehending adherence to fixed procedures. While related to later evolutionary notions of trial and error, themselves joined by a winding genealogy to early modern arithmetic,[13] Recorde's dialogical didactic form should be understood in the context of customs and assumptions from early modern technical and philosophical education, informed in part by interpretations of Classical dialogues like Plato's *Meno* where mathematical lessons are imparted by reasoning past initial errors. This early modern pedagogical history suggests a long view of the relationship between algorithms and learning, linked by judgment and reason, rather than opposed in fantasies of mechanism that evacuate expertise from algorithms' operation.

So long a view, it would appear, that in some ways it is circling back on itself: one of the most significant new areas of algorithmic thinking today—data-intensive machine learning—revolves around coaxing digital computers into something very much like what the Master wishes of the Scholar in Recorde's 1543 arithmetic. Eschewing principled, determinate, deductive rules in favor of adaptive and often inscrutable webs of associations,[14] machine-learning systems must be trained on a ground set of prepared data and are expected, at least in the beginning, to make basic, sometimes embarrassing, mistakes. If machine learning is the most striking (and perhaps urgent) current parallel to Recorde's didactics, it is hardly the only one. Like the early modern "algorism" that preceded it, more recent modern algorithms derive from habits and principles of procedural learning and practice whose customary naturalization can obscure the essential operation of human agency and error. Notwithstanding modern ambitions for error-free computing through formally rigorous hierarchical design, to code is to tinker, to debug.[15]

Such agency and error are central, even foundational, to Recorde's text. Here, mistakes are generative elements of pedagogy necessary not just to learning but to systematic reasoning. Each error carried specific lessons about reckoning well and general lessons about how to think. Making errors central and productive, rather than exceptional and regrettable, gives a striking and fruitful analytic purchase on algorithms past and present.

Names, Reason, and Authority

The Master and Scholar's exchange about names comes at a pivot in Recorde's opening dialogue. Having convinced the Scholar of the merit of numbering, the Master exchanges with the Scholar a bit of methodological verse.

Recorde's books often break into verse, and almost as often—as in the present example—the publisher composed the text as undifferentiated prose. Listen, rather, for the rhyme:

> S. And I to your auctoritie, my wyttes do subdewe: what so ever you say, I take it for trewe. M. That is to moch, and mete for no man, to be beleved in al thynges, without shewynge of reason. Though I myght of my Scoler some credence requyre, yet except I shewe reason, I do not it desyre. (5ᵛ)

Recorde's dialogues hinge on the Master's exercise of authority, but the Master's authority must be properly derived. The Master's experience becomes evident in response to the Scholar's ignorance, discernment in response to error. The Scholar has to get it wrong before the Master can show how to get it right. No precision without misprision.

The Master's reasoning begins with naming. When the Scholar first resists the Master's question, the latter insists, "For greate rebuke it were, a science to have studied, and yet can not tell how it is named" (6ʳ). That is, one must know what one studies. Such knowledge fits one's learning into a systemic whole: here, the lesson on names continues, through naming, into a presentation of the "kyndes" or "partes of nombrying" (6ᵛ)—addition, subtraction, and so on—which serves as both a conceptual survey and an outline of the book. Before learning rules and procedures, one must be able to see what they do and how they relate to others. In the Master's words, "Fyrste you muste knowe what the thinge is, and then after learne the use of the same" (7ʳ). Without that context, arithmetic and algorism lack independent meaning, inscrutable as farts or mumbles.

Readers of Recorde and other early print vernacularizers learn to look closely at word couplings, which typically pair a coinage derived from another language with a more familiar near synonym.[16] The pattern holds for "arithmetike," which according to the Master comes from "arithmos in greke," meaning "nomber" (6ʳ). The pair in the usual scheme to arithmetic would not be algorism but numbering, which Recorde places at the end of the sentence defining both arithmetic and algorism. Matching arithmetic instead to algorism in the vernacular pair, Recorde did more than just indicate the presence of multiple equivalent terms with distinct etymologies. Arabian algorism's brief mention in *The Ground of Artes* serves at least two specific expository purposes, asserting numbers' novelty and explaining their oddity.

In the chapter on numbering, Recorde suggests both purposes, writing (as Master): "I mought here shewe you, who were the fyrste inventours of this arte, and the reasons of all these thinges that I have taught you, as why you shuld recke[n] your order of places backewarde, I meane from the ryght syde towardes the lefte, with many other thynges, touchynge the causes and reasons of it" (13ᵛ–14ʳ). Naming the art of numbering with the Arabic algorism, Recorde establishes its place outside of and postdating the Classical sciences. This genealogy gives reason to numbering's peculiar terminology and conventions, such as placing the "first" [smallest] part of a number on the right. At the same time, the attribution situates numbering's users in a distinctly modern relation to Classical knowledge.

The Master never gives the promised explanation of the art's first inventors and the reason for reckoning backward in the first edition of the text. However, the 1558 and subsequent editions include an extended attribution of algorism to "the Chaldays," who "did set these figures as thei set all their letters. For they wryte backwarde as you tearme it, and so doo they reade."[17] The same right-to-left order, the Master notes, "may appeare in all Hebrewe, Chaldaye and Arabike bookes." This satisfies the Scholar, but the Master continues by saying that "the Caldays and Hebrues do not so use their owne numbres," implying (but not stating) that Arabic algorism is the necessary explanation for numbers' ordering. Moreover, the Master adds, English pronunciation from left to right, starting with the largest number, puts the numbers in the same order as the Arabic writing from smallest to largest, right to left. This coincidence settles the matter, and the dialogue continues to the next chapter, Addition.

The work of misprision did not end with the Scholar's initial invocation of awgryme. A second instance reinforces the corrupted term's connection with both the process of learning through correction and the problem of place value in numerical notation. This latter misprision takes place in the chapter on numbering, after the Master asserts that the numeral figure 9, when unaccompanied by other figures, always represents the number nine. The Scholar correctly affirms "Then 9 without more figures of Awgrym[18], betokeneth .ix.," but immediately shows his ignorance when, asked to interpret "this example 3679," he matches the figure 7 to the number .vii. (rather than .lxx.) and so on (9ᵛ–10ʳ).

Where the Scholar uses the wrong name, it reinforces that the Scholar has also mistaken the concept and usage. As commodious as the figures of algorism will prove in *The Ground of Artes*, they harbored their share of pitfalls and counterintuitive uses. The Scholar's misunderstanding of place value, a problem distinctive to the new notation under discussion, lets Recorde portray the dangers of unthinking number-work. In a very direct sense, the figure's context (in a sequence of figures) determines its meaning (as a numerical value), just as contexts will determine mathematical and other kinds of meaning for the rest of the text. The error shows that even the most apparently transparent instructions are not inevitably applied correctly, that not just their mechanics but also their contexts and rationales must be learned and practiced.

Putting Division in Its Place

Long division has long been an iconic, even notorious, part of elementary mathematics curricula. To the extent Recorde's book marked a starting point for such curricula in the English vernacular, it has been there from the start. The procedure's laborious and difficult-to-motivate litany of rules and prescriptions about how to align and manipulate numbers vex and bore students and teachers alike. In recent years, long division has been a lightning rod of sorts for debates about how to teach children to perform arithmetic, think algorithmically, or develop mathematical intuition or creativity.[19]

Recorde's method was not quite the same one most readers of this chapter will have learned in primary school, but it bears a strong family resemblance. Working from the greatest place values to the least, one finds the quotient digit-by-digit by comparing the divisor to the greatest remaining place values of the dividend, subtracting the largest possible round multiple of the divisor from the dividend at each step. The procedure uses the properties of the place value system to break one large division problem into a sequence of simpler problems of division, multiplication, and subtraction.

As in primary school lessons today, this general strategy and its mathematical justification are not spelled out as such in Recorde's text. Algorism was an efficient symbolic means for finding the right answer, not explaining why it was right. Justifications and philosophical principles—including for numerical questions—were matters for geometry, not arithmetic. In Recorde's arithmetical context, understanding was a matter of learning the correct circumstantial application of the method. This, in turn, meant attempting and then learning to avoid incorrect misapplications, and these errors and misprisions show Recorde's sixteenth-century perspective on a centuries-old procedure.

The text's chapter on division follows a set pattern established previously for other basic operations of arithmetic. The dialogue begins with a demonstration by the Master of a simple example, spurred along by queries from the Scholar. The Master marks each step—where to look, what to calculate, what to write, where to write it, what to call it, and so on—and refers to a sequence of inset figures showing the calculation in progress. Then the Scholar works a more difficult example with interjections from the Master to rectify typical misunderstandings. After these, the sequence repeats with the same numerical examples for a method of checking the outcome of the calculation.

The first error, in fact, tells us more about the text's composition for print than about its author or pedagogical philosophy. The Master begins by explaining how to align the numerals of the dividend and divisor in order to begin the procedure, with the last (leftmost) figures aligned except when the divisor's leftmost figure exceeds that of the dividend, in which case the divisor should be offset by one position to the right. The Master describes the difference by contrasting the quotient of 365 (the number of days of a year) by 28 (the days in a "co[m]men moneth"), by 52 (the weeks in a year), and by 4 (the number of quarters of a year) (62^r). Following the just-described rule, the "2" of 28 would be written under the "3" of 365, whereas the "5" of 52 and the 4 would be written instead under the "6" of 365. These examples involve both of the two possibilities for alignment while opening considerations of possible misunderstandings—for instance, mistakenly applying the rule to the "2" from 52 instead of the "5," mistakenly expecting the "8" of 28 to be smaller than the number above it, or in the shifted case mistakenly expecting 4 to be greater than the number above it after the shift (the "6" of 365). These considerations are not spelled out in the text, however, and the accompanying figures show the 28 erroneously lined up just like the 52, with the "2" under the "6" of 365. Coming in the Master's voice and without comment, this misalignment seems clearly to be a production error (one that remained uncorrected in the 1558 edition) rather than a pedagogical intervention. The alignment appears correctly and without comment on the next page and in the remainder of the calculation (62^v *et seq.*).

The first deliberate misprision comes in the second stage of the Master's example of dividing 365 by 28. Having established 1 as the first digit of the quotient in the first stage, the Master instructs the Scholar to align the divisor of 28 directly under the remainder of 85 from the first stage, so that the "2" of 28 sits under the "8" of 85. Naively following the rule just introduced, the Scholar announces that this 2 in the divisor may be taken 4 times from the 8 in the dividend. "Truthe it is," the Master replies, "that you maye fynde 2 foure tymes in 8," but (turning to the next digits) one cannot likewise find 8 four times in 5 (63^v). Thus, the second digit of the quotient should be 3, not 4, and the Master completes the calculation accordingly. The lesson here: one must look ahead in the calculation and adapt earlier steps accordingly.

Misunderstandings also allow the Master to rule certain considerations as out of bounds or premature. Finding a remainder of 1 after dividing 365 by 28, the Scholar suggests that one might "parte the 1 that remayneth into 28 partes" (64^r). As fractions and proportions had their place in a later part of the book, the Master here replies "That is well sayd, and so must we do in suche cases whe[n] there remayneth any thynge, but I wyll lette that passe nowe, and wyll make you parfecte in hole Divisyon, and wyll here after teache you peculyarly of broken nombre callyd fractions" (64^{r-v}). Proposing to continue the division beyond where the Master would stop, the Scholar shows a misunderstanding not of mathematical principles but of the proper stopping point for a procedure in its conceptual and pedagogical context. The Scholar's misstep previews and begins to justify a later element of Recorde's arithmetic while creating an opening to explain the appropriate place of the current element. The Master's correction, accordingly, asserts a distinction between whole division and fractional arithmetic, reserving the latter for a subsequent exposition.

At this point, the Master hurries the Scholar into the next example. But on recapitulating the result of 365 divided by 28 afterward, another problem surfaces. The Master summarizes the conclusion, "wherby I knowe that in a yere (which contayneth 365 dayes) there are 13 monethes, reckenyng 28 dayes (or 4 wekes) just to a moneth, and 1 day more" (66^{v}). "Why then," responds the Scholar, "do we call a yere but 12 monethes?" To this, the Master replies only that "now it is not convenient to entangle your mynde w[ith] other thynges the[n] do dyrectly pertayne to your mater," and so rules the question outside the scope of the lesson (66^{v}–67^{r}). The Scholar's insight that the arithmetical calendar invoked in the example contradicts the usual calendar becomes, here, a different kind of error: expecting a mathematical reason for an anomaly with non-mathematical causes (history, astronomy, religion, and so on). When learning arithmetic, the Master instructs the Scholar, one must keep to things that are arithmetical, and recognize when the answer you seek might require other kinds of knowledge and reasoning.

Rules without Practice

The second worked example of long division, with the Scholar at the helm, offers more opportunities for mistakes. The Master sets the problem of dividing 136,280 by 452, and the Scholar commits a first mistake from the very moment of transcribing the numbers in question. Having set the "4" of 452 under the "1" of the dividend, the Scholar—when prompted by the Master—admits that "I had forgotten" to "sette the dyvisor one place more forward towarde the ryght hande," since 4 exceeds 1 (64^{v}). Following the earlier printing error, this is the first purposeful mistake regarding alignment, signaling this as a notable sort of error for the proper use of place value. The

Scholar admits to forgetting so that the Master can insist on the importance of assessing each step and remembering adjustments and exceptions in place-valued arithmetic.

The Scholar completes the first stage of the division without further incident, but encounters a hitherto untreated exception when setting up the next stage. By pedagogical design, the Scholar finds upon shifting the divisor by one place to start the next stage that all the numerals above the leading digit of the divisor have already been canceled out in the first stage. The Master explains that in such situations one must "write in the quotie[n]t a cyphre 0" and adds "The reason of this wyll I showe you hereafter" (65^v). Though the explanation is deferred, two lessons are clear. First, when setting up a stage of the division one does not simply move incrementally to the right each time; rather, one must continue to look at the value remaining from the dividend. Second, as in other operations of arithmetic, ciphers index shifts in place value, a point the Master reinforces by instructing the Scholar to place a cipher for every time such a shift is needed.

The Scholar continues the calculation, eliciting a brief suggestion from the Master that taking 5 from 28 to leave 23 could just as well have been described as taking 5 from 8 to leave 3, with the "2" in the higher place unchanged. Stuck at the last step, which amounts to subtracting 2 from 230, the Scholar asks the Master "what shal I nowe do?" The answer is to "do as you learned in Subtraction in a lyke case" and borrow 1 from the 3 in the second place (66^r). Having successfully applied the principles of one kind of arithmetic to another, the Scholar completes the division and the Master summarizes the result (66^v).

One successful quotient alone did not teach the method, however. The Master instructs the Scholar that "you have lerned a shorte maner of dyvisyon, whiche I wolde have you often to practyse, so that you may be parfecte in it" (67^r). Here, the Master evokes an earlier invocation from the chapter on Numeration:

> I wyll yet exhorte you now, to reme[m]ber both this, that I have said, and all that I shall saye, & to exercyse your selfe in [the] practise of it: for rules without practise, is but a lyght knowledge, and practise it is, that maketh men p[er]fecte and prompte in all thynges. (14^v)

It lacks the pith of the adage that practice makes perfect, but the message is unmistakable. In *The Ground of Artes*, just knowing the rules without having perfected their application through repeated practice can only be a light, superficial knowledge. The exposition bears this out repeatedly: simply following a rule, failing to place it in its conceptual and procedural context, leads again and again to mistakes and misunderstandings. Algorithmic thinking required the discerning execution of methods learned and practiced at length, with skill and judgment.

With rule-based calculation so evidently error-prone, Recorde has the Scholar worry about how to "examyne and trye my worke, whether I have done well or no," without an expert on hand to offer corrections (67^r). "Sume men (yea and co[m]menly) do trye that," the Master offers, "by the rule of nine," suitably adapted to division. The rule of nine first appears in the chapter on Addition, when the Master instructs the Scholar that "it is often practise that maketh a man quyke and rype in all thynges. But bycause of suche greate summes, in whiche there may chau[n]ce to be some errour, I wyll teache you, how you shall prove whether you have done well or no" (23^v). (Here and elsewhere, "prove" and "proof" have the sense of a trial or test—as in "the proof of the pudding is in the eating" or "proof-reading"—not, as typically used in mathematics, of a rigorous demonstration of certain validity.) The procedure involved treating all the numerals in the calculation "as though they were all unities," ignoring place values and adding them together as units. Whenever the sum "encreaseth above 9, cast away 9" and continue. If the summands and the sum yield the same result after such a calculation, "then have you well done, but yf they be unlyke, then have you myssed" (24^r).

A brief digression into twentieth-century mathematics may help account for the method's principles and limitations. In anachronistic terms, the procedure applies the canonical surjection from the ring of integers to the ring of integers modulo 9. Since this is a ring homomorphism and $10 \equiv 1 \bmod 9$, this amounts to adding the digits on both sides of the equation and reducing modulo 9 as one goes. The method allows one to catch an error of addition provided that the net of all errors in all the digits of the calculation is not itself a multiple of 9. The method is not as uniformly effective for division problems since there are more ways for errors to compound as multiples of 9—at one extreme, the method says nothing about the quotient when the divisor is itself divisible by 9—but it was certainly quicker than the alternative of checking directly by multiplication, and mostly reliable. Such considerations, however, did not enter into the dialogue in *The Ground of Artes*.

To check division with the rule of nine, the Master revives the cross, a device introduced in the Addition chapter for checking sums involving multiple denominations (e.g., of currency; 34^v–35^r).[20] Angled in an "x" shape, the cross displays numerals as two pairs—left and right, top and bottom—for ready manipulation and comparison. Here, the cross organizes the four constituents of division: the dividend, divisor, quotient, and remainder, the latter implicitly by way of an intermediate calculation. Keeping to a third-person description of what "sume men … do trye," the Master describes first casting out nines from the quotient and divisor, noting the results on the left and right of the cross respectively. The Master then multiplies the results together and adds the remainder from the division, casts out nines again, and places the result at the top of the cross. Upon casting nines from the dividend and noting the result at the bottom of the cross, one passes the test if the top and bottom of the cross agree. Significantly, after the Master demonstrates the method by checking the result of 365 divided by 28, the Scholar then performs the corresponding check for the calculation of 136280 divided by 452 without mistake.

Or rather, without computational mistake. When the Scholar proudly declares that with the rule of nine "I knowe that the division was well wrought," the Master admonishes that the rule is only "the co[m]men profe" and that "the more certayne workying" would require inverting the calculation with a full process of multiplication (68ᵛ). The Scholar's error, here, is one of overconfidence in common tests in place of those that, while more cumbersome, are also more certain. Even when one has followed rules correctly, one must know what those rules mean and what they imply for subsequent assessments and calculations.

Farting for Modernity

Which brings us back to arses. That is, to the relationship between naming, mistaking, and algorithmic thinking in early modernity. Recorde presented algorism in 1543 as a fruitful and dynamic approach to solving mathematical problems, one requiring dedicated practice and experience to execute well. Learning to think algorithmically required acquiring habits of rule following premised on each rule's sensitive dependence on context and judgment. Context and judgment, in turn, hinged on naming, of knowing whereof one spoke. That is why, in *The Ground of Artes*, the commodities and names of the various parts of arithmetic came before the practice of those parts.

Before getting even arithmetical names right, it was necessary to get them wrong. Algorithmic reasoning was creative, alive with possibility. Misprision marked that creativity pedagogically by demonstrating the possibility of miscreating while validating the importance of experience and offering a means to attain it. Through mistakes, adaptations, practice, and discernment, one could learn to recognize and resolve the multifarious configurations of numerical arithmetic. Getting words wrong was thus a prelude to getting numbers and procedures wrong, and all of these kinds of mistakes showed how to get arithmetic right. Recorde's dialogical exposition proceeded by error and correction because this was the essence of arithmetic, its source of power.

Writing in dialogue, Recorde eschewed the other dominant form of early modern mathematical exegesis: the orderly and methodical deduction associated with Euclidean geometry. Indeed, Recorde's own 1551 book of Euclidean geometry, *Pathway to Knowledg*, kept to the Euclidean mode of sequential constructive exposition even as it subverted other elements of the genre.[21] For most early modern authors, Classical geometry stood, to varying degrees of explicitness, in constant contrast to the newer Arabic mathematical arts of algebra and algorithm.[22] Arithmetic, associated not just with post-Classical civilization but also the worldly preoccupations of commerce and money,[23] was always overtly or implicitly a departure from the stale intellections of Ancient minds.

Pedagogical dialogues, by their very construction, are polyphonous texts, bringing together multiple forms of speech. Valerie Allen indicates that the broken, profane, multisensory expression of farting was very much a part of late medieval and early modern conceptions of language.[24] Farts linked to laughter, puns, incantations, and other linguistic productions that sat opposite the reasoned forms of trivial grammar and rhetoric. There were, in the period's rhetorical, medical, and alchemical theory and discourse, multiple meaningful ways to expel air.[25] In its hybridity, excess, and alienation, the boundary-troubling experience of farting marks, in Allen's analysis, a mode of transition for modernity itself, one she compares to the transition from trapped wind to passed gas via the anal threshold and bifurcated cheeks.[26]

Chaucer's ars-metrike sat on this kind of epochal precipice, parodying number-obsessed early modernizers like the so-called Oxford calculators of Merton College.[27] The ontology and epistemology of geometrical partition in the Summoner's Tale—situated in the devil-inspired construction problem "That every man sholde have yliche his part / As of the soun or savour of a fart?"[28]—implicated fundamental questions about the nature and divisibility of numbers and measures.[29] A close reading of Chaucer's ars-metrike, in this light, places this farting arithmetic at the cusp between an older model that sought to resolve these questions through Classical and canonical authority and an emerging model that favored experiment and experience.[30]

This understanding of Chaucer's ars-metrike and its more explicit and better studied mass of allusions allows a more capacious interpretation of Recorde's Scholar's misprision of arithmetic as arsmetrike. If Classical geometry united head and hand—reasoning and drawing—in a robust authoritative tradition, early modern arithmetic found its most honest expression in a baser organ, with its characteristically modern manner of speech. Recorde makes a virtue of the Scholar's persistent penchant for mistaking, forgetting, and misapplying—not quite speaking in farts, but perhaps speaking after the manner of farting. These misprisions give an experiential basis for knowledge that can penetrate beyond the "light knowledge" of rote procedure.

In the century after Recorde, playwright Ben Jonson gives the vain and oft-mistaken Sir Glorious Tipto in *The New Inne* (1631) the line:

> Fart upon *Euclide,* he is stale, & antique,
> Gi'me the modernes[31]

If the great mathematical minds of antiquity spoke in geometry, modernity belonged to the arses that farted algorism. Only in the seventeenth century, when the calculator-mad likes of Leibniz enrolled arithmetic in their fantasies of mechanical decidability, did algorithms begin to lose the scent of creativity and indeterminacy. Even then, philosophers and artisans entertained a tremendous variety of views on the relationship between mechanism and creativity.[32]

Across this history, formally deterministic mechanical rationality has always been the stuff of principles rather than practices. It is one thing to have rules and a system for counting to a billion, and quite another to carry it off with fingers and toes, words and tallies.[33] Discharge the task to a machine—say, the prolifically counting electronic computers that drive twenty-first century digital economies—and the result is more mediation and more credence in the unseen operations of counters that further defer human accountability. One way to understand modernity is in terms of a widening and deepening willingness to credit in-principle abstractions over context-laden practices.

Recorde's early modernity, by contrast, was one where principles were intimately bound up in practices and contexts. If algorithms seem now to evacuate judgment rather than demand it, that is only because the early modern embrace of algorism has been joined by a later modern move to write misprision out of calculation's essential operation. Even where context and contingency have been embraced as essential to the practice of reasoning, formalist fantasies have sterilized such practices of their error-bound roots. Nicolas Bourbaki, modern prophet of rigorous logical structures, called for mathematicians to exercise "experience and mathematical flair" in place of tedious idealized deterministic deductions.[34] For Bourbaki, a high modern confidence in the structural guarantees of in-principle calculation justified informal reasoning's predominance for working mathematicians, with careful delineations of context ensuring the rigorous validity of playful reasoning.[35]

As modernity's critics have recognized under many different rubrics, such a divorce of error from understanding, of mistake from calculation, was no small feat.[36] Calculation is an enterprise replete with misprision, from the schoolroom to the electronic computer database and beyond. Where Recorde's narrative performs the imbrication of mistake and reason, later pedagogies relegate mistake to unwritten disciplinary byways of training, debugging, accidents, and exceptions. But that imbrication is precisely what makes modernity's reason fruitful. Fart upon stale and antique philosophies and historiographies of deterministic algorithmic thinking! In early modern arithmetic, we can recognize that algorithmic modernity itself called for new, unruly ways of speaking, and different organs of reason.

3
The Orderly Universe

How the Calculus Became an Algorithm

Amir Alexander

The Geniuses

Back when I was in high school in Jerusalem I was part of a group of mostly male students known among our classmates as "the geniuses"—"ha-geʾonim" in Hebrew. Being called a "genius" was not exactly a compliment. A genius was certainly not cool, and didn't have much social cachet, but was not exactly a bottom-feeder either. One could fall much lower. Like its American counterpart, the "nerd," the term managed to combine a degree of social contempt with a dose of grudging respect. Yet in my high school, quite apart from the social opprobrium it carried, the assignation also had a specific technical meaning: the "geniuses" were the students who chose to study "5 units mathematics," the most advanced level of mathematics offered in Israeli schools.

In "5 units mathematics" my fellow geniuses and I studied many difficult things—or so it seemed to us at the time (later experience caused me to revise that assessment). We studied algebra, analytic geometry, trigonometry, and combinatorics, and other fields from the mathematical canon. Yet there was one field that was set apart from all the others, a field that was considered so advanced and challenging that we could not touch it until our final year of high school: It was, of course, the calculus.

To math students in tenth or eleventh grade, who had only the vaguest notion of what it was about, the calculus possessed a near-magical aura. It was secret knowledge, accessible only to the lucky few, and wielded by all-powerful magus-mathematicians. Our teachers would drop hints about it on occasion, when dealing with other questions: "of course this can be done far more effectively with the calculus," they would say, and smile knowingly. A few of our classmates, who perhaps had a better than average claim to the "genius" label, were held in awe because they had actually studied it on their own, and were counted among those in the know. As for the rest of us, we simply had to wait patiently for the day when we too would be initiated into the mysteries of the calculus.

Amir Alexander, *The Orderly Universe* In: *Algorithmic Modernity*. Edited by: Morgan G. Ames and Massimo Mazzotti, Oxford University Press. © Oxford University Press 2023. DOI: 10.1093/oso/9780197502426.003.0004

But when in our senior year the day finally came, we were in for a surprise. The purpose of the calculus, we learned, is to find the slopes of curves and the areas they enclose. Given the curve of a function $f(x)$, we were told, the derivative $f'(x)$ is the function describing the slope of the curve, and the integral $\int f(x)dx$ is the function describing the area between the curve and the x axis. Expecting a grand revelation that would transform our view of mathematics, if not the universe, we were treated to a series of mechanical procedures that seemed rather like a cookbook, listing different recipes for different occasions. After the years of buildup, this was bound to be something of a disappointment.

Here's roughly what we learned about derivatives (Table 3.1):

Table 3.1 Simple Derivatives

If f is a	then $f(x)=$	and $f'(x) =$
Constant	c	0
Line	x	1
	ax	a
Power function	x^n	$n\,x^{n-1}$
Square root	$\sqrt{x}$	$(\tfrac{1}{2})x^{-\frac{1}{2}}$
Exponential	e^x	e^x
	a^x	$\ln(a)\,a^x$
Logarithm	$\ln(x)$	$1/x$
	$\log_a(x)$	$1/(x\ln(a))$
Trigonometric function	$\sin(x)$	$\cos(x)$
	$\cos(x)$	$-\sin(x)$
	$\tan(x)$	$\sec^2(x)$

To facilitate more complicated cases, we were provided with a set of rules for combining two functions, $f(x)$ and $g(x)$ (Table 3.2):

Table 3.2 Rules of Differentiation

Rules	if the function is	then the derivative is
Multiplication by constant	cf	cf'
Sum rule	f + g	f' + g'
Difference rule	f − g	f' − g'
Product rule	fg	fg' + f' g
Quotient rule	f/g	(f' g − g' f)/g²
Reciprocal rule	1/f	−f'/f²
Chain rule (using ')	f(g(x))	f'(g(x))g'(x)

And here's what we learned about integrals ... (Table 3.3)

Table 3.3 Simple Integrals

If f is a	then f(x)=	and ∫f(x)dx=
Constant	a	$ax + C$
Variable	x	$x^2/2 + C$
Power function	X^n	$X^{n+1}/n+1 + C$
Reciprocal	$(1/x)$	$\ln|x| + C$
Exponential	e^x	$e^x + C$
	a^x	$a^x/\ln(a) + C$
	$\ln(x)$	$x\ln(x) - x + C$
Trigonometric function	$\cos(x)$	$\sin(x) + C$
	$\sin(x)$	$-\cos(x) + C$

... and their accompanying rules (Table 3.4):

Table 3.4 Rules of Integration

Rules	if the function is	then the integral is
Multiplication by constant	$cf(x)$	$c\int f(x)\,dx$
Sum rule	$(f + g)$	$\int f\,dx + \int g\,dx$
Difference rule	$(f - g)$	$\int f\,dx - \int g\,dx$

The calculus, it turned out, was not a magical secret that would resolve all mysteries, as we had been led to believe. It was simply an algorithm, a series of predetermined procedures one must follow in order to achieve the desired results.

The calculus is, to be sure, an imperfect algorithm. It requires different procedures for different cases, which makes it a collection of algorithms rather than a single unified one. Furthermore, even all those procedures together do not cover all possible cases, leaving many functions out in the cold. The problem is manageable in the case of differentiation (or finding of the derivative), but is inescapable when it comes to integration. Here the standard methods, such as integration by substitution and integration by parts, are just a collection of mathematical tricks designed to discover the function whose derivative is the starting function. As often as not, even these tricks fail, making integration more of an intuitive art than a rule-bound algorithm. In high school we were only dimly aware of these difficulties, since the problems presented to us were chosen to be solvable with the tools we had. We no doubt overestimated the consistency and reliability of the procedures we had been taught, yet fundamentally we were right: the calculus was designed to be a reliable algorithm, simple and easy to use, and in many cases it was just that.

From Method to Algorithm

When it came to the calculus, my experience as a high school genius fell well short of the intellectual revelation we had been promised. And yet, in a deeper sense, it was absolutely true to what the calculus was designed to be. From its origin in the work of Newton and Leibniz, the calculus was meant to be an algorithm—a prescribed series of mechanical steps aimed at achieving the desired results. The very term by which the procedures became known suggests this: unlike other fields, such as trigonometry, analytic geometry, combinatorics, or any other, the term "calculus" gives no hint about the subject matter of the field or the type of questions it addresses. All it implies is the existence of a series of predetermined procedures, literally—a "calculus." The subject matter or the mathematical grounding hardly seems to matter: possibly alone among mathematical fields, what makes the calculus what it is, is not what it studies, but simply its algorithm.

This is true not only conceptually, but also historically, because it was the algorithmic nature of the calculus that separated it from related techniques that preceded it. Before there was the calculus, there was the "method of indivisibles," a mathematical approach whose roots go back to Archimedes, but which was revived by Galileo's followers in the seventeenth century. Like its progeny, the calculus, the method of indivisibles dealt with the slopes of geometrical curves and the areas, or volumes, that they enclosed. Like the calculus, it approached the problems by considering continuous magnitudes as made up of infinitesimal components—lines of points, surfaces of lines, volumes of surfaces. Unlike the calculus, however, the method of indivisibles was *not* an algorithm.[1]

For a taste of the method of indivisibles, consider this simple example from Bonaventura Cavalieri (1598–1647), professor of geometry at the University of Bologna and the method's founding father:

If in a parallelogram a diagonal is drawn, the parallelogram is double each of the triangles constituted by the diagonal (Figure 3.1).

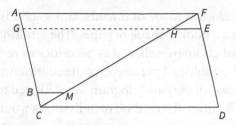

Figure 3.1 Cavalieri's proof that the area is double each of the triangles constituted by a diagonal.
Source: Exercitationes Geometricae Sex (1647), Book 1, prop. 19.

In the diagram this means that if a diagonal FC is drawn for the parallelogram AFDC, then the area of the parallelogram is double the area of each of the triangles FAC and CDF. Cavalieri proceeded:

> Let equal segments FE and CB be marked off from points F and C along the sides FD and CA respectively. And from the points E and B mark segments EH and BM, parallel CD, which cross the diagonal FC at points H and M respectively.

Cavalieri then shows that the small triangles FEH and CBM are congruent, because the sides BC and FE are equal, angle BCM is equal to EFH, and angle MBC is equal to FEH. It follows that the lines EH and BM are equal. The same can be shown for all other parallels to CD, namely that those that are marked in equal distances from the point F and C along the sides FD and AC, are equal between themselves, as are the extremes, AF and CD. Therefore all the lines of the triangle CAF are equal to all the lines of the triangle FDC. Since "all the lines" of one triangle are equal to "all the lines" of the other, Cavalieri concludes, their areas are equal, and the parallelogram is double each of them. Q.E.D.[2]

One might wonder why Cavalieri would go to such lengths to prove such a simple, even trivial proposition. A quick glance at the diagram is enough to reveal that the triangles FAC and CDF are congruent, and their areas are consequently equal. Cavalieri, needless to say, knew this, but his purpose here was not to prove a novel theorem; it was, rather, to demonstrate the effectiveness and reliability of his new method. And for that purpose the demonstration is revealing indeed.

First and foremost, for Cavalieri, the demonstration shows how to successfully manipulate the elusive objects called "indivisibles" in order to achieve reliable results. The concept, as Cavalieri explains in his introduction, is founded on the intuitive perception that a surface is made of lines in the same way that a piece of cloth is made of threads, and a solid is made of surfaces in the same way that a book is made of pages. The intuition is, no doubt, appealing, but as Cavalieri knew well, and as his critics were quick to point out, it runs afoul of well-established paradoxes of the continuum, known since antiquity. The simple case of the parallelogram was designed to convince readers that, despite the difficulties, the method of indivisibles would reliably produce true results, as long as its procedures were handled correctly.[3]

But the example also shows something else, which Cavalieri and his associates likely took for granted. The method of indivisibles was a "method" in the sense that it suggested a broad strategy for resolving questions of areas and volumes (or "quadratures," in the language of the time) and questions of tangents. Such questions, according to the "method," should be addressed by dividing the curve (or surface, or solid) into its infinitesimal component parts, that is, points (or lines, or surfaces). By summing up all the indivisible components one could determine the length (or area, or volume) of a figure, and by subtracting an indivisible from its neighbor one could calculate the tangent at that point.

The method was not, however, and never aspired to be, an algorithm. It did not supply a series of predetermined steps that applied in every, or even most cases. Instead, it made use of specific features that are different for each problem: in this example it relied on the geometric properties of the parallelogram in order to divide it into straight, parallel lines, and on the geometric features of the component triangles to ensure that each line in one triangle had its equivalent in the other. But when the problems change, the mathematician's approach must change along with them. To determine the area enclosed inside an Archimedean spiral, for example, Cavalieri divided it into circular indivisibles, which he then laid side by side to form a parabola. When Torricelli sought to calculate the slope of an "infinite parabola" (a family of curves we would characterize as $y^m = kx^n$), he did so by calculating the relative "thickness" of the horizontal and vertical indivisibles that intersect with the curve.[4]

Despite the fact that the method of indivisibles was the progenitor of what became the calculus, its practice could hardly be more different. To a mathematician the method suggested a general approach to certain problems, but no more: the actual resolution of each case was left to the mathematician's ingenuity, aided perhaps by a growing body of examples of comparable problems resolved by his colleagues. The calculus, in contrast, strove to be a series of fixed rules that could be applied uniformly and unthinkingly to any question of a given type, and guaranteed to produce the correct results. Admittedly, the calculus was not always successful in producing such uniform rules. But that was nevertheless its aim, and the key feature that distinguished it from the method of indivisibles: the method of indivisibles was a mathematical approach; the calculus was an algorithm.

Making an Algorithm

Let us consider for a moment what it takes to establish a method as an algorithm. First of all, an algorithm cannot be just a random sequence of operations. It has to produce a result, and that result has to be verifiable as correct. This means that the algorithm has to refer to some universe, outside of itself, that provides a check on its results. If, for example, we were asked to divide 456 by 8, we could follow any number of step sequences and arrive at different results, but only the steps of the division algorithm (or something like it) would result in the correct answer, 57. How do we know that 57 is the correct result? One answer is that out there in the world, if we obtain 456 objects and divide them into 8 equal parts, each of these parts will include 57 objects. Another answer might refer not to the "real" world but to the characteristics of a formal mathematical system, demonstrating that 57 is the only answer that is consistent with its assumptions. Either way, if an algorithm is to function, its operations and results must refer to some universe—real, material, formal, theoretical, or other—that exists outside the algorithm itself.

This rather minimal requirement is not, to be sure, limited to algorithms, but is shared with any mathematical system. How else, after all, can one decide whether a mathematical result is correct or not? The method of indivisibles, for example, clearly refers to the broader world of geometrical objects, and its results must be consistent with, and preferably provable by, the traditional methods of geometry. Even the highly abstract theorems of modern mathematics, which are explicitly divorced from any material or quasi-material world, nevertheless refer to a purely rational mathematical universe in which all truths are strictly rigorous and consistent.

The second requirement for an effective algorithm is more demanding and not so easily satisfied. If an algorithm is to be successful, the world to which it refers must be regular, orderly, and predictable. An algorithm, after all, is a predetermined series of steps that should be applicable in every case. If each case requires a different procedure, then an algorithm would obviously be useless. The division algorithm, for example, is just as effective for dividing 1127.3 by −73.49 as it is for dividing 456 by 8. If each of these cases had required a different procedure, then there would be no algorithm. Yet the method of indivisibles makes no such demands on the world. Every problem, for the method, is indeed different and requires a different approach, and there is no reason to believe that what had worked in one case will also work in another. The geometrical world, accordingly, may be orderly or chaotic; there is no way of telling in advance. All that can be done is to approach each case as it presents itself, and attempt to resolve it in accordance with its unique characteristics and circumstances.

Not so the calculus, which presents a rigid set of steps that are applied routinely. By making use of its procedures, we express confidence that the same steps will produce correct results not only to a particular problem we have studied, but to all problems, including those we haven't even considered. This can only be the case if the geometrical world of the calculus is a remarkably orderly one, operating according to strict, regular, and predictable rules that extend to each and every case. If this were not so, then it would be pointless to attempt to decipher its secrets with an algorithm.

It follows that the differences between the method of invisibles and the calculus that emerged from it are more than a matter of refinement and systematization. At their core, the two methods differ in the kind of world in which they exist and operate—irregular and unpredictable for the method, regular and predictable for the calculus. The transition from one to the other required not just mathematical sophistication, but a transformation in the way the mathematical world is seen and understood. As we shall see, it is precisely this kind of transformation that accompanied Newton's development of his version of the calculus, which he came to call the "Method of Fluxions."

The Disordered Universe

Newton began working on the "Method of Fluxions" in the mid-1660s but modified it repeatedly over the next four decades. His earliest speculations on the subject can be found in the "Trinity Notebook" that he kept between 1663 and 1665, which he labeled "Quaestiones quaedam philosophicae," or "Certain Philosophical Questions."[5] Nothing in the speculations contained in the pages of the Notebook suggests that by the end of 1665 Newton would be in possession of a mathematical tool that surpasses anything developed by the leading mathematicians of the time. But the Notebook does indicate an early interest in the nature of mathematics and its relation to the physical world.

What becomes clear from the pages of the Notebook is that Newton made almost no distinction between mathematical and physical objects. That mathematical objects such as points and lines have a real presence in the world is evident on the very first lines of the very first passage of the Notebook.

Inquiring over the nature "of the first matter," Newton asks "[w]hether it be mathematical points, or mathematical points in parts, or a simple entity before division indistinct, or individuals, i.e. atoms."[6] Conversely, objects in the world can be perfectly described by analogy to mathematical objects: "I shall all along draw a similitude from numbers," Newton writes a few pages later, "comparing mathematical points to ciphers, indivisible extension to units . . . a multitude of atoms to a multitude of units."[7] Mathematical objects, then, are manifested in matter, while all matter is structured mathematically. It follows that the world, as Newton sees it, is composed of mathematical objects, and Newton's investigation is aimed at determining which mathematical objects are involved and in what way.

We might be excused at this point if we concluded that since Newton's world is mathematical, it must be mathematically ordered, that is, logically and systematically like Euclid's geometrical universe. In the seventeenth century this was a widely held view, expressed most elegantly by Galileo when he wrote in 1623 that the universe is written in the language of mathematics.[8] That Newton has something quite different in mind becomes clear in a section of the Notebook entitled "Of Quantity," which follows closely on the heels of his discussion of first matter and atoms. Instead of seeking out elegant mathematical harmonies in the manner of Galileo, or Kepler, Newton heads straight to the convoluted heart of the concept of the continuum. "All superficies" (i.e. surfaces), he writes, "bear the same proportion to a line, yet one superficies may be greater than another." This, he continues, means that "though all infinite extensions bear the same proportion to a finite one, yet one infinite extension may be greater than another."[9] Drawing the mathematical implications from this paradoxical state of affairs, Newton concludes that although $\frac{2}{0}$ and $\frac{1}{0}$ are both infinite with respect to any finite magnitude, nevertheless $\frac{2}{0}$ is two times $\frac{1}{0}$.

Newton quickly follows up this mind-bender with others. The angle of contact, he writes, is to another angle as a point is to a line, because the "crookedness" of a circle—which is "four right angles"—is composed of an infinite number of such angles of contact. A point, he continues, is to a line as a line is to a surface, and a surface is to a solid. And finally this:

> It is indefinite, that is undetermined, how great a sphere may be made, how great a number may be reckoned, how far matter is divisible, how much time or extension we can fancy . . . a/0 exceeds all number and is so great that there can be no greater, but infinite number is called indefinite in respect of a greater.[10]

There is nothing particularly novel in all this. Newton's probing of the continuum, with its paradoxes and contradiction, follows in the footsteps of ancient and medieval scholars going back to Zeno the Eleatic. And yet, the paradoxical mathematical world of the Notebook is significant when it comes to Newton's view of the world in his early years at Cambridge. While the universe, in Newton's view, was indeed deeply mathematical, this fact made it neither orderly, coherent, or predictable. It was, rather, riddled with paradoxes and pitfalls, was fundamentally indeterminate and largely incomprehensible. Mathematics, as Newton showed, could model these contradictions, but it neither resolved them nor imposed upon them a discernable logical order. To reach reliable mathematical results in such treacherous terrain one must tread carefully, avoid pitfalls and paradoxes, and seek out those parts of the mathematical world where order prevails.

Such a view of mathematics does not encourage the development of algorithms. The world of reference for mathematics, as Newton saw it, is the physical material world, a view to which he held fast throughout his long career. Yet this world was disorderly and unpredictable, and full of unhappy surprises that could not be known until encountered. This mathematical universe might be successfully investigated through the method of indivisibles, which is flexible enough to adjust to differing circumstances. But it could not support an algorithm, which requires a regular and predictable universe to be of any use. And indeed, as we shall see, Newton's early efforts with calculating infinitesimal magnitudes were far from the consistent calculus algorithm they would ultimately become.

On May 20, 1665, Newton penned a brief calculation that may have been his first step on the road to his method of fluxions.[11] Unlike in the Trinity Notebook, which was concerned with broad philosophical questions, Newton here is attempting to resolve a specific mathematical question: given the equation $ax + xx = yy$ (which describes a hyperbola), find its subnormal at any given point. The question is a typical one for the method of indivisibles, and so is Newton's approach: seeking the subnormal $bd = v$ at the point b (see Figure 3.2), Newton adds a minute increment "$bc = o$" to b along the x axis. He then applies the original equation to the point $c = b + o$, whose abscissa intersects the curve at point f. Assuming that for a very small "o" $cd = fd$ (both are radii of the circle of curvature), he formulates the relationship between the point b, the sought-for subnormal at that point (i.e., v), and the increment "o." Finally, Newton declares that o "vanisheth into nothing," leaving him with the correct determination of the subnormal for the original point b. For a given point x, he concludes correctly, the subnormal of the curve is $x + \dfrac{a}{2}$.

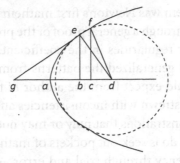

Figure 3.2 Newton's calculation of the subnormal of a hyperbola in his 1665 treatise.

The facility Newton displays with the geometrical characteristics of the hyperbola, as well as with the practices of the method of indivisibles, is impressive. Though a simple student at Trinity College, unknown and unrecognized, he is nevertheless investigating questions and producing results on a par with the leading mathematicians of his day. Yet if we are looking for hints of the calculus algorithm for which he became famous in later years, then there is little enough to see. Newton does, to be sure, make use of infinitesimal methods, which would become the foundation of the calculus. But as is always the case with the method of indivisibles, he applies the approach to a specific case, making use of the unique characteristics of the given curve. Newton's solution could very well be a helpful example for someone attempting to resolve a similar question and a related curve, such as a different hyperbola, parabola, or ellipse. But it is very far from being an algorithm—a fixed procedure that can be applied to any given case. In the irregular world described in the Notebook, there was no room for algorithms.

In the years that followed, Newton discovered the binomial theorem, a method to expand any function of the form $f(x)=(a+x)^{m/n}$ into a power series of the form

$$a^{m/n}+C_1a^{\frac{m}{n}-1}x+C_2a^{\frac{m}{n}-2}x^2+C_3a^{\frac{m}{n}-3}x^3+\ldots$$

where C_i are constant coefficients that can be calculated through Newton's method. The function $f(x)=(1+x)^{1/2}$, for example, which is another way of writing $\sqrt{1+x}$, can be expanded into

$$1+\frac{1}{2}x-\frac{1}{8}x^2+\frac{1}{16}x^3-\frac{15}{128}x^4+\frac{7}{256}x^5-\ldots$$

The binomial theorem was Newton's first mathematical breakthrough, but he did not arrive at it through a general proof of the properties of polynomials. Instead, he looked for regularities in the coefficients of simple cases, when m/n is an integer, and generalized the patterns from there.[12] This is exactly the approach one would expect from the author of the Notebook: since the world is irregular and strewn with inconsistencies and paradoxes, there is no point in seeking demonstrations that may or may not apply in any particular case. The best one can do is seek out pockets of mathematical regularity, and establish their consistency through trial and error. As long as the regularity proves true, and produces correct results, no further proof is required—or likely possible.

Armed with his new discovery, Newton set out to expand on his earlier use of infinitesimals. In *De Analysi per aequationes infinitas*, his first systematic exposition of his infinitesimal method, he showed how powerful and effective the binomial theorem could be for determining the quadrature of curves (i.e., the area between the curve and the x axis).[13] But by 1669, when he sat down to write the treatise, his outlook had evolved substantially from the days of the Notebook and the binomial theorem. For unlike in his earlier discussions of infinitesimals, in *De Analysi* Newton is clearly attempting to reduce his approach to a systematic rule-bound method.

Newton begins *De Analysi* by laying down three rules, two of which would ultimately be incorporated into the standard calculus algorithm. First, he notes, if $ax^{\frac{m}{n}} = y$ describes a curve, then $\left(\dfrac{n}{m+n}\right)ax^{(m+n)/n}$ is equal to the area enclosed between the curve and the x axis. Second, he notes, if y is a compound of not one but several terms $ax^{\frac{m}{n}}$, then each term should be treated separately as prescribed by the first rule, and all results then combined to give the area enclosed by the compound curve. In modern (Leibnizian) terms we would say that rules 1 and 2 prescribe how to integrate a polynomial. The third rule is not about integration as such, but rather about how to reduce more complex functions into the form of a power series, so that they can be treated in accordance with rules 1 and 2.

As compared to previous iterations of Newton's infinitesimal method, *De Analysi* represents a major step toward generalization in at least two ways. First, by laying down rules (numbers 1 and 2) that can be simply and mechanically applied to any function that is presented as a power series; second, by offering a variety of methods (rule 3) that would convert as broad a range of functions as possible into the desired form. For the first time, Newton is showing a desire to provide a standard and general method that can be broadly—perhaps even universally—applied.

Yet even as *De Analysi* represents a step toward the systematization of Newton's method, it is, nonetheless, a limited one. Rules 1 and 2 provide regular procedures with broad application, but Newton does not prove them as such: instead he merely states them, then offers a range of examples to show how they should be applied in specific cases. When, toward the end of the treatise, he ultimately comes around to demonstrating rule 1, he does so almost as an afterthought, and once again does not provide a general proof. Instead, he proves his case for a specific function, $z = \frac{2}{3}x^{3/2}$, implying that other cases are likely similar.[14] Rule 3 is not really a rule at all, but rather a collection of mathematical tricks that might prove useful in reducing certain functions to the form of a power series. When these fail—as they often do—Newton turns to numerical methods for approximating the roots of what he calls "affected equations," and uses them to construct the power series.

In sum, despite Newton's effort to regularize his approach, *De Analysi* falls far short of a systematic method, not to mention an algorithm. Its "rules" ultimately amount to little more than a series of useful recipes, to be tried out when conditions seem promising. The treatise, without a doubt, presented powerful methods for integrating a broad range of functions, and in that respect was superior to anything possessed by other mathematicians at the time. Yet the mathematical world of *De Analysi* was simply not regular and predictable enough to support a reliable uniform algorithm.

The Deep Science of Motion

For a practicing mathematician dealing with questions of tangents and areas, *De Analysi* provided a powerful set of tools. So powerful, in fact, that for the remainder of his long career Newton did little to improve on the utility of his method. *De Quadratura Curvarum*, considered Newton's most comprehensive treatment of the calculus, and published in 1704 as an appendix to the *Opticks*, adds little in the way of finding tangents and areas to what was already present in the 1669 treatise. It follows that when Newton returned to the topic repeatedly in the final decades of the seventeenth century, his goal was not to uncover some hidden mathematical secrets that had evaded him in his youth, but something else: to transform a practical heuristic approach into a regular systematic method, grounded in proper demonstrations. If he succeeded, the haphazard and irregular practices of *De Analysi* would become regular, reliable, and routinely applicable. They would become an algorithm.

The change in tone is already evident in *De Methodis Serierum et Fluxionum*, a treatise Newton composed less than two years after *De Analysi*, but which remained unpublished for decades. Here Newton makes clear the true subject matter of his method, an issue that had remained ambiguous in his previous formulations: The "Method of Fluxions and Fluents," as he now calls it, deals with flow through time, or, in other words, with motion. The instantaneous speeds of the flow are "fluxions," the quantities generated by them are "fluents," and the "infinitely small additions by which those quantities increase during each infinitely small interval of time" are called "moments." The problems dealt with in the method of fluxions are in line with this subject matter: First, "given the length of space continuously (that is at every time), to find the speed of motion at any time proposed; Second, given the speed of the motion continuously, to find the length of the space described at any time proposed." The first problem, in other words, is to move from the fluent to the fluxion; the second problem is to move from the fluxion to the fluent.[15]

This definition of the issues already indicates a more systematic and regular approach. For one thing, the method is now anchored in a well-defined phenomenon—motion. It seems natural to expect that such a well-known and regular phenomenon will be described by equally regular functions and procedures. For another, the two key problems dealt with are identified as inverse to each other, pointing to deep a hidden symmetry within the method. Newton, to be sure, was already aware of the inverse relationship between problems of tangents and areas when he composed *De Analysi* some years before. But it is only in *De Methodis* that the relationship becomes key to his procedures, as he moves smoothly back and forth between problems of fluxions and fluents.

Since the overall problem in *De Methodis* is defined in more regular and general terms, it follows that the methods utilized in the treatise are also more general and systematic. *De Methodis* relies more on general demonstrations than *De Analysi* and less on specific examples (though these, to be sure, are still very much present). Furthermore, the treatise supplies a comprehensive catalogue of curves for mathematicians to refer to when seeking a fluxion or fluent. All this does not make Newton's method of fluxions in *De Methodis* into a regular algorithm. But there is little doubt that such an algorithm is precisely what Newton was ultimately seeking.[16]

Newton reached the highest level of generality and consistency for his method of fluxions in *De Quadratura Curvarum*, a treatise originally written around 1691 but revised and published in 1704.[17] At first glance, it seems that not much had changed: the utility of the method in *De Quadratura* is little improved since the time of *De Analysi* decades before, and like *De Methodis* it is founded on the notion that the method of fluxions describes continuous motion. Furthermore, as Newton makes clear, these motions take place in the physical world. All mathematical magnitudes, he explains in the opening, are created through motion—a line by the movement of a point, a surface by the motion of a line, and so on. "These geneses," he continues, "take place in the reality of physical nature and are daily witnessed in the motion of bodies."[18]

But some things are clearly different. Newton is no longer satisfied with the notion of the infinitesimal "moment" that he used to account for this motion three decades before. Whether viewed as an increment of a line, as in *De Analysi*, or an increment of time, as it is implicitly in *De Methodis*, the concept of moment saddles infinitesimal methods with all the well-known conundrums of the continuum, going back to the paradoxes of Zeno and the problem of incommensurability. Back in his student days, when he was jotting down his thoughts in the Notebook, Newton was little troubled by these logical difficulties. In fact, he seemed happy to point out paradoxes of infinity, such as his conclusion although $\frac{2}{0}$ and $\frac{1}{0}$ are both infinite, $\frac{2}{0}$ is nonetheless two times $\frac{1}{0}$. This approach was what one might expect in a disordered world, where order and regularity cannot be assumed, but must be carefully and laboriously sought out. But four decades later, when he wrote and then published *De Quadratura*, Newton perceived the world as orderly and rule-bound, just as he had described it in the *Principia*. In this universe there was no room for troublesome entities like the infinitesimal "moment."

And so, Newton came up with a new interpretation of his Method of Fluxions: instead of "moments," his method would be based on what he called "first and ultimate ratios." A fluxion is no longer an infinitesimal increment of a curve, produced in an infinitesimal "moment," but the limit of the ratios between increments to the abscissa and ordinate, as they approach zero.[19] To illustrate his approach he offers a proof for a general and simple case:

Let the quantity x flow uniformly and the fluxion of the quantity x^n need to be found. In the time that the quantity x comes in its flux to be $x + o$, the quantity x^n will come to be $(x + o)^n$, that is (when expanded) by the method of infinite series

$$x^n + nox^{n-1} + \frac{1}{2}(n^2 - n)o^2 x^{n-2} + \dots;$$

The ratio between the increment to the ordinate and the increment to the abscissa is o to $nox^{n-1} + \frac{1}{2}(n^2 - n)o^2x^{n-2} +$, or 1 to $nx^{n-1} + \frac{1}{2}(n^2 - n)ox^{n-2} + \ldots$. Now, Newton continues, "let those augments come to vanish," and "their last ratio will be 1 to nx^{n-1}." Consequently, he concludes, "the fluxion of the quantity x is to the fluxion of the quantity x^n as 1 to nx^{n-1}."[20]

What had Newton gained by introducing his method of "first and ultimate ratios?" Certainly no problem-solving advantage. The result that the fluxion of x^n is nx^{n-1}, for example, would be self-evident to anyone who had studied *De Analysi*. What the method of first and ultimate ratios did provide, however, is greater logical consistency, as the hard moments are replaced with soft increments that trend toward nothingness and finally vanish, leaving behind their last ratio. It's not that everyone was convinced that Newton had indeed solved the problem that had bedeviled infinitesimal mathematics since antiquity: many were still critical, and Bishop George Berkeley, writing some decades later, famously called Newton's increments "the ghosts of departed quantities." But the important thing is not whether Newton was successful or not, but what he was trying to do: establish an orderly and logically consistent method of fluxions, suitable for the orderly and consistent world envisioned by the author of the *Principia*.[21]

In addition to providing what he considered sound mathematical foundations, Newton's attempt to provide general and rigorous proofs for his method in *De Quadratura* accomplished something else: they extended the reach of the method of fluxions algorithm beyond the particular cases discussed in earlier tracts, to every curve and every magnitude. Here, too, Newton was not entirely successful, as certain functions remain even today beyond the reach of the standard algorithms of the calculus. But by providing general proofs of the procedures of the method of fluxions he sought to extend his method as far as possible into the mathematical—and thereby the physical—world. Ideally, the standard algorithms connecting fluxions and fluents would describe the world as reliably and consistently as the universal law of gravitation.[22]

Logically consistent and universally applicable, the method of fluxions of *De Quadratura* had come a long way since Newton's early investigations into infinitesimal methods in his student days at Cambridge. A rough and ready heuristic, intuitively appealing while boldly incorporating well-known paradoxes, had become a rigorous mathematical system, founded on axioms and worked out through deductive proofs. A series of useful examples, intended to suggest a promising approach to mathematicians confronted with similar problems, became a standard list of rules. And all in all, a flexible approach, which must be specifically adjusted in each and every case, had become a set of standard procedures to be applied to any specific case (or at least to the broadest possible range of cases). Between "A Method for Finding Theorems" of 1665 and the final version of *De Quadratura* in 1704, the broad mathematical approach known as the method of indivisibles had become a standard algorithm. Newton called it the Method of Fluxions; we know it as the calculus.

God, King, and Algorithm

How does one account for this transformation, which produced Newton's calculus algorithm? The first thing to note is that from beginning to end, the subject matter of Newton's mathematics was the physical material world. This is as clear in the Trinity Notebook, where he models mathematical magnitudes on an atomistic conception of nature, as it is forty years later, in the *De Quadratura*, where he insists that the motions analyzed in the method of fluxions "take place in the reality of physical nature."[23] Newton's most eloquent statement of this position is in the preface to the *Principia* of 1687, where he states that "geometry ... is nothing other than that part of *universal mechanics* which reduces the art of measuring to exact propositions and demonstrations."[24] It follows that since mathematics is inseparable from physical reality, the kind of order found in mathematics is also the order to be found in the physical world. More specifically, a change in Newton's conception of mathematics entails a change in his view of the world, and vice versa.

The mathematics of the Trinity Notebook (1663–1665), of "A Method for Finding Theorems" (1665), and even of *De Analysi* (1669) requires little of the world in the way of order or predictability. Unconcerned with rigorous demonstration or general results, and founded on deeply problematic intuitions regarding the composition of the continuum, Newton's early mathematics seeks only to resolve certain questions when such resolutions are possible. It offers a practical method for producing reliable results in some cases, but makes no claims about others. Even the cases it does resolve retain a certain inexplicable mystery about them, since the resolution itself is founded on paradoxical assumptions. The physical world entailed in Newton's mathematics of the 1660s, in other words, is irregular, unpredictable, and ultimately unknowable. Certain regularities can, to be sure, be found, and Newton makes the best of them by calculating, for example, the tangents and areas of power series curves. But such regularities are only partial, and cannot be predicted or relied upon. All mathematicians can do, in such a world, is to try to apply their craft, and see what works.

But by the time Newton wrote down his mature views on the method of fluxions decades later, his views of the order that prevailed in the universe had changed substantially. The world of *De Quadratura* (1691, 1704) is a regular and orderly one, rational and predictable. Only an orderly world would support the operation of a standard algorithm, and the general and rigorous proofs that Newton provides are aimed at establishing that it is indeed so. Even the paradoxes of the continuum and the infinitely small, which inserted an element of mystery and unpredictability into the early versions of the method of fluxions, are accounted for here—at least to Newton's satisfaction. It is a world governed by strict, all-pervasive, and universal mathematical laws, which apply to everything, everywhere, and always. It is, in other words, the world made familiar by Newton's most influential work, the *Principia Mathematica* of 1687, in which a few simple mathematical rules account for everything from the fall of an apple to the motions of planets. For Newton, the laws of the calculus, as demonstrated in the *De Quadratura*, are no different from the three laws of motion or the law of universal gravity expounded in the *Principia*: at their core, the rules of the calculus are laws of nature, simple mathematical principles that govern and describe everything in the physical universe.

The profound shift in Newton's view of the world, and the kind of order one might find within it, was, in part, a philosophical journey: in the 1660s he had come into contact with the views of the Cambridge Platonists, and in particular Henry More, whose writings were one of the key influences on the *Notebook*.[25] More, Ralph Cudworth, and their fellow Platonists argued that the world is pervaded by a vital "hylarchic principle" or "spirit of nature" and cannot be fully explained by mechanical principles.[26] It is also likely that in those years Newton shared in the Baconian ethos of the early Royal Society. Entailed in this view was an entrenched suspicion of overarching mathematical schemes in the manner of Descartes and Hobbes, and a strong preference for empirical observation and experimentation.[27] By the time he wrote the *Principia* and *De Quadratura*, however, Newton had abandoned this early empiricism, replacing it with a vision of a mathematically governed world that seems more in line with Galileo than with Bacon.[28]

In part, the transformation may also have involved a religious evolution for Newton, who never doubted that God's place in the world was an integral part of natural philosophy.[29] In the 1660s, Newton, like the Cambridge Platonists and leading fellows of the Royal Society, seemed to have believed in a God that was closely involved in the operations of the world. For the most part, God operated according regular principles known as the laws of nature, but on rare occasions He could also deviate from them, producing miracles.[30] No law of nature, in such a world, could ever be universal and unqualified, and a mathematician seeking regularities would be wise to tread humbly and cautiously. By his later years, however, Newton abandoned this view in favor of a more distant God, who rules the world from on high as an omnipotent king. God, as Newton pronounced in the "General Scholium" to the second edition of the *Principia*, "governs all things, not as the soul of the world, but as Lord over all."[31] A world ruled over by a distant all-powerful God who lays down universal laws was bound to be more orderly and rule-bound than one pervaded by a capricious spirit of the world.

Finally, the transformation in Newton's conception of the universe may also have carried political overtones. Newton's early views, of a world imbued by a divine spirit, bore more than a passing resemblance to the teachings of the enthusiastic sects of the English Revolution. To radicals and revolutionaries, the notion that God pervaded the world implied that there was no need for an organized church, and that they had as much of a claim for a place in God's kingdom as their social betters. In the decades that followed the Restoration of 1660, such views came to be reviled by the ruling classes, who considered them subversive to their social position and to the state as a whole. In contrast, the world implied in Newton's later writings— orderly, rule-bound, and governed by a distant and all-powerful God—was far more reassuring. To the ruling classes, as well as to the still-shaky monarchy, such a world carried the promise of an orderly and divinely sanctioned social and political order.[32]

These philosophical, religious, and political developments are closely intertwined, so much so as that making distinctions between them seems almost arbitrary. All, however, likely played a part in the transformation of Newton's view of the world from his early Cambridge years to the writing of the *Principia* and beyond. And all, consequently, had a role in recasting the method of indivisibles as an algorithm, and thereby creating what we know as the calculus. The so-called invention of the calculus, it follows, cannot stand alone as a purely mathematical development, hatched in the minds of Newton and Leibniz with support from a few other leading mathematicians. To the contrary, formulating the calculus required a profound transformation in the ways the world as a whole and the order that governs it are perceived. It is a transformation whose roots and implications extend far beyond the boundaries of higher mathematics.

And what is true of the calculus is true as well of algorithms in general, whether in mathematics, computers, or other fields. By their very nature, algorithms make steep demands on the world in which they operate, demanding that it be highly regular, consistent, and predictable. Inevitably, such regularity requires support from fields far beyond mathematics—from science and philosophy, or even religion and politics. Simply put, an algorithm is not simply a mathematical creation: it is a cultural one.

Back in my high school days, of course, I and my fellow geniuses had no inkling of the web of cultural currents that came together in the algorithm we were taught. We dutifully mastered its procedures and learned to apply them reliably to different cases in order to reach correct results. We may have found the process dry and mechanical, but in retrospect we were probably wrong about that. There is indeed magic here: in Cambridge in the late 1600s, Newton dreamed of a perfectly regular, predictable and orderly world; 300 years later in Jerusalem, the students of "5 units mathematics" were still working in Newton's dreamworld.

4
The Algorithmic Enlightenment

J. B. Shank

Ever since the category "the Enlightenment" burst onto the scene as a master label for thinking historically about the intellectual dynamics of modernity, the term has undergone a variety of semantic transformations. Defined with a definitive article to mark it as an era of world-historical development, the Enlightenment has further become a theater of historiographical combat where disputants contest the normative value of the eighteenth century as a moment of modernizing change. For those who celebrate the Enlightenment, it was a time when positive ideals like human rights and humanitarianism, religious toleration and secularism, popular sovereignty and democracy, reasoned debate and science, and the values of individual liberty and personal self-actualization began to be championed as both universal human virtues and political ideals appropriate for the entire planet. For those who decry the Enlightenment, these imagined values are mere illusions that mask a time when new and dehumanizing systems of techno-scientific domination emerged together with rationalized apparatuses of illiberal oppression in a historical shift that laid the groundwork for the social and political pathologies of the modern era.

In a century that gave the world both Jefferson's opening paragraphs of the United States' Declaration of Independence and the early foundations of modern racial science, aligning the normative compass of the Enlightenment is no simple task. Connoting at first a temporal stage in Western philosophical development, the label is now used just as frequently to describe historical processes like secularization, professionalization, urbanization, and new understandings of publicity. Also marked by it are the new media regimes and literary practices of the eighteenth century, along with its changing patterns of economic circulation and commodification, and the newly global dynamics of European travel, trade, colonization, and empire. Recent thinking about the Enlightenment also tends to prefer distilled reductions of its essential features to complex accounts of its actual historical character. Epigrammatic analyses of the modernization brought by "Enlightenment critique," "the

J. B. Shank, *The Algorithmic Enlightenment* In: *Algorithmic Modernity*. Edited by: Morgan G. Ames and Massimo Mazzotti, Oxford University Press. © Oxford University Press 2023. DOI: 10.1093/oso/9780197502426.003.0005

Enlightenment public sphere," or "the Radical Enlightenment" have become historiographical staples, as have accounts of the Enlightenment defined by its imagined conceptual DNA: "rationalization," "scientization," "discipline and surveillance," or "a quantifying spirit," to name some of the contenders for genetic pride of place. Still other accounts define the Enlightenment as a historical period characterized by certain value-establishing events: "The Age of Liberty and Democratic Revolution," or the time of "The Consumer Revolution" or the "Age of Sensibility," to name only three. The Enlightenment is also captured through the innovations that are said to have made eighteenth-century culture historically new and forward-looking, currents such as naturalism, materialism, secularism, and the new awareness of the freedom of the self-possessed individual. In a more negative vein, the era has also been described as one of Enlightenment sexism, racism, Eurocentrism, and as a period that gave us new Enlightenment regimes of police, surveillance, management, and governmental control.

Given the porous fluidity of the term the Enlightenment today and the contests that surround it, it would be folly to expect to use the algorithm as an analytical key unlocking the essence of the Enlightenment in all of its complexity. My title, therefore, is not conceived with this grand project in mind. Rather, in the spirit of this volume, I will isolate through this one concept a characteristic thread of the Enlightenment when viewed as a moment in European intellectual history, one that serves as a tie joining many of the larger entanglements associated with the eighteenth-century Enlightenment overall. At the core of my discussion will be the singular fact of the rather sudden emergence of the algorithm in its fully modern form in the century of light, and the way that the changes initiated in relation to this new concept served as a significant marker of eighteenth-century historical development overall. While avoiding all claims to the algorithm as the quintessential ingredient defining the Enlightenment as a distinct era, I will locate the algorithm at the heart of eighteenth-century European thought and culture, and argue for its centrality in a number of different areas where forward-looking thinking was an agent for modernizing, world-shaping transformation.

The Enlightenment Birth of the Modern Algorithm

As Michael Barany shows in this volume, "only in the seventeenth century . . . did algorithms begin to lose the scent of creativity and indeterminacy" that they had retained throughout their early development. His title, with its emphasis upon the linguistic instability of the term during this fifteenth- and sixteenth-century period of modern European genesis, also points to the epistemological flexibility of the algorithm as at once a simple neologism describing the ancient art of reckoning with numbers and as a marker of something new, a different method perhaps, or a new discipline, but something that distinguishes the algorithm as an innovative kind of quantitative reasoning. Its scent of creativity was found in this pluralism, and in the way the word was deployed to mark the user of algorithms as an innovator. But only later did the term acquire its iron authority as an impersonal instrumental determinant of epistemologically inviolable outcomes.

All the historical dictionaries of the European vernaculars record this youthful changeableness, and they all offer the same basic history of change, locating the origins of the word "algorithm" in Arabic thought, especially mathematics.[1] The word shares in its opening "al" the same pedigree as other terms of Arabic descent—algebra, alchemy, and alembic, but also alcohol, alfalfa, and alcove. Like algebra, the word algorithm may also be the result of a bastardized mispronunciation of an Arabic name in medieval European discourse. Algorithm was at first "algorism," and even "algourism," a spelling that points to the role of Spain as a crucial site of its European importation, and in its earliest uses it described mathematical operations performed with Arabic numerals, another important Islamicate import.

For centuries, the word held firm under this meaning, which connected it to the manipulation of symbols in the service of quantitative calculation. In Edmond Huguet's nineteenth-century dictionary of sixteenth-century French, for example, no entry for "algorithme" is offered, but "algorisme" and "algoriste" are present, denoting respectively "arithmetic, calculation" and "he who is a good and just calculator."[2] In this vein, Jean de Fontenay was listed as the author of a "book on agorisme," that was "otherwise called a book on numbers (*chiffres*)."[3] Similarly in English, the *Oxford English Dictionary* lists the oldest definition of "algorithm" as "the Arabic system of numbering, characterized by a zero." It also notes its early synonymous connection with what the dictionary calls "the now archaic word algorism," and with the word "calculation." The *OED* also lists as the earliest example of the use of the word "algortithm" in Edward Phillips's 1656 dictionary, *The New World of English Words*, which defined "algorithme" as "the art of reckoning by cyphers."[4]

The *Philosophical Transactions of the Royal Society of London* showed that this same symbol-based meaning was still customary in 1699 when it defined "the algorithm" to mean the "numeral figures now in use." Likewise, Dr. Samuel Johnson still treated algorism and algorithm as synonyms, capturing this commonality by defining each together in his 1755 *Dictionary of the English Language* as "Arabick words, which are used to imply the six operations of arithmetik, or the science of numbers."[5] In Italian, "algoritmo" was also preceded by "algorismo," which the *Grande Dizionario Italiano dell'Uso* traces to the medieval importation of the "Arabic mathematician al-Xwarizmu, man of Corasmia, a historical region divided today by Turkmenistan and Uzbekistan." In this premodern Italian understanding, algorithm meant "ciphers that express quantity, and their arithmetic calculation" and "the system of calculating through Arabic numerals" and "rules for the resolution of a numerical calculation."[6]

Already in these understandings we see the important connection between the algorithm and the abstract representation of quantitative reckoning through symbols. We also see the beginning of the word's connection with the derivation of scientific knowledge from systematic, rule-bound symbolic mathematics. These remain central to the algorithm today. But what is absent from these early definitions is a separation between the algorithm as a simple denotation of symbolic arithmetic calculation and the idea of it as a general instrument capable of performing abstract reasoning and universal knowledge making. The power attributed to the algorithm today derives from its generalization as a conceptual tool applicable to any sort of inquiry whatsoever, and the precise historical break into this more modern understanding occurred in the eighteenth century.

Amir Alexander traces this shift in his contribution to this volume. "It was the algorithmic nature of the calculus that separated it from related [mathematical] techniques," he writes. Before there was *the Calculus*—the definitive article, which we still use today, points to the special nature of this mathematics—there were many different mathematical methods, rules, and operations for dealing with indivisible, infinitesimal, and incommensurable magnitudes. The many terms for the unstable object that the calculus resolves also points to the older pluralisms and flexibilities of its methodologies, and with the emergence of the precise rules, or algorithm, for differentiating and integrating evanescent quantities, a new higher-order conceptual and abstract rigor was added to the meaning of the word algorithm. Earlier methods of resolving imponderables "suggested a general approach to certain problems, but no more," Alexander writes. "The actual resolution of each case was left to the mathematician's ingenuity." But the calculus was different. Its aim was to produce an unfailingly successful rule for accomplishing this task in every case without exception, and this was "the feature that distinguished it from [earlier methods]." "The calculus," argues Alexander, "was an algorithm."

While the new math aspired for a new kind of abstract universalism of procedure, the actual calculus developed simultaneously and independently by Isaac Newton and Gottfried Leibniz in the 1670s was not always successful in realizing this dream. Alexander notes Newton's vigorous work to establish the algorithmic rules for his method of fluxions, but while this left him assured of his method, others across Europe were not as confident about the procedural clarity of this mathematics. Rather than receiving universal assent, the new calculus provoked intense epistemological debates focused on the very legitimacy of the new analytical science as its defenders called it.[7] Only after a century of fierce intellectual struggle were the deep logical and epistemological conundrums of the new calculus fully clarified, and in the interim the eighteenth century witnessed both the rapid dissemination and use of the new algorithmic mathematics and a persistent call for its eradication as a logically suspect and fallacious mathematical science.

The historical record of the European vernacular languages again documents this history in detail. The *OED* points unconsciously to the key events when it lists three definitions of the noun "algorithm," noting a development from the word's initial archaic connection with the Arabic system of numbering to its most recent meaning as "a procedure or set of rules used in calculation and problem-solving." "In later specialized use," the dictionary states, an algorithm is "a precisely defined set of mathematical or logical operations for the performance of a particular task." The entry also offers a third definition of very recent vintage (2006), one specific to medicine. There an algorithm is defined as "a step-by-step protocol used to reach a clinical diagnosis or decision." This understanding is very closely related to the second definition noted above, which spoke of a precisely defined set of logical operations used in the performance of a task. While noting this generalizing semantic change over time, the dictionary entries also offer an implicit historical account of these shifts that point to the eighteenth-century dynamics that brought about these changes.[8]

Under the first, archaic definition that speaks of algorithm in terms of the Arabic numbers used in calculation, the entry quotes the English botanist John Wilson, writing to Sir Isaac Newton in 1721. In the letter, he explains to him that "it does not appear from the *Commercium* that you were acquainted with the true Characteristics and Algorithm of Fluxions or Differences, before their Master."[9] This is followed by the mathematician Nicholas Saunderson, writing in 1739 in his *Elements of Algebra*, who speaks of "the War that was lately waged among Mathematicians concerning the Algorithm or Principle of Fluxions."[10]

These *OED* illustrations are offered under the second, more modern definition of algorithm, the one that spoke of it as "a precisely defined set of mathematical or logical operations for the performance of a particular task."[11] As such, they point to the historical episode—the eighteenth-century debate about the foundations of the calculus—that was decisive in pushing the archaic understanding toward the more modern conceptualization. In the same vein, the dictionary quoted as its first example of the new, more generalized definition of the algorithm an 1811 article in a learned periodical called the

Eclectic Review. The article stated that "the calculus of variations wants a new algorithm, a compendious method by which the theorems may be established without ambiguity and circumlocution." The entry ends with a very recent, and much less mathematical, example of its usage taken from a 2006 article that spoke of "the Bee Algorithm," "a calculation based on the way honeybees lead each other to nectar." The algorithm here is a general analytical tool that helps companies "develop more efficient industrial practices and allows businesses to test their manufacturing processes against 3000 variables."[12]

Present in all these entries, if nowhere explicitly articulated, is a conceptual history of the algorithm that traces its genealogy in three steps: (1) the algorithm's ancient origins in the premodern European reception of Arabic mathematics, particularly arithmetic and algebra; (2) the decisive transformation of the algorithm as a word and concept through its participation in the innovations associated with the new infinitesimal calculus and the mathematical analysis built from it; and (3) finally, the growth and expansion of the algorithm in both conceptual power and influence after 1800, driven by the expansion and increasing influence of mathematical analysis to science overall. This last transformation was catalyzed by the birth of the electronic computer in the twentieth century, and taken together, these changes in the last two centuries produced our understanding of the algorithm as a general conceptual apparatus and instrumental tool perceived by many to be universally applicable in a vast array of different domains.

The *OED* citation from Wilson's 1721 letter to Newton points to the way that the eighteenth-century debates about the infinitesimal calculus served as the precise historical crucible from which this modern understanding of the algorithm was born. The *Commercium*, which he referenced, was the 1712 *Commercium epistolicum D. Johannis Collins, et aliorum De analysi promota, jussu Societatis Regiæ in lucem editum*, the Royal Society of London's official indictment and prosecutorial brief directed against Gottfried Wilhelm von Leibniz, accusing him of both stealing and bowdlerizing Isaac Newton's previously developed infinitesimal calculus.[13] The document lit the fire of what would become the calculus priority dispute, a titanic polemic that pitted the defenders of Newton against those of Leibniz in a bitter public battle fought over the right to claim priority in the invention of this new mathematics.[14] Saunderson had this polemic in mind when he referenced the "war among mathematicians," and by describing it as he did in terms of a battle "concerning the Algorithm or Principle of Fluxions," he was pointing to the way that the algorithmic nature of the new infinitesimal calculus, or, more precisely, the claim that the new mathematics should be understood algorithmically as a new instrumental tool of universal mathematical science, was in many respects the most important contention in the whole struggle.

While the right to claim priority in the discovery of what Leibniz called "the infinitesimal calculus," and Newton "the method of Fluxions," was ostensibly the major bone of contention, a deep set of epistemological differences also drove the argument. These made the battle as much a struggle over correct mathematical practice as it was an argument over who first discovered the calculus. Newton and Leibniz ultimately held very different understandings of the nature and value of the new mathematics, which they both in fact arrived at more or less simultaneously. And while it has ever after been called the "Priority Dispute," a better name for it might be the "Battle over Analytical Mathematical Fundamentals." This was because the battle was waged at the precise moment when symbolic, algebraic mathematical analysis was moving to the center of European mathematical practice, and the struggle over the calculus was a major theater in the larger fight over the role and value of this kind of mathematical work overall.[15]

In his letter, Wilson addresses Newton as a Leibnizian, and thus as a critic, and when he cites the *Commercium* by stating that "it does not appear ... that you were acquainted with the true Characteristics and Algorithm of Fluxions or Differences, before their Master," he was both privileging Leibniz's prior arrival at the truths of the calculus, and giving him a superior understanding of their fundamentals.[16] When he invented his own calculus, Newton was not likely acquainted with Leibniz's differential algorithm, the essential core of his infinitesimal calculus. Yet Wilson spoke with many at the time who recognized in Newton's work an explicit parting of ways with Leibniz because he disdained more generally the analytical, algorithmic approach to mathematics that Leibniz championed.

As "a tool for simply finding out," to use the language of the *Commercium*, there was nothing wrong with instrumental algorithms like the one developed by Leibniz for rectifying curves algebraically. Newton also conceived of his own fluxional method as a convenient tool useful for similar work. He further used it in his own mathematics, albeit as a private tool that he never presented publicly as formal mathematics.[17] However, to declare instrumental mathematics of this sort to be a new kind of universal mathematical science, as Leibniz did, was to confuse non-rigorous tools with real demonstrative mathematics, and Newton made clear his own devotion to traditional standards and his hostility toward Leibniz's innovating theorizations. The problem with Leibniz's calculus, Newton and his defenders argued, was its claim that instrumental algorithmic reasoning was a new kind of universal mathematical science even though the demonstrative rigor supporting this claim was not available. From Newton's traditionalist position, Euclidean demonstrative rigor was the foundation that secured true mathematics, and from this perspective Leibniz's new "analytical mathematical science" was nothing more than a reckless erasure of mathematical first principles in the name of innovation and economy.

Newton likewise developed his own idiosyncratic geometric method of first and last ratios, and then used it in his *Principia mathematica*, so that he could accomplish geometrically and synthetically what Leibniz accomplished algebraically and analytically.[18] This difference sat at the heart of the "Priority Dispute," along with the larger natural philosophical debate about gravity that erupted out of it. While the Newtonians won the battle over the correct physics of the cosmos, it was Leibniz who won the mathematical war since it was his algorithmic and analytical calculus that became the foundation for the classical mechanics and mathematical physics that was built in Newton's name after 1730.[19]

Speaking historically, what is today called "classical Newtonian mechanics" should rather be understood as the result of a dual transformation driven by the new Newtonian physics and mechanics and the Leibnizian mathematics that, when fused with it, laid the foundations for modern mathematical physics. This dual understanding is especially important with respect to the history of the algorithm since it was Leibniz's algorithmic algebraic and analytical understanding of the calculus that gave the new Newtonian physics its wider instrumental orientation and impact.

Knowledgeable commentators at the time understood this, even if they often confused the Leibnizian source for the algorithmic understanding of the calculus with Newton's very different geometric and non-algorithmic work in the *Principia*. As the French astronomer and royal academician Jean Sylvain Bailly, who was also a Freemason and mayor of Paris during the revolution, put it in his *History of Modern Astronomy*: "Geometry, of which [Newton] was a master and possessed in all its detail, received from him a new form. It became in his hands a more subtle instrument, one more suited to profound research."[20] The key innovation, said Bailly, was Newton's "*calcul de fluxions*." "This invention caused a revolution in the exact sciences," he declared. "Like the application of lenses to instruments and the invention of micrometers in the practical sciences, instruments that gave men organs that permitted them to penetrate into the knowledge of causes, the calculus of fluxions, or differential calculus, serves as a sort of micrometer that the mind (*esprit*) uses to see in a more intimate and certain manner the relationships (*rapports*) between things."[21] Bailly's colleague at the Académie Royale des Sciences, Joseph-Louis Lagrange, echoed this assessment with respect to the history of physics. "Mechanics," Lagrange asserted, "became a new science in Newton's hands, and his *Principia*, which appeared for the first time in 1687, was the agent of this revolution. Ultimately, the invention of the infinitesimal calculus put mathematicians in a position to reduce the laws of moving bodies to analytical

equations, and since that time the study of the resulting moving forces has become the principal object of their work."[22]

Pulling these assessments together, what they point to is the way that
Enlightenment mathematical physics acquired a very instrumental understanding of mathematics and its usefulness for science overall (first astronomy, physics, and mechanics, and then others) through the Continental
European fusion after 1700 of Newton's physico-mathematical explanations of
the cosmos with the very algorithmic conception of the calculus central to the
new mathematical science of algebraic analysis. Since this fusion remained the
heart and soul of Enlightenment mathematical physics throughout the eighteenth century and beyond, the often-noted influence of Enlightenment mathematical physics as a model for science overall in this period also carried with
it a very instrumental and algorithmic orientation. Euler's development of the
function concept after 1730 crystallized the algorithmic nature of the calculus
as Leibniz conceived it, and when astronomers and physicists deployed this
instrumental conception of mathematics to explain a variety of concrete empirical phenomena, a paradigm was established that lay the groundwork for
an Algorithmic Enlightenment overall.[23] The rules still taught to introductory
students of the calculus today, like "Rolle's Theorem" after Michel Rolle (1652–
1719) and "l'Hôpital's Rule" after Guillaume François Antoine, Marquis de
l'Hôpital (as well as Marquis de Sainte-Mesme, Comte d'Entremont, and
Seigneur d'Ouques-la-Chaise [1661–1704]), further point to the eighteenth
century as the key moment when the calculus, and with it physical science,
became algorithmic.

In this guise, the calculus became the basis for a new understanding of
mathematical science whereby physical problems were first reduced to quantifiable problems, manipulatable through the algorithmic resolution of abstract
equations, and then solved mathematically as a prelude to physical explanation. Quantifying empirical phenomena so as to make it susceptible to algorithmic mathematical analysis also became, after the model of calculus-based
Newtonian physics, a new paradigm for natural science itself. In this way, an
algorithmic understanding of the calculus became a paradigm for a newly instrumental approach to the sciences overall, or, to put it in Bailly's terms, the
calculus, treated as a scientific algorithm, became a new instrument, akin to
the micrometer, telescope, and other precision graduated measuring devices
useful for knowing things in a newly useful and universal way.

What began in astronomy and physics with Clairaut's prediction of the return of Halley's comet to within three days in 1758 reached one climax with
Laplace's *La mécanique celeste* around 1800, which claimed to leave no astronomical motion unaccounted for. These achievements were soon mirrored in
other sciences ranging from Lavoisier's new chemistry and the astonishingly

rapid quantification and then instrumental mastery of electricity. What characterizes all these Enlightenment sciences was a move toward precision quantification and explanation through the dual employment of instrumental tools of quantified observation and measurement and instrumentally conceived tools of analytical mathematics, especially the calculus.[24] At the center of all of this was Leibniz's algorithmic conception of the calculus, which broke decisively with older notions of demonstrative, geometric mathematics in creating a newly instrumental and utilitarian-oriented mathematical method capable of rendering complex problems explainable. In this sense, the calculus was the progenitor of the algorithmic Enlightenment, not because all Enlightenment thought, even scientific thought, became mathematical, but because the instrumental model of mathematics as a universal tool for empirical analysis offered by calculus-based mathematical physics became the model for new instrumental scientific approaches in a wide array of eighteenth-century domains.

From Mathematical Analysis to the Algorithmic Enlightenment

The historical dictionaries of the European vernaculars record with remarkable clarity the stunning nature of this eighteenth-century transformation. One marker is the invention at this moment of the adjectival form "algorithmic," a linguistic innovation that marks the generalization of the algorithm from a word denoting arithmetical operations with Arabic numerals to its more generalized modern understanding as an instrumental routine of systematic symbolic analysis overall. The *OED* illustrates the shift nicely by noting that in 1799 the *Monthly Magazine* could speak of "the algorithmic notation (universally applicable since the invention of logarithms)" as "inexpressibly superior to all other modes." In this formulation, the word "algorithm" still denotes mathematical operations with numerical symbols, but a new abstraction and awareness of such operations as one freely chosen analytical mode among many is also present. The entry further rounds out the contours of the change by illustrating the origin of the adjectival term "algorithmic" in calculus-based physics. For as William Gardener declared in his translation of J. H. M Wronski's 1820 address to the British Board of Longitudes: "In this fixation of the conditions of the more or less great perfection of celestial mechanics, we have not examined those which depend on a greater or less perfection of algorithmic methods, with which are determined the heavenly motions."[25] Here the use of the word "algorithmic" to describe the new approaches to celestial mechanics, not to mention mathematical physics more generally, is evident.

The French case is very similar. The fourth edition of the *Dictionnaire de l'Académie Française* published in 1762 still defines algorithm in terms of arithmetical calculation, calling it a "didactic term," and defining it as "the art of calculation" and using the phrase "the algorithm of fractions" to illustrate its usage.[26] By the publication of the sixth edition in 1835, however, the algorithm has become a term "in algebra" describing "a process of calculation." It is also listed as a "particular genre of notations," and the example given for it is "the differential algorithm," a direct reference to the calculus.[27] *Le Grand Robert de la langue française*, the dictionary that is the French equivalent to what the OED is for English, likewise notes an early meaning of "*algorithme*" rooted in the rules of arithmetical calculation with Arabic numerals ("*l'algorithme de la division*," for example). But a later meaning, born of the eighteenth century, is also offered, one that defines the word more generally as "an ensemble of operational rules proper to a calculation." The illustrative example offered is "*algorithmes du calcul integral*," again directly linking the change to calculus-based science in the eighteenth century. *Le Grand Robert* also dates the birth of the adjectival form "algorithmique" to 1815, noting that it was a term developed in the sciences to describe "something which uses algorithms."[28]

Il Grande dizionario Italiano dell'Uso offers a similar history, with a medieval understanding listed first that makes "algoritmo" a variation of the older "algorismo" describing operations with Arabic numerals, and then a modern understanding, developed after 1800, that treats it generally as "a method or systematic operation for resolving a problem." Google NGram also points to the same shift through its measurement of the frequency of the use of words in its corpus of digitized books.[29] An NGram scan picks up no discernible use of the word "algorithm" before 1750, a fact that we know to be erroneous given the evidence of the historical dictionaries. But after this date, its usage spikes up significantly as it becomes a frequently used word in its precise scientific but also newly generalized meaning. After 1930 an even more precipitous increase occurs as the algorithm starts to become both a frequently used concept among those doing quantitative work in the sciences, and a wider description of routinized, operational instrumental thinking in all sorts of domains.

Returning to the shifts of the Algorithmic Enlightenment, what we should be attentive to is not the linguistic shift per se in the eighteenth century, rooted in new deployments of the term "algorithm" to describe new knowledges and knowledge-making overall. That's what happened with the concept in the twentieth century. Instead, the Algorithmic Enlightenment is about first the clarification of the algorithmic method as a paradigm for reasoning in mathematics, and then its deployment as a paradigm for the new Enlightenment mathematical physical sciences, which were the first to incubate this new instrumental manner of thinking. Second, the Algorithmic Enlightenment is about the stabilization of this paradigm in the practice of quantified natural science, first in astronomy and mechanics, and then into adjacent fields. Finally, it is about the wider transformation of Enlightenment thinking overall through the semi-conscious application of algorithmic conceptualizations as new pathways for innovation within knowledge work.

The rise of the algorithmic paradigm itself can be connected back to the central problem addressed by the new differential and integral calculus when using it to pursue natural science. The operation that gave birth to the modern understanding of the algorithm was Leibniz's use of what he called an infinitesimal, which was a magnitude smaller than any other magnitude yet still greater than nothing, to conceive of the continuous arc of a curve as a series of discrete infinitesimally small instants, or points. This was the physical principle behind using the calculus to map physical motion in nature. Mathematically, it allowed the continuous curve to be rectified, giving it a discrete magnitude that could then be used to derive precise quantitative solutions. Physically it allowed phenomena like planetary motion or the movement of electrical charge to be explained mathematically through recourse to mathematical equations. Leibniz also created an algebraic notation for all this, and established a set of rules, or routines of arithmetical algebraic calculation, to accomplish this work. Here the older understanding of algorithm as arithmetical operations with Arabic numerals began to acquire a new and specific association with scientific explanation, one that also worked to universalize its meaning into a general instrumental logical-symbolic tool.

This algorithmic paradigm radically changed the perceived capacity for mathematics to accomplish useful physical explanations, as both Bailly and Lagrange emphasized. And as Euler and others stabilized this algorithmic approach through the mathematics of the function, allowing abstract differential equations to become universal instruments for a new quantified celestial and terrestrial physics, and as astronomers like Clairaut, d'Alembert, Laplace, and others used these new instrumental mathematical tools to determine the precise quantitative relationships operative in the observed empirical phenomena of the heavens and on Earth, a powerful fusion occurred that secured the algorithm as a paradigmatic scientific tool and model for using quantitative algorithmic reasoning as a foundation for modern science itself.

The Algorithmic Enlightenment that followed in the wake of this transformation was on the one hand grounded in the continuing development of this brand of quantified mathematical physical science, and its establishment as a paradigm for natural science overall, one that could then be extended into the life sciences and the new social and human sciences born of the eighteenth century. Writing in 1972, Jean Piaget articulated well this outcome when he said in his *Épistemologie des sciences de l'homme* that "a problem remains only philosophical as long as it is only treated in a speculative manner. ... It becomes scientific as soon as one succeeds in sufficiently delimiting it in a manner where the methods of experimental, statistical, and algorithmic verification can allow for the achievement of solutions with a certain intellectual affinity and synergy (*un certain accord des esprits par convergence*), not one made from opinions or beliefs, but through fairly precise technical analyses (*recherches*)."[30] Here we see the algorithm as a new universal tool for scientific understanding overall, one that secures human claims about natural experience precisely through a reduction of all questions to relations of quantifiable variables, followed by a determination of all legitimate answers through an instrumental process of rule-bound, algorithmic determination.

The Algorithmic Enlightenment was certainly the moment when the algorithm as a specific epistemological tool was developed and put into the center of modern science overall. To take one example, it was in relation to their wider work in calculus-based mathematical physics that Euler, Daniel Bernoulli, and Abraham de Moivre (to name only a few) came upon the algorithmic rule, later called the Binet Formula, that allows one to determine the nth number in a Fibonacci series through a general equation.[31] Like the universal concept of the function, these instrumental understandings of mathematical science were simply the modus operandi of mathematical work in the century of light.

But the Algorithmic Enlightenment was more than the moment when algorithmic understandings became central to natural science. It was also about the eighteenth-century extension of this generalized instrument, the quantifying algorithm, to knowledge-making in domains that were not so much mathematized and quantified as instrumentalized in a newly rigorous, systematic, and rational and calculating way. In this broader Algorithmic Enlightenment, a space where a broader and less direct influence of the new algorithmic mathematical physics was felt, the algorithm served more as metaphor, or guiding spirit, than as an explicit model for mathematicized knowledge work. Here the Algorithmic Enlightenment exerted its influence beyond direct transformations in the nature of mathematical scientific thinking and more through new methods of instrumental knowledge-making and rational control.

The Algorithmic Enlightenment at Large in the Eighteenth Century

Take as a first example Enlightenment political economy, which was directly influenced by the eighteenth-century algorithmic transformation even if the result was not a newly mathematized approach to the economy in any robust sense. Unlike other Enlightenment sciences—the science of electricity is a good counterexample—political economy did not take from the new mathematical physics either a new inclination toward quantitative theorization of the flow of goods and material resources, or an urge to mathematize the empirical outcomes of economic life. Economic science would need another century before it would become an algorithm deploying quantitative mathematical science.[32] Instead, Enlightenment political economy modernized by first isolating a new and discrete sphere called "the economy," or to be more historically precise "society," a zone that later, through even more specialization, became the economy at the heart of modern economic science. In what has been called "the invention of society," eighteenth-century thinkers began to recognize a new zone of interrelations, material, cultural and social, that operated, like other natural systems, in a regular and lawful way. Once conceived in this way, society could then be studied in ways that deployed algorithmic tools to both comprehend, and, more important, manipulate these natural process for utilitarian purposes.[33]

Michel Foucault has described this transformation under the category of "governmentality," a new socio-intellectual formation that combined a new awareness of the self as an isolated sphere to manage and care for, along with a new notion of society composed of individuals whose self-management became directly related to new conceptions of the state, society (newly distinct), sovereignty, justice, and normative calculations of moral and social propriety. In Foucault's formulation, the new governmentalized self, enacted within a new Enlightenment relation between the sovereign state and "society," was attended by a new regime of "*sciences et technés*," or knowledge formations, that sustained the new configuration through rational calculation and control.[34] It was here that the algorithm as an innovative conceptual tool participated after 1750 in the rise and establishment of Foucault's new governmental Enlightenment.

At the center of the shift was a new set of "governmental sciences" that attempted to quantify so as to manage the realities of social life. Political economy was at the center of these changes, along with related sciences like demography, public health and hygiene, and police—the neologism derives from the same source as the word "polite," connoting a new idea of a human being managed, or "polished," into a proper moral subject. Also central was the new mathematics of "state numbers," or statistics, along with the algorithmic tools of probability theory and stochastic calculation that were also born at the same time.[35]

Algorithms understood in the eighteenth century as general quantitative rules for guiding calculation were important to each of these new sciences.

To illustrate, consider what many have taken to be the founding document of Enlightenment political economy, François Quesnay's 1759 *Tableau Économique*.

Tableau Économique

Objets à considérer. 1°. Trois sortes de dépenses; 2°. leur source; 3°. leurs avances; 4°. leur distribution; 5°. leurs effets; 6°. leur reproduction; 7°. leurs rapports entr'elles; 8°. leurs rapports avec la population; 9°. avec l'Agriculture; 10°. avec l'industrie; 11°. avec le commerce; 12°. avec la masse des richesses d'une Nation.

DÉPENSES PRODUCTIVES relatives à l'Agriculture, &c.	DÉPENSES DU REVENU, l'Impôt prélevé, se partagei aux Dépenses productives et aux Dépenses stériles.	DÉPENSES STÉRILES relatives à l'industrie, &c.
Avances annuelles pour produire un revenu de 600ᵗᵗ *sont* 600ᵗᵗ	*Revenu annuel de*	*Avances annuelles pour les Ouvrages des Dépenses stériles, sont*
600ᵗᵗ *produisent net*	600ᵗᵗ	300ᵗᵗ

Productions ... *moitié passe icy* ... *Ouvrages, &c.*

300ᵗᵗ *reproduisent net* .... 300ᵗᵗ 300ᵗᵗ

moitié ...

150 *reproduisent net* 150 150

moitié, &c. ... *passe icy* ...

75 *reproduisent net* 75. 75

37. 10ˢ *reproduisent net* .... 37. 10. .... 37.10

18. 15 *reproduisent net* .... 18.15. .... 18.15

9...7...6ᵈ *reproduisent net* .... 9...7...6 .... 9...7...6ᵈ

4.13...9 *reproduisent net* .... 4.13...9. .... 4.13...9

2...6.10 *reproduisent net* .... 2...6.10. .... 2...6.10

1...3...5 *reproduisent net* .... 1 3 5. .... 1...3...5

0.11...8 *reproduisent net* .... 0.11...8. .... 0.11...8

0...5.10 *reproduisent net* .... 0...5.10. .... 0...5.10

0...2.11 *reproduisent net* .... 0...2.11. .... 0...2.11

0...1...5 *reproduisent net* .... 0...1...5. .... 0...1...5

&c.

REPRODUIT TOTAL...... 600 *ll de revenu; de plus, les frais annuels de 600 ll et les intérêts des avances primitives du Laboureur, de 300 ll que la terre restitue. Ainsi la reproduction est de 1500 ll compris le revenu de 600 ll qui est la base du calcul, abstraction faite de l'impôt prélevé, et des avances qu'exige sa reproduction annuelle, &c. Voyez l'Explication à la page suivante.*

François Quesnay, *Tableau économique* (Paris, 1759)

Quesnay was a physician in service to the royal court, especially the official royal mistress Madame de Pompadour. His work, with its attempt to quantify the ebb and flow of resources within the national economy, was once imagined as the founding document of modern quantitative macroeconomics.[36] More recent scholarship, however, has restored its premodern character, situating it in one recent study as a Baroque court spectacle, akin to other wondrous contrivances that revealed magical secrets through the deft manipulation of pulleys, levers, and spring doors. The truth is somewhere in the middle, and the modernizing aspect places Quesnay's *Tableau* at the heart of the new algorithmic approach to political economy emergent after 1750.[37]

What Quesnay envisioned was a quantitative machine that could reduce the flow of resources in the economy to a regular system, one that could be used to visualize, so as to understand and manage usefully, material economic life. In brief, the *Tableau* traced the circulation of real wealth produced from agriculture into money, showing how it passed through landlords, merchants, and artisans and then back into consumable goods that in turn were sold to regenerate the cycle all over again. More precisely, Quesnay's numbers, which were by and large arbitrary, made the system appear to be a harmonious mechanism that generated prosperity for all involved. Developed through his connections with royal fiscal administrators inside the French monarchy, Quesnay's *Tableau* demonstrated publicly a new quantitative and instrumental approach to royal fiscal management that was then becoming ascendant within the French state.

Among the officials associated with these changes was Vincent de Gournay, the first person to use the phrase "*laissez faire, laissez passer*" to describe prudent economic policy. What Quesnay's *Tableau* revealed to thinkers like these was the percolation within the French state and in its environs of a new and more algorithmic approach to royal financial management that used the discourses that we now call "liberal economic science" as tools for a new kind of governmental management of the kingdom. At the heart of these emergent trends was the new understanding of the economy as a lawful, natural system, one that if properly conceived and comprehended could be manipulated through the proper application of precise algorithmic rules. It would take over a century for all of this to become anything like today's supposedly predictive quantitative mathematical science of macro- and microeconomics, but already in the Enlightenment the seeds of these new sciences were beginning to be planted.

Quesnay was the founder of a movemen that self-identified as the "Physiocrats," a name that they drew from their founder's medical training. The label referenced an economy that was conceived as an organic and potentially harmonious natural system, and critics called Quesnay and his followers

"*Les Économistes*," a term that was at first charged with dispersion. To these critics, ideas about "the lawful rules and science of the economy" smacked of empirical quackery rather than physiological science. But connected as they were to the royal administration in charge of French fiscal policy, the Physiocrats began to exert an important influence on French political life.[38]

Most consequential, and notorious, were the efforts made in the 1760s and 1770s to change the French royal policies concerning the circulation of grain. The Bourbon monarchy had secured absolute power after 1550 through a compact with the French people that guaranteed a sufficient supply of bread to every royal subject in exchange for obedience to the sovereign. By the 1760s, the adequate daily supply of high-quality grain had become arguably the most important public measure of effective of royal rule, and given the centrality of this political bond in Old Regime France, nothing was more destabilizing than grain shortages and famines. These were moments where the king was perceived to have abdicated his basic responsibility as provider for his royal family, and even moderate disruptions in the royal provisioning system could be highly destabilizing. Accordingly, when algorithmically minded French administrators began to experiment after 1760 with Physiocratric models regarding the free-market management of grain distribution, the widespread social consequences were enormous.[39]

For Physiocratic administrators, letting the market decide where grain would go was to let the natural, lawful systems revealed in Quesnay's *Tableau* operate, ensuring over time the greatest prosperity for all. What ordinary Frenchmen saw, however, was grain going places where it was not necessarily needed, and shortages in places where famine was threatening to decimate populations. In the wake of this disconnect between algorithmically conceived royal economic policy and the socio-political dislocations it caused within the traditional economy of bread, a new kind of governmental state management produced a new kind of social and political unrest, one that was further fueled by the widespread incomprehension among ordinary subjects about the rationale for these new arrangements. At the elite level, an intense intellectual battle over the validity and prudence of Physiocratic economic theory ensued. Among the uneducated populace—French literacy rates in the eighteenth century were still well under 30 percent—more imprecise responses became the norm. Steven L. Kaplan has documented one widespread outcome: the rise of what he has called "The Famine-Plot Persuasion in Eighteenth-Century France," or the popular myth that grain shortages were the direct result of a cabal of French ministers working nefariously under the cover of "scientific economic theory" to hoard grain so as to subvert the crown. Under this persuasion, the king was a good father trying to fulfill his obligations to his people, but undermining him were sinister royal officials bent on subverting his efforts for their own selfish and subversive gain. Accordingly, when well-stocked grain wagons were seen leaving town with the excuse that better prices could be found elsewhere, peasants read this as evidence of deceitful administrators working to subvert the monarchy by denying the populace its royal right to accessible grain. Adding credibility to this imaginative fiction was Physiocratic discourse, which claimed, incredibly to many, that the shortages were not real, but merely local dislocations that would eventually be worked out naturally and systematically for the greater good of all.[40]

The social unrest and dislocations caused by the grain-trade battles of the 1760s and 1770s were among the most important sources for the popular revolutionary upheaval that erupted in France after 1789. In this case, if not in every other, revolutionary agitation was a direct consequence of the new algorithmic understanding of the economy and of economic science emergent after 1750. It was also an outcome shaped by the new governmental conception of state administration and public social management that the new algorithmic governmental thinking fostered. The wider emergence of political economy as a new Enlightenment social and human science was further shaped by the same historical influences. The footnotes in Adam Smith's epochal 1776 *Inquiry Concerning the Wealth of Nations* are teeming with references to Physiocratic writings, and if Smith is seen, as he is by many, as the father of liberal, free-market economic science, then his work should rightly be seen as a continuation and extension of the French political economy emergent after 1750.

The Influence of the Algorithmic Enlightenment on the wider development of economic science, and the Enlightenment social and human sciences overall, is also profound. Some of Smith's most famous economic concepts, in fact, are perfect examples of instrumental algorithmic thinking applied to the new sphere of economic life. His famous argument for the existence of an "invisible hand," which magically regulates markets for the greater good of all, is highly resonant, for example, with the arguments about the mysterious power of the infinitesimal to transform curvilinear magnitudes into discrete numbers so as to make the complex motions of bodies apparent and instrumentally reducible to scientific explanation. That Smith was an expert student of eighteenth-century mathematical physics—he authored a history of modern astronomy—only adds credence to this analogy.

Many scholars of Smith's new economic science have also noted its ties to the new mathematical physics of the Enlightenment.[41] Also algorithmic in spirit and substance is Smith's equally famous analysis of the natural benefits of the division of labor. Using the manufacture of pins as the "system" with which he proposes that we think instrumentally and algorithmically, he develops a loosely quantified account of how ten individuals doing separate related tasks can produce a quantitatively greater number of pins than the same number working individually to make single pins alone. The economic import of the demonstration is derived directly from applying quantitative, instrumental algorithmic logics to the analysis of this complex empirical phenomenon.[42]

Other milestones in the history of liberal economic science were likewise products of the Algorithmic Enlightenment. David Ricardo's infamous laws, which became foundational for the new of "Dismal Science" of economics in the early nineteenth century, illustrate the point. His "Iron Law of Wages," for example, claims a natural quantitative relation between the reduction of worker pay to a bare subsistence level and the establishment of economic equilibrium. His "Law of Comparative Advantage" in trade likewise shows the quantitative material advantage that individuals gain by producing what comes most easily and trading for the rest rather than trying to produce every needed good individually. Each of these laws is an instrumental, quantitative rule designed to explain precisely the empirical realities of economic social life. Each is also, in turn, an instrumental tool for using this newly specialized knowledge to manage human beings and their social interactions. Malthus's population principle, with its ineradicable natural and instrumentally quantitative conflict between population growth and the availability of the material resources necessary to sustain human life, is likewise emblematic of the same algorithmic instrumentality.[43]

Also indicative of the Algorithmic Enlightenment was the governmental deployment of algorithmically conceived "*sciences et technés*" in the management and control of political and social life. Malthus's population principle, for example, was one of several algorithmic laws born of the new eighteenth-century sciences of demography and population studies that led directly to new regimes of urban planning and social and political management. Related algorithmic discoveries and calculations rooted in other "natural systems" such as the "semen economy" and the "system of the female passions" were also features of the widespread influence of the Algorithmic Enlightenment. In this precise case, "the semen economy," which appeared to scientifically account for and explain changing birthrates and fertility patterns, became the site for algorithmic arguments about moral propriety, along with new prohibitions against masturbation and male sexual deviancy. The idea of an "economy of female passions," linked instrumentally to everything from monstrous births and female infertility to the moral character and conduct of woman, was likewise made into an algorithmic ground productive of new calculations about how to manage society governmentally in ways that were said to improve the body politic as a whole.

The new public hygiene regimes emergent in the eighteenth century were similarly dependent on the Algorithmic Enlightenment in this way, and the confluence of all of these new governmental sciences at the heart of eighteenth-century cultural life is perhaps best seen by looking at the exuberant work of

Enlightenment social planners, who sought to maximize this new potential quickly and decisively through imaginative conceptions of radical urban reform. One example traces its roots back to the culmination by mid-century of the precise mapping projects begun a century earlier, projects that produced a newly precise quantitative grid from which to locate people, places, and things in France and around the world.[44] Brought to one moment of completion in 1739 by Michel-Étienne Turgot, father of the Physiocrat and Encyclopedist Anne-Marie Jacques Turgot, who marked the high water mark of Physiocratic political influence when he became First Minister and Controller-General of France under King Louis XVI in 1776, the year Smith's *Wealth of Nations* was published, the so-called Plan de Turgot was a precisely regimented bird's-eye view map of the city of Paris. It illustrates well the new possibilities that this kind of Enlightenment algorithmization of the empirical world made possible.

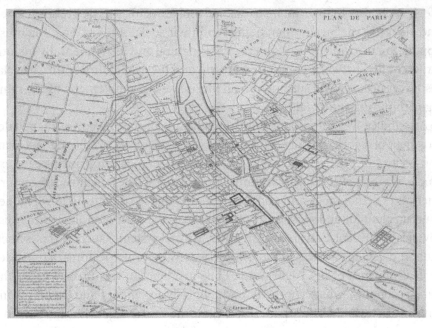

Plan de Paris, commencé l'anneé 1734: dessiné et gravé sous les ordres de Messire Michel Étienne Turgot, Marquis de Sousmons Paris, 1736[45]

The map was based on a geometrical grid that located every physical location in Paris in a newly precise and quantified way. A new efficiency for the city's postal system was one advantage of the new instrumental cartographic rationality that the map unleashed. Projects in algorithmic governmental science like census-taking and the public management and delivery of freshwater and the removal of sewage were also enhanced by this new mathematically ordered city space. New "police" capacities, as in the governmental capacity to manage people and society toward moral virtue, were also made possible by the new political-social grid. In 1749, the Parisian chief of police Jacques-François Guillauté illustrated the new potential, laying out a comprehensive program for governmental reform in his *Mémoire sur la réformation de la police, soumis au roi*.[46] Guillauté built his program on a new and precise algorithmic deployment of the urban management capacities enabled by the Plan de Turgot, turning the map into a machine for the instrumental policing of the city. In the treatise, the engraver Gabriel de Saint-Aubin offered an imaginative rendering of the new instruments of rational management and control that sat at the heart of Guillauté's plan.

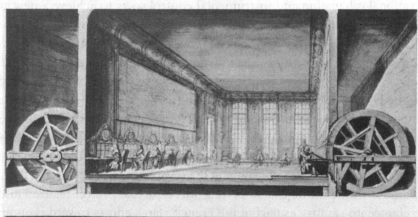

"The Paperholder," drawing by Gabriel de Saint-Aubin (1749), in M. Guillauté, *Mémoire sur la Réformation de la Police, soumis au roi en 1749*

In this imaginary command-and-control center for urban policing, an elaborate machine serves to transform each node on the grid of the Turgot map into an individual dossier alphabetically organized to contain and make available the necessary information pertaining to every urban resident and location. Supervised and scrutinized by a silk-wearing officer, the apparatus concentrates the Algorithmic Enlightenment into a single concrete fantasy of life improved through instrumental rational control. The vast wheels that would link one desk and depository with another, facilitating instrumentally the general management of the system overall, further suggest the algorithmic vision of the device overall, a vast machine making empirical realities available for rational calculation and instrumental control so as to achieve a new and more precise law and order as a consequence of systemic algorithmic efficiency.

This vision of the rationally managed modern city illustrates well the way that instrumental algorithms were influential far beyond the analytical mathematics and sciences from which they were born. Claude Nicolas Ledoux's imagined city to be built around the royal saltworks at Arc et Senans in eastern France is yet one more illustration.[47] A direct product of Physiocracy as it met royal fiscal administration (in this case the focus was on salt rather than grain, another basic nutritional necessity managed in new ways by the crown in the eighteenth century), the *Saline* built at the edge of the *forêt* de Chaux (hence its name as the Saline de Chaux as well) was an actual project realized in the late eighteenth century as part of the French state's efforts toward economic modernization. Ledoux the architect designed the buildings for the salt factory, which still stand today on their original site.

The Saline de Chaux at Arc et Senans. Architect Claude Nicolas Ledoux, ca. 1780

His design for the factory was indicative of the Algorithmic Enlightenment in its careful attention to the instrumental refinement of the production process, one where highly salinated water was brought by aqueducts to the factory and then reduced to crystalline salt through wood-fired boiling. Buildings were designed and spaced to allow for aeration and fire protection, and the site was chosen for its location in the heart of a royal forest that would provide the necessary fuel. Worker dormitories were also provided, and arranged strategically, to provide for the workforce necessary to sustain this remote industrial outpost.

Recognizing the wider social issues that such an innovative industrial factory posed, Ledoux also attempted to engineer an appropriate life-world suitable to this machine of royal economic government. His worker dormitories, for example, included leisure spaces and accommodations for women and families, and he also conceived of a police system for the manufactory centered in the two Greek-columned edifices at the axial center of the edifice. Here the royal officials of the General Tax Farm, the royal officers responsible for the management of the *gabelle*, or royal salt tax, would live and work together with other expert royal administrators, industrial engineers, chemists, and factory foremen in the surveillance and management of the entire operation.

Only the buildings seen in the photo above were actually built, but during his imprisonment during the revolution Ledoux extended his vision for Arc et Senans, imagining an ideal city to be built around the factory that would extend his rational algorithmic planning of industrial salt production to the moral management of the community overall.

Ideal View of the City of Chaux, from Ledoux, *L'architecture considérée sous le rapport de l'art, des moeurs et de la legislation*

Extending the axial alignment into a full circle, Ledoux imagined a series of buildings ringing the factory that would transform it into an ideal society. Enlightenment notions of hygiene and bodily management were one vector of influence, and among the buildings imagined by Ledoux was a public brothel, to be built in the shape of male genitals, where the semen economy and the female passions could be managed together and thus more rationally and effectively regulated. A massive library would serve the intellectual needs of the community, and public arenas and theaters suitable for edifying entertainments and assemblies were also imagined. In its broadest conception, every aspect of social life would be anticipated and algorithmically planned for, transforming through instrumental architectural planning everyday life into happiness and utility for all. As a fitting tribute to the guiding spirit animating the whole complex, Ledoux chose to include a monumental cenotaph celebrating Isaac Newton as a public monument and temple.[48]

Not lost in the move to grandiose utopian planning was the governmental management of the entire community contained in the original factory building. Although extended now across an entire circle, the central administrative axis from the original plan was preserved as the regulating center of governmental administration, surveillance, and control. In his various designs for the ideal city, which he left unpublished in a dossier discarded in his prison cell, Ledoux offered a schematic image of his plan, one that in its abstract geometric character isolated the algorithmic instrumentality of his thinking.

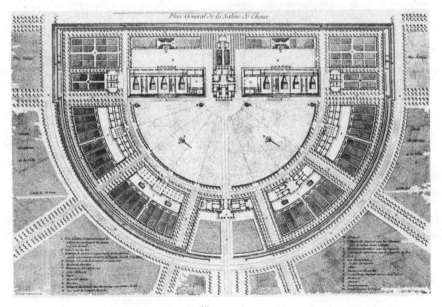

From Ledoux, *L'architecture considérée sous le rapport de l'art, des moeurs et de la legislation*

Those familiar with Foucault's work will recognize in this image a link to Jeremy Bentham's "panopticon." Foucault in fact deployed this precise image in his seminal 1975 book *Surveiller et punir*, translated into English as *Discipline and Punish*.[49] The pairing of Ledoux's architectural fantasy with Foucault's understanding of Enlightenment governmental rationality illustrates perfectly what I have been pointing to as the larger influence of the Algorithmic Enlightenment. Ledoux's instrumental rational planning, like Bentham's imagined prison reform, is not noteworthy for its attention to quantitative mathematical theorization. The algorithms deployed in each case are not, therefore, instruments for turning quantitative data into empirical knowledge even if the quantifying spirit is not irrelevant to either project. Yet, as I have been arguing, they are thoroughly algorithmic nevertheless because conceptually each is imagined as a way for human passions, conceived as continuous material flows, to be made observable empirically so that they can then be reduced to causal relations that can be rationalized and managed through regimes of instrumental calculation. Conceived, as it often is, through language that talks of Bentham's "calculus" of pleasure and pain, and his instrumental rules for the correct maximization of happiness and the minimization of suffering, his utilitarian ethics is thoroughly algorithmic in its attempt to treat moral conduct in terms of instrumental calculations bound by universal rules, laws, and principles. His "panopticon," which is indeed implicit in Ledoux's ideal plan for Arc et Senans, is similarly orchestrated. Positioning the prisoner as an individual body in need of rational management and control, the panopticon serves as an algorithmic instrument designed to simplify and universalize surveillance in ways that maximize carceral efficiency while minimizing effort, cost, and the physical pain of the inmate. It is nothing less than an algorithm for efficient carceral government, and without specifying this aspect explicitly, Ledoux likewise draws on the same algorithmic conception when imagining his axial system of surveillance and control in the salt factory as an architecture of managerial control and utilitarian social planning.

What for Bentham is an algorithmic conception of effective incarceration is for Ledoux an algorithmic conception of efficient industrial production and labor management. Each, in other words, is emblematic of the Algorithmic Enlightenment even though neither is a highly precise mathematical project or quantitative in orientation. Let me use one more example to conclude this chapter by driving home the same point: the quantitative voting theory of the Marquis de Condorcet. Condorcet literally reduced voters and votes to quantitative data points, and then used complex mathematical tools to develop a complex algorithmic theory for understanding and predicting democratic election outcomes.[50] The fact that Condorcet has exerted a direct and powerful influence on twentieth-century rational choice theory, not least in the work of Kenneth Arrow, further supports this understanding. The algorithmic link between Condorcet and modern quantitative economics and political theory is direct, but I would argue that the Algorithmic Enlightenment can also be used to understand other eighteenth-century thinkers who, like Bentham and Ledoux, were not expert mathematicians and did not manifest their connection to the algorithm through inclinations toward quantification and mathematization.

A case in point is the political theory of Jean-Jacques Rousseau. Anything but a mathematizer and quantifier, Rousseau's famous principle of the General Will, which was widely influential in the development of the French revolutionary democratic tradition, is, I would argue, another illustration of the Algorithmic Enlightenment despite its absolutely unmathematical terms and foundation. In brief, and in opposition to figures like Condorcet, Rousseau argues that a rational calculus of interests within a democracy that leads to rule by the majority is an abdication of the real foundation of democratic justice. If an outcome is just, then the will of the largest number of people is irrelevant to that, and is nothing more than an arbitrary principle based on the whims of opinion and fashion. True justice, Rousseau argues, must be grounded in universal truth assented to by everyone, and accordingly he demands unanimous consensus to be the measure of any truly just outcome in a democratic deliberation.

Mathematical calculation is therefore of little use in rationalizing such processes, and yet Rousseau's insistence upon the geometric rigor of his deduction turns it into an instrumental certainty that cannot be contested. When turned, as it was during the revolution, into a constitutional principle governing democratic process and deliberation, Rousseau's doctrine was made into a kind of universal algorithm for the instrumental government of the French polity. The point to see here is that while the history of the algorithm is about the increasing quantification of modern life, and the ever-expanding influence of instrumental mathematical science in all manner of human being and doing, its import today also extends far beyond mere mathematization.

In their famous critique of Enlightenment, Theodor Adorno and Max Horkheimer captured this link in their description of what they took to be the pathological aspect of Enlightenment, the one that led not to democracy and human rights, but directly to fascism and the gas chambers. The illness they saw lay in the idea of a thoughtless "factual mentality" that "separated thought from the business (of knowing) and adjusting actuality." "Factuality wins the day," they lamented when looking at eighteenth-century thought; "cogitation is restricted to its repetition. ... The world is reduced to a gigantic analytical judgement."[51] The authors stressed the role of mathematization in this, writing that: "For the Enlightenment, whatever does not conform to the rule of computation and utility is suspect."[52] Yet what is most powerful in their analysis is not mathematization itself, but its role in a new kind of automated, reflection-less, routinized, and blindly calculating instrumentality. It is this reduction of complex life and being to machine-like, rule-bound regularity that makes the Enlightenment pathological, Adorno and Horkheimer contend, and in linking the algorithm to Enlightenment conceived this way, my point is to encourage us to look beyond computational mathematics when seeking the limits and dangers of our own algorithm-centered contemporaneity.

5

Capitalism by Algorithm

Numbers for the Innumerate in the Eighteenth- and Nineteenth-Century Atlantic World

Caitlin C. Rosenthal[1]

Anxieties about algorithms arise from many sources: the computers we carry in our pockets tell us where to go and how to get there, search bars offer suggestions about everything from what to eat to whom to love, and specialized software tells police departments and armies what and whom they should target. Popular new books alternate between praising "algorithms to live by" and warning that algorithms are stealing our jobs, controlling our money, and taking over the world. As they increase in complexity and the data sets they draw on grow ever larger, algorithms seem to be slipping beyond individual comprehension and intuition. They are getting more important and simultaneously harder to see. We are living in what one author has called a "black box society," where algorithms processed at lightning speed seem to control us though we can no longer control—or even understand—them.[2]

These anxieties may be well founded, but blaming them on scale, speed, and complexity misses part of the story. This essay is about the very simplest of algorithms—a set of mathematical steps called the "rule of three." Though it has earlier origins, the simple procedure became exceptionally popular in the eighteenth- and nineteenth-century Atlantic world, spreading with the expansion of markets, commerce, and capitalism. It was both empowering and ominous, and as it spread it changed the process of negotiation, sometimes in invisible but critically important ways. In short: it previewed the promise and the peril of modern algorithms, but with paper and human memory as its conduits.

In his 1828 dictionary, Noah Webster called the rule of three "the Golden Rule,"[3] and a brief tour of eighteenth- and nineteenth-century mathematics textbooks reveals its ubiquity. Most introductory American and British textbooks, or "arithmeticks," presented the rule as the culmination of a basic mathematical education that began with numeration (learning to count and

Caitlin C. Rosenthal, *Capitalism by Algorithm* In: *Algorithmic Modernity*. Edited by: Morgan G. Ames and Massimo Mazzotti, Oxford University Press. © Oxford University Press 2023. DOI: 10.1093/oso/9780197502426.003.0006 to interpret digits); progressed through addition, subtraction, multiplication, and division; and ended with the rule of three, the "most important of all arithmetical rules."[4] We still learn to count, add, subtract, multiply, and divide, but the rule of three has long since disappeared from mathematical education in the United States.

What was this "precious Gem in Arithmetick"?[5] And why was it described in such fantastic terms? The name refers to its most common formulation: a way of using "three numbers given to find a fourth."[6] Emmor Kimber's *Arithmetic Made Easy for Children* instructed readers to "multiply the second and third terms together, and divide the product by the first; the quotient will be the fourth term, or answer." The obscure directions become clearer from practice exercises. Kimber asked: "given that 8 yards of cloth cost 24s., how much will 96 yards come to?" Following the instructions yielded the answer: the second number was 24 shillings (the cost of 8 yards). That number was multiplied by 96 yards (the third number), and their product was divided by 8 yards (the first number). The result was 288 shillings, or 14 pounds and 8 shillings. Or, in algebraic terms, $(24 \times 96)/8 = 288$. To apply the rule of three was simply to convert fractions.[7]

Comparing simple arithmetical processes to flashy twenty-first-century algorithms run on massive databases may seem strange to modern readers. But to many of their users, arithmetical rules and table books were black boxes—opaque systems of rules that produced answers they could not verify by hand or by intuition. Though conversions of currencies and calculations of wages seem simple today, to the eighteenth- and nineteenth-century workmen and shopkeepers who depended on them, they could be as opaque as modern algorithms are to the uninitiated. To many of their users, techniques like the rule of three were black boxes no easier to open than modern algorithms run at scale.

In the expanding market economy of early America, anyone who had to work, buy, or sell needed access to the language of numbers. Increasingly, this was everyone. The rule of three offered a measure of influence to the innumerate—to women, wage laborers, farmers, and even enslaved persons. Once they could speak the language of numbers, they could convert measures and currencies, compare prices, and calculate wages, all of which gave them power in the expanding world of commerce. But this process of democratization also had consequences: the rule privileged and legitimated certain varieties of exchange. As with more conventional technologies, it simultaneously reflected and remade the market in ways that were important but easy to overlook. The transactions it enabled came to seem not just correct, but fair and even moral. And, even as the rule placed the power to negotiate in different hands, it also bounded the terrain of negotiation—often in ways that were invisible to its users.

The Rule of Three

The ability to set a quantity of one thing (be it currency or a commodity) equivalent to some amount of another was an entry-level competency for commercial exchange in early America and the Atlantic world. Instructors like Emmor Kimber offered a sampling of the seemingly infinite variety of situations where the technique could be used: examples involved calculating the cost of different amounts of sugar, butter, tea, coffee, salt, and iron. He also showed students how to use the rule to calculate interest and exchange currencies.[8] Kimber demonstrated how to change Massachusetts currency into dollars, how to turn South Carolina currency into New Jersey currency, and how to equate New Hampshire currency with federal money. The advanced student could learn the "double rule of three," which enabled the discovery of a sixth number when five were known ("if 3 men in 4 days eat 5lb. of bread; how much will suffice 6 men for 12 days?" Ans: 30lb).[9]

Almost every transaction required an application of the rule, and contemporaries saw it as fundamental to economic independence. Indentures frequently specified that apprentices learn it before they became masters.[10] Business handbooks recommended that someone taking on an apprentice "agree to teach and instruct the said apprentice ... to read and write and cypher as far as the Rule of Three."[11] These recommendations appear to have been widely used and sometimes even mandated by law. In 1792, the state of Virginia required that indentured orphans receive instruction through the rule of three, and a Kentucky act allowed poor boys to be bound out by the state, so long as the "the person to whom they shall be bound to teach them some art trade or business ... and, if a boy, common arithmetic including the rule of three."[12]

Success in applying the rule did not require any conceptual knowledge of fractions or ratios. Rather, it depended on arranging the numbers in the correct order. As *Cocker's Arithmetic* explained, discovering "which is the first, which is the second, and which the third" was "the greatest difficulty" of using the rule.[13] Students learned this by rote, occasionally with the aid of a rhyme. Stokes's *Rapid Arithmetic* offered the following:

By Rule of Three to answer find,
Third and fourth terms make of same kind.

If Answer more than third term makes
The less term then the first place takes
If answer LESS than third term shows
The greater term in first place goes.
Second and third together 'ply
And by the first division try.
The quotient will the answer be,
In the same name as third term see....[14]

A better poet might have rendered memorization easier, but further training in mathematics was unnecessary. Though the rule did not rely on print, it operated like a set of written instructions. These were portable from person to person, place to place, and problem to problem.

The rule of three was, in essence, an algorithm designed to be run by a user who lacked training in mathematics. Education in the rule "deliberately relied on memory, not on understanding."[15] Indeed, those who pursued further instruction did not even need the rule. One student who had already studied fractions reflected on his father's mistaken aspiration that he be "pushed on in arithmetic ... through the rule of three." His knowledge of fractions rendered the rule redundant. But his father, "ignorant as he was of arithmetic," believed that the rule of three was the end of "a competent education for his class of persons." Interestingly, the calculative methods of the less numerate were not necessarily inefficient. As the student remarked, his father "had a method of his own by which he could calculate values by his head more rapidly and quite as correctly as I could do by my arithmetical process."[16]

By the mid-nineteenth century, the rule of three had come under fire from educational reformers. As one lamented, the "sorry learner attempts to fix in his memory a series of words, whose meaning he knows nothing about." Perhaps the rule worked "mechanically" to help students solve sample questions, but how often would it help them solve the real problems of business? American educator and author David Perkins Page recalled his own harrowing "passage through the rule of three." Page asked his teacher, "why sir, do I multiply as the rule says?" In reply his teacher repeated the rule, "Why, because 'more requires more and less requires less,'—see the rule says so." When he replied that he wished to understand more, the teacher looked "as if idiocy itself trembled before him." So Page "shrunk away" to his seat and proceeded to "follow the rule because 'it said so.'"[17]

A contributor to the *Southern Literary Journal* critiqued the methods of instruction along the same lines. Writing in 1839, he lamented that arithmetic "as commonly taught" offered "no improvement whatsoever to the mind." The problem was that "scholars are made to commit certain rules to memory" and to "solve problems mechanically by these rules, without ever seeing the necessity or the reason of the rule." Asked why they found the answer in the method that they did, he claimed that students replied "*the rule says so.*" Asked if they saw any reason behind the rules, they answered "*No sir! The rule says I must work so and so, and they answer will come. But why, I don't know.*"[18]

But this shortcoming was also the rule's main advantage. Teaching the golden rule as a series of mechanical instructions enabled men and women with limited access to formal education to master and use the rule themselves. The technique appears in handwritten ciphering books and in account books, sometimes showing how individuals worked to teach themselves in the absence of more formal education. In early New England, farmer Thaddeus Fish puzzled over profits to be earned in the sale of eggs, literally demanding of the rule that it help him learn whether a woman had "lost or gained" by her sales.[19] Abraham Lincoln's copybook also includes a page full of neat calculations titled "The Single Rule of Three" (Figure 5.1).[20] Lincoln had received a simple education and explained that he was able to "read, write and cipher to the rule of three. But that was all." What little he learned, he practiced, and his copybook shows him at work on the rule of three.[21] The

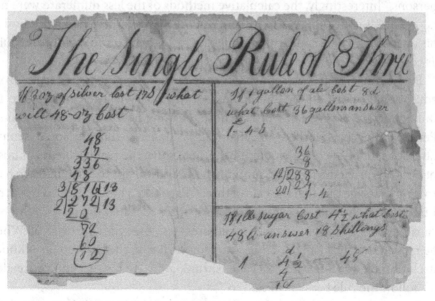

Figure 5.1 Though Abraham Lincoln had little formal education, his copybook shows him practicing "The Single Rule of Three" in 1826.
Source: Original in the John Hay Library at Brown University.

rule of three was not just a curious moment in the history of math education. Rather, it was a flexible technology that suited the everyday needs of an expanding commercial society.

Practices like the golden rule of three offered new opportunities for all kinds of people to negotiate effectively in an increasingly global economy. Even enslaved people sometimes put the rule to work: Olaudah Equiano describes how a captain's clerk "gave me a smattering of arithmetic as far as the rule of three." His future owner Robert King allowed him to conduct minor trade on his own account, probably using the rule. Through these exchanges, Equiano saved money for the purchase of his freedom.[22] Other slave narratives and biographies also mention the rule: G. W. Offley writes of trading food for lessons from a young boy who did not have enough to eat. The boy taught him to "cypher to the single rule of three."[23] James Pennington also "learned to read, write and cipher, as far as the single rule of three."[24] As historian Thomas Wickman writes about Equiano, the enslaved used numbers to "reason about what was fair as they crossed boundaries."[25] In the expanding Atlantic economy, everyone could benefit from being able to compare currencies, convert weights and measures, and evaluate potential exchanges.

The rule of three was a technology of commensuration—a tool for making comparisons and setting the unequal equal. Wendy Espeland and Mitchell Stevens define commensuration as measurement of "characteristics normally represented by different units into a common metric," and they argue that commensuration is always an embedded social and political processes.[26] Many scholars have described the social implications of more complex calculative practices like cost-benefit analysis, commodity grading, accounting, and household labor.[27] However, little attention has been paid to the most basic practices—techniques that appear more like arithmetical properties than social processes.

It is correct to say that ten times two is twenty, or three times one is three. But it does not follow that wages for twenty hours will always be twice those for ten, or that the price of three items should be three times the price of one. Indeed, in large negotiations—between big businesses—such linear relationships rarely hold. But with some exceptions, these kinds of relationships often do hold for wage workers and individual consumers. The history of the rule of three shows that both the knowledge of how to perform even these simple computations—and, perhaps more important, the belief that they were fair and valid—had to be learned. In an economy where many buyers and sellers were illiterate, innumerate, or both, the spread of such learning relied on the rule of three and a range of other simple calculative technologies.

Paper Technologies

Technologies like the rule of three did not develop in isolation. Like today's algorithms, they depended on cutting-edge networks and data sets. In the case of the rule of three, however, information was transmitted and stored not over wires and on computers but through books and newspapers, which were newly accessible to larger and larger audiences. Everyday activities like buying goods or being paid for labor sometimes required several calculations. Such calculations are easy to overlook in isolated sources like account books, the contents of which have already been conveniently converted into a single currency or unit of measure. But the paper trail of commensuration and its democratization survives in tables of weights and measures, scrawled calculations, and prices currents. These often-invisible processes were ubiquitous in a world where currencies and measurements changed from place to place.

Printed tables giving the ratios between different metrics could be found in all kinds of manuals and advice books. These tables reflect the complexity of commerce in eighteenth- and nineteenth-century Europe and America. As local markets connected to the global economy, many units and currencies circulated side by side. Even weights and measures varied dramatically by location. Joseph Blunt's *Merchant's and Shipmaster's Assistant*, published in 1822, instructed readers on how to convert among many systems of measurement. For example, he included a table showing how to convert between the long measures of various European cities. This table (Figure 5.2) listed fourteen different units, including the yard, five different kinds of ells, three types of canes, three varieties of vares, as well as Portuguese covedoes and brasses of Florence. As Blunt advised below the table, a simple application of "the common rule of three, or proportion" would translate his ratios into "any quantity required."[28]

Meaningful differences separated even units bearing the same names: 100 English or Scottish yards equaled 133⅓ ells of Holland and Amsterdam, but only 131⅔ ells of Antwerp and Brussels (Figure 5.2). Elsewhere Blunt specified that 10¼ English bushels of corn equaled 83 Winchester bushels, which amounted to various numbers of sacks, muddes, cahizes, setiers, and hoedts. He also gave a table for weights that paralleled his table of long measures. This chart listed conversion ratios for changing twenty-three different kinds of pounds into one another. The smallest of these "pounds" weighed less than 30 percent of the largest, so mistaking one for the other could turn a profitable exchange into a heavy loss.[29]

Tables of weights and measures could be used in combination with prices currents to track changes in value over both space and time. From the

A TABLE

Representing the conformity which the Long Measures of the principal trading cities of Europe have with each other.

The Ells of Amsterdam, Haarlem, Leyden, the Hague, Rotterdam, and other cities of Holland, as well as the Ell of Nuremberg, are equal among themselves, and are comprehended under the Ell of Amsterdam, and the Ell of Berne and Basle under that of Hamburg, Frankfort, and Leipsic.

MEASURES.		Yards of England, Scotland, and Ireland.	Ells of Holland & Amsterdam.	Ells of Antwerp and Brussels.	Ells of Hamb. Frankf. Leipsic and Cologne.	Ells of Bergen & Drontheim.	Ells of Sweden and Stockholm.	Canes of Marseilles & Montpellier.	Canes of Toulon Albi, and Castres.	Canes of Genoa of 9 palmes.	Vares of Castile and Biscay.	Vares of Cadiz and Andalusia.	Vares of Portugal or Lisbon.	Covedoes of Portugal or Lisbon.	Brasses of Florence, Leghorn, and Lucca.
100 yards of England, Scotland, and Irel.		100	133⅓	131⅔	160	146⅔	154	46⅔	50	40⅔	107	109⅕	81⅓	133⅓	154⅔
100 ells of Holland and Amsterdam....		75	100	98¾	120	110	114½	35	37½	30½	80	81⅘	61	100	116½
100 ells of Antwerp and Brussels......		76	101¼	100	121½	111⅛	116	35½	38	30½	81	84	61½	101¼	118
100 ells of Hamburg, Frankfort, &c....		62½	83⅓	82¼	100	91¾	95½	29¼	31¼	25⅛	65⅝	68½	50⅓	83⅓	97
100 ells of Bergen and Drontheim....		67½	90	89	108	100	103	31½	33¾	27½	72	74¾	55	90	105
100 ells of Sweden and Stockholm.....		65½	87½	86½	105	96½	100	30½	32¾	26¾	70	71½	53½	87½	102
100 canes of Marseilles and Montpellier	make	214½	286	282½	343⅓	314⅔	327½	100	107¼	87½	228¾	234	174¾	286	333⅓
100 canes of Toul. and Upper Languedoc		199½	266⅔	263⅓	320	193⅓	304	93⅓	100	81½	213⅓	213	162¾	266⅔	309⅛
100 canes of Genoa, of 9 palmes......		245½	327	323	392⅔	359⅔	374½	114½	122⅔	100	261¾	268½	199½	327	381
100 vares of Castile and Biscay.......		93¾	125	123⅓	150	137½	143¼	43¾	46⅔	38½	100	102¼	76¼	125	145⅔
100 vares of Cadiz and Andalusia.....		91¾	122¼	119	146⅔	134½	139¾	32¾	45	37¼	97⅔	100	74⅔	122¼	142
100 vares of Portugal or Lisbon......		123	164	162	196½	180½	187⅔	57½	61½	50	131½	134	100	164	191
100 covedoes of Portugal or Lisbon...		74	100	98⅚	120	110	114½	35	37½	30½	80	81½	61	100	116½
100 brasses of Florence, Leghorn, &c..		65⅓	85⅓	84⅓	102⅔	94	98	30	32	26½	68¼	70⅓	32½	95½	100

N. B. By means of this Table, the reader may please to observe, that 100 yards of England make 133⅓ ells of Holland; and in like manner he will find how the measures of other places in the table correspond with each other. By the common rule of three, or proportion, he will easily make his computations for any quantity required.

Figure 5.2 Anyone engaged in the Atlantic economy had to grapple with wildly varying units of measurement. Joseph Blunt's *Merchant's and Shipmaster's Assistant* (1832) offered help converting among the many different "Long Measures" of the "principal trading cities of Europe."

Source: Original in the John Carter Brown Library at Brown University.

fourteenth century forward, these circulars made prevailing commodity prices available to anyone who could pay for a subscription. While tables of weights and measures varied by place and remained stable across time, prices currents were specific to place but changed over time. Armed with both of these and the golden rule of three, a trader could calculate the value of any good in any port. Prices currents were expensive—for centuries they were used only by merchants and traders. But, in the late eighteenth century falling prices made them more accessible, and by the mid-nineteenth century they appeared regularly in general newspapers.[30]

With the help of tables and prices currents, numbers traveled well. They retained validity across time and space and could be brought up to date (or place) with a simple application of the rule of three. Fragmented currency and measurement systems presented real barriers to exchange, and paper technologies of calculation bridged gaps, connecting local systems of payment and measurement to global trade. Nowhere is this clearer than in the conversion of currencies. For example, in Jamaica, the "Grand Mart" of British America, traders had to negotiate among a wide array of currencies, even if they restricted their business to gold coin.[31] Blunt's *Assistant* offered a special table for traders there (Figure 5.3). He listed the legal weight and value of five kinds of Portuguese coins, four kinds of Spanish coins, and three kinds of English coins. Of course, coins could be clipped, so the relationship between currency

The following are the GOLD COINS which circulate at JAMAICA, with their legal Weight and Value :

		Dwt.	Gr.	Currency.
Portuguese	Joannes, or Joe . . .	18	12	£5 10 0
	Half-Joannes, or Joe .	9	6	2 15 0
	Quarter do.	4	15	1 7 6
	Moidore . . . : .	6	22	2 0 0
	Half do.	3	11	1 0 0
Spanish	Doubloons	17	18	5 0 0
	Double Pistole . . .	8	16	2 10 0
	Pistole	4	8	1 5 0
	Half do.	2	4	0 12 6
English	Guineas	5	8	1 12 6
	Half do.	2	16	0 16 3
	Seven-Shilling Piece .	1	19	0 10 10

If gold coin be lighter than the weight expressed above, 3 pence are deducted for every grain deficient.

Figure 5.3 Conversions between currencies and conversions between weights and measures sometimes overlapped. This table describing the many kinds of gold coins circulating in Jamaica shows the complexity of converting currencies, including adjusting for variation in weight on clipped or worn coins.

Source: From Joseph Blunt's *Merchant's and Shipmaster's Assistant* (1832). Original in the John Carter Brown Library at Brown University.

and weight blurred. Blunt appended instructions on how to adjust the value of coins that were lighter than their minted weight. And this was only one of several related currency tables—Blunt also included ratios for converting between European monies and the different varieties of American dollars.[32]

Many table books and prices currents assumed that their users already knew how to read tables. But some authors and printers supplied directions, formulated as a series of simple, algorithm-like procedures that resembled the language used in the rule of three. In 1805, Nathaniel Allen drew or copied a table into his cyphering book to assist with the reduction of "the currencies of the several United States; also, Canada, Novascotia, and Sterling, each to the par of all the others." He inscribed instructions at the top of the chart, telling users to "see the given currency in the left hand column, and then cast your eye to the right hand till you come under that required, and you will have the rule."[33] Another handwritten table that belonged to Sarah Pollock gave similar advice. At the top of the document the copyist wrote: "seek the given Coin in the left hand column, then cast your eye to the right hand till you come under the required Coin and you will have the Rule."[34] Users did not need any mathematical background to use tables like these, nor did they need to have any general fluency in consulting tables. They merely had to follow instructions. Just as the innumerate could memorize the "rule of three," anyone who could read, multiply, and divide could use these tables to convert among currencies.

The use of basic multiplication tables could render even the need to multiply unnecessary. *Cocker's Arithmetic* offered a simple multiplication table with procedural, algorithm-like instructions that could replace the need to memorize multiplication tables—as well as the arithmetical intuition that generally accompanied such learning. The lengthy prose instructions under Cocker's simple table reflect the complexity of instructing someone who could neither multiply nor read tables in how to find a product (Figure 5.4). As he explained, "if you would know the Product of any two single Numbers multiplied by one another, look for one of them (which you please) in the up-permost Column, and for the other in the side Column, and running your Eye from each Figure along their Respective Columns in the common Angle (or Place) where the two columns meet, there is the Product required."[35]

Before using a table like Cocker's to multiply, students of course had to learn to count. But even this process—called "numeration" in arithmeticks and other manuals—could be explained using a table and prose instructions. Daniel Fenning's *British Youth's Instructor* offered a table and directions for numeration similar to that found in other late eighteenth- and early nineteenth-century copy books and manuals (Figure 5.5). Fenning advised students that "this table you ought to get by heart, at least, to understand the nature of it." But like Cocker he advised them to use it as a series of physical procedures, telling students where to cast their eyes, beginning at the top of the chart. The title of Fenning's book explained the reason for this mechan-ical process: the book was intended to help those who could not do math to find the answers they needed to conduct their business. As the full title adver-tised, by using Fenning's volume, "any Person may, of himself, (in a short Time) become acquainted with every Thing necessary to the Knowledge of Business." It aimed to help "Country Youths in particular" with "the whole" work having been "designed for such as been hitherto neglected," especially those who "have not had an Opportunity of becoming acquainted with Figures."[36]

Tables and table books used algorithmic procedures to turn basic literacy into practical numeracy. Common laborers could consult simple table books called "ready reckoners" to verify their wages. Calculating payments could be complex, especially in the heat of a negotiation or the brief moment when a worker reached the front of a payroll line. Determining your due often re-quired both multiplication of fractions and conversion between currencies. *Arnold's Ready Reckoner*, a text published in Providence in 1842, makes this potential difficulty clear. A worker who had agreed to be paid "4s 6d" or 4 shillings and 6 pennies per week and had worked 5⅛ days could find the amount he was due in dollars and cents simply by looking at the index, turning to page 10, and finding the intersection of his wage rate (4s 6d,) with the time worked (5⅛ in the left-hand column). The answer provided by the text was 64 cents.[37]

$MULTIPLICATION\ TABLE$

1	2	3	4	5	6	7	8	9
2	4	6	8	10	12	14	16	18
3	6	9	12	15	18	21	24	27
4	8	12	16	20	24	28	32	36
5	10	15	20	25	30	35	40	45
6	12	18	24	30	36	42	48	54
7	14	21	28	35	42	49	56	63
8	16	24	32	40	48	56	64	72
9	18	27	36	45	54	63	72	81

The ufe of the preceeding table is this. In the uppermoft line or column you have expreffed all the digits
from 1 to 9; and likewife beginning at 1, and going
downwards in the fide column, you have the fame; fo
that if you would know the product of any two fingle
numbers multiplied by one another, look for one of
them, which you pleafe, in the uppermoft column, and
for the other in the fide column ; and running your eye
from each figure along the refpective columns, in the
common angle, or place where thefe two columns meet
there is the product required. As for example : 1 would
know how much is 8 times 7. Firft, I look for 8 in
the uppermoft column, and 7 in the fide-column ; then
do I caft my eye from 8, along the column downward
from the fame, and likewife from 7 in the fide-column
I caft my eye from thence toward the right hand, and
find it to meet with the firft column at 56 ; fo that I
conclude 56 to be the product required. It would have

Figure 5.4 Lengthy prose instructions telling users how to look at a table often
accompanied even basic tables like this one from Cocker's *Arithmetic*.
Source: Edward Cocker, *Cocker's Arithmetic* (Belfast: Printed by and for Henry and Robert Joy, 1756).

Though savvy calculation alone did not win negotiations, the inability to
cypher would almost certainly lose them. Knowing how much you were owed
or whether your wage would cover your room and board was no small matter.
Some ready reckoners were marketed directly to workmen. The *Mechanics'
and Laborers' Ready Reckoner* aspired to help laborers to "find … at a glance,
the amount due them for labor, for any length of time, from one quarter of a

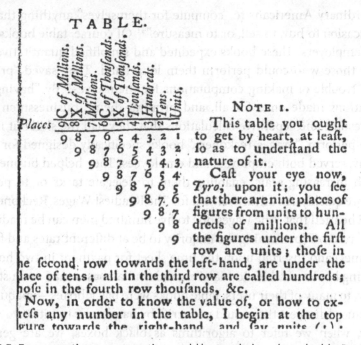

TABLE.

Places	C of Millions.	X of Millions.	Millions.	C of Thousands.	X of Thousands.	Thousands.	Hundreds.	Tens.	Units.
	9	8	7	6	5	4	3	2	1
		9	8	7	6	5	4	3	2
			9	8	7	6	5	4	3
				9	8	7	6	5	4
					9	8	7	6	5
						9	8	7	6
							9	8	7
								9	8
									9

NOTE 1.

This table you ought to get by heart, at leaſt, ſo as to underſtand the nature of it.

Caſt your eye now, Tyro, upon it; you ſee that there are nine places of figures from units to hundreds of millions. All the figures under the firſt row are units; thoſe in he ſecond row towards the left-hand, are under the place of tens; all in the third row are called hundreds; hoſe in the fourth row thouſands, &c.

Now, in order to know the value of, or how to ex-reſs any number in the table, I begin at the top ure towards the right-hand, and ſay units (1),

Figure 5.5 Even counting or numeration could be taught by rule and table. Daniel Fenning's *British Youth's Instructor* told students to cast their eyes to a table of numeration that explained the "nine places of figures from units to hundreds of millions." Such an approach to counting appears unnecessarily complex to modern readers, but it offered numeration by rote to the uninitiated.
Source: D. Fenning, *The British Youth's Instructor: or, A New and Easy Guide to Practical Arithmetic* (1767).

day to one year," all "without reference to figures."[38] These books shaped negotiations over wages and working conditions, even helping workers understand their legal rights. Following the passage of the 1874 Factory Act in England, John Calvert of Manchester published two guides to assist with calculating wages for the new 56½-hour workweek. One volume, priced at two shillings, was intended "for office use." The other was sized "for the pocket" and cost a mere three pence, "within the reach of working men."[39] Those who did not own a copy could also demand reference to their employers' books. Another wages reckoner suggested that it "be consulted by both the Employer and the Employed to settle any difference that may arise between them."[40]

Differences in measurement and currency had to be overcome by a wide array of individuals who were caught up in the new, more calculating commercial world. In a 1790 report on the standardization of weights and measures, Thomas Jefferson explained his fear that the "present complicated and difficult ratios" placed commercial calculation "beyond" the "comprehension" of many Americans. A standard system of measurements would enable ordinary Americans to "compute for themselves" anything that they "have occasion to buy, to sell, or to measure."[41] Of course, table books could also aid employers. These books expedited and simplified quantitative tasks even for those who could perform them long hand. They saved "practical men the trouble of making computations for themselves." By "having such computations made once for all, and printing them," businessmen could avoid tiresome and repetitive calculations that they had the skill but not the desire to perform. "Wages-reckoners," books specifically designed for calculating pay, served both employers and employees. They helped businessmen deal with increasing scale, enabling them to delegate tasks or to perform them more quickly. An advertisement for The Readiest Wages-Reckoner Ever Invented boasted that "The wages due to five hundred men can be readily and correctly read off (presuming the majority to be at different rates and for different times) and inserted upon the time sheet for payment in two hours."[42] Comparing books of tables to massive modern databases may seem strange. However, to many of their users, table books were black boxes—opaque tools that produced answers they could not verify by hand or by intuition.

Today, when we refer to algorithms as black boxes, we are generally describing either their overwhelming complexity or the fact that they are proprietary. With modern algorithms and machine learning, not only do users not understand how they work—sometimes designers do not know either. And when designers do grasp their logic, aspects may be proprietary or difficult to communicate. By contrast, early modern algorithms were neither complex nor secret. Nonetheless, though they could be understood, most of their users did not have the necessary training. Nor did they seek it or even judge it important. They were black boxes not because users *could not* look inside but because they generally *did not* look inside. The users of such tools simultaneously judged them to be fundamentally important and also politically neutral. They democratized access to arithmetic as an essential tool for negotiation, but with this broadened access came a set of assumptions about the correctness of the answers such tools provided.

To do ideological work, algorithms need not be impenetrable: their politics simply have to be overlooked. The democratization of conversion and proportion through the rule of three was not without costs. In *Capital*, Karl Marx famously described the dual nature of commodities as possessing both "use value" and "exchange value." He explained this shift in perception through what could be described as a series of applications of the rule of three—a string of conversions from one good or unit to another. Eventually all goods are converted into a single "money-commodity"—gold, in Marx's analysis. He argues that this process of transformation—from a thing known qualitatively to a commodity perceived quantitatively—mystifies and obscures social relations.

More specifically, "the rise of the commodity form erases the value of labor, hiding it under the apparent fluctuations in the relative values of commodities." In Marx's phrase, use value asks, "how and what?" while exchange value asks merely, "how much?"[43] The rule of three enabled a far wider range of people to answer the second of these questions. In doing so, it offered them a measure of power. However, at the same time, the rule helped to shift the terrain of negotiation from the first question to the second.

As the terrain of negotiation shifted to "how much?," quantity was held up as a goal in itself. Peddlers of counting tools boasted about the sheer number of calculations they contained, as if the quantity of calculations could somehow help to order the world around them. The cover of Warne's *Model Ready Reckoner*, a pocket-sized volume, touted 40,000 calculations, and William Chadwicks's much larger *Combined Number and Weight Calculator* advertised that it contained "250,000 direct calculations," which "by a single addition to each" would produce "a combination of over 20 million calculations!"[44] As one historian has written about Victorian Britain, such tables displayed the "limitless fecundity of numbers," transforming them "into a commodity that would bring the power of calculation within the reach of the ordinary citizen."[45]

Early calculating machines were quite literally closed boxes that concealed mechanisms. In 1858, an English writer described "Baranowski's Ready Reckoner, lately invented in America." This "simple machine" was designed for various "questions in which sums of money are concerned; such as days' wages at so much per day, prices at so much per lb., or interest at so much per cent." A version of this machine designed for wages would be housed in "an upright box, with a handle at the bottom, rows of figures up the front, and a number of small slides moved by studs." The device was literally a black box. Concealed "within the box" was "a paper on which rates of wages are printed." By turning the handle the appropriate wage rate was "brought to the opening," and the machine will show "the amount of wages earned in any fractional number of days and hours, at that rate." Those who were quick at mathematics, sometimes themselves referred to as "ready reckoners," might not find the machine particularly useful. But for the innumerate and the minimally numerate, machines and table books greatly increased speed and accuracy.[46]

Like table books, calculating devices were lauded for the sheer size of the numbers they manipulated—they gave their users access to unprecedented and seemingly incomprehensible amounts of information. A New York magazine praised a French calculating machine called an arithmometer in particularly grand terms. The wonders of this practical device, already used in "many great financial establishments," were best expressed the size of its sums.

It could "furnish in a few seconds products amounting to 999,999,999,999,99 9,999,999,999,999,999," a "marvelous number" that compared to "the infinite multitude of stars which stud the firmament, or the particles of dust which float in the atmosphere."[47] As the question "how much?" came to dominate markets, questions of quantity replaced more complex questions, with large numbers signifying more than math.

Capitalism by Algorithm

On the rare occasions that historians train their sights on units of measure, it is usually with the intention of adjusting them away. But leaving out the messy process of adjustment misses the complexity of everyday commerce. The process of shifting between currencies, units, and quantities was both difficult to master and fundamental to survival in the markets of the eighteenth- and nineteenth-century Atlantic world. The rule of three enabled the otherwise innumerate to engage in the increasingly global economy. Alongside a wide array of paper technologies like ready reckoners and table books, the rule made certain kinds of conversion possible without a deeper understanding of mathematics. It was easily learned and easily transported from one location to another. The rule turned basic literacy into practical numeracy. It could also be memorized, enabling even the illiterate to speak the language of mathematics.

Market expansion required standard methods for commensuration to be learned and accepted, not just in counting houses but by the many other individuals connected to the growing economy. The expansion of the rule of three could make such transitions appear as neutral and natural as arithmetic. When students were taught that a calculation was correct because "the rule says so," it might also suggest that wages or prices were correct because a rule said so. This simple algorithm might even discourage its users from questioning the social relations that made the rule necessary in the first place.[48]

So too with algorithms today. With supercomputers in our pockets, we all stand to benefit: algorithms offer ever-larger numbers of people access to new ways of managing everything from money to love. But these "democratized" technologies can also change the terms of negotiation in ways that are easy to overlook. They let us shop for the lowest prices, but they rarely highlight the political circumstances that produced different prices and different products. Nor do they reveal the social relations that make different prices available to different people. And they never question the structures and values that sent us shopping in the first place. Likewise, they can help us find and build relationships, but they rarely tell us why these relationships are appealing or what kinds of racial and economic divisions they might both reflect and reinforce. Such dangers are magnified because of what Ruha Benjamin has dubbed the "progressive narratives that surround technology"—in her analysis the "assumption that technology is race-neutral" when critical analysis of such technologies often reveals the opposite.[49]

And yet, the story of the rule of three also contains the opposite lesson offered by the landslide of recent work on data and algorithms: that algorithms can put new tools in different and sometimes unexpected hands. Even as quantified, capitalized language gained authority, the rule of three reminds us that individuals were using arithmetic to negotiate on the ground. They encountered mathematics not as an abstract discourse that swept away the texture of daily life, but as an absolutely essential tool for making sense of it. Armed with a table book, calculating individuals could independently confirm or dispute the costs of doing business. A worker could argue for increased compensation. An enslaved person could contest the price of goods in an otherwise unequal marketplace. The stakes of this knowledge were high: for many, being overcharged or shorted on wages was a matter of survival. And, for middling craftsmen and shopkeepers, care in calculations separated profit from loss.

In a sense, the "rule of three" and other early modern "apps" offered workers and enslaved people a new tool to audit, but it was a narrow tool indeed. The power granted by this "app" mostly failed to direct attention to questions of equity, fairness, or justice. Perhaps it even distracted people from these political questions. As Safiya Noble has written, "social inequality will not be solved by an app ... we will not sort out social inequality lying in bed staring at smartphones."[50] To make algorithms into truly abolitionist tools will require the cultivation of "unflinching accountability" that goes far beyond techno-optimistic design thinking and extends to the democratization of data.[51]

Speed and size are changing the ways we encounter algorithms. But speed and scale are not the only or the most important reasons that algorithms become black boxes. In a sense, algorithms have always had the potential to obscure and redirect: they are recipes and procedures that simplify more complicated tasks and thus make them more easily accessible but simultaneously more invisible. This was as true in the eighteenth- and nineteenth-century Atlantic world as it is today. Then and now, black-box technologies could be democratic. But these tools could also cover over power and politics. Algorithms can help us to navigate and negotiate the world around us, but they can simultaneously define and narrow the terms of our negotiations, often in ways that are easy to overlook.

6

Material Mathematics

British Algebra as Algorithmic Mathematics

Kevin Lambert

In 1844, mathematician George Boole published a paper in the *Philosophical Transactions* that would win him the prestigious Royal Society's gold medal for mathematics (generally known as the Royal Medal).[1] The theoretical practice central to Boole's paper, known as the "method of separation of symbols," directly informed all of Boole's work, including his work on logic.[2] It also substantially extended the reach and power of a research program closely associated with the authors and readers of the recently founded *Cambridge Mathematical Journal*. The "method of separation of symbols" involved separating symbols of operation from the quantities on which they operated and then manipulating those symbols as if they were symbols of quantities.[3] This procedure or algorithm is now a standard practice for students of the differential calculus, and Boole's paper was a landmark in what Uta Merzbach and Carl B. Boyer have described as the way in which "British mathematicians in the second half of the nineteenth century were thus again becoming leaders in algorithmic analysis."[4]

The designation "algorithmic analysis" is, however, faint praise. Brian Rotman has described a hierarchical division in Western thought that values *arithmetica* (numbers contemplated philosophically; ideal, perfect objects) over *logistica* (numbers as empirical objects calculated with by slaves). That hierarchy also appears to inform historian of mathematics Jeremy Gray's recent contention that we "should not look at the story of pure mathematics in Britain in the 19th century as a success story, but as a particular kind of failure."[5] Victorian algebra fails for Gray because it lacks a profundity and rigor he finds in nineteenth-century continental mathematicians such as Augustin-Louis Cauchy. Cauchy's mathematics, for instance his work to provide a rigorous foundation for differential and integral calculus, is an almost perfect example of the philosophically ideal and more perfect *arithmetica* in Rotman's hierarchy.[6] And it would seem that British algebra did not fully succeed, at least as far as Gray is concerned, because it did not transform

Kevin Lambert, *Material Mathematics* In: *Algorithmic Modernity*. Edited by: Morgan G. Ames and Massimo Mazzotti, Oxford University Press. © Oxford University Press 2023. DOI: 10.1093/oso/9780197502426.003.0007 eighteenth-century mathematics into rigorous nineteenth-century analysis in the way that Cauchy's did.

But just as art historians have begun to appreciate that they should not dismiss Victorian painting because it was not French Impressionism, it may be time to begin taking nineteenth-century British algorithmic mathematics more seriously.[7] There are some interesting parallels between the arguments made by historians such as Gray, who see Victorian algebra as a kind of failure and resistance to the use of computers in late twentieth and early twenty-first century mathematics research. In a book published in 2003, Donald MacKenzie writes that: "The areas where 'proof' is predominantly computer proof, and where computer proofs are positively welcomed, indeed more highly valued than human proofs have … to date remained mainly outside of pure mathematics."[8] Initiatives such as the journal *Experimental Mathematics* have been working to make visible "mathematical experiments … carried out on computers," experiments that will "have a theoretical impact."[9] And in an effort "to raise hackles," philosopher Ian Hacking has written that "Experimental mathematics provides the best argument for 'Platonism' in mathematics: that is, the idea that mathematics is just 'out there,' a given. We explore it with many tools, including pencil and paper, and now computers. But contrary to many philosophers, this does not leave everything the same, not to worry."[10]

For Rotman, Hacking, and MacKenzie, attention to the increasing use of computers in mathematics research makes it look less like an ideal science, art of necessary truth, or *arithmetica* and more like *logistica*, and suggests that the experimental or material practices, with which mathematicians have always been involved, need greater attention. In this chapter, then, I propose embracing the label "algorithmic mathematics" as an accurate description of Victorian mathematics.[11] In so doing I wish to follow Rotman's interest in what he calls *arithmetica* or nomadic mathematics and the way it calls attention to the importance of the materiality of mathematics. One of the things I will show is that the emphasis on empiricism in nineteenth-century British mathematics, whatever its failures, allowed it to stay open to possibilities that continental rigor closed down. That openness, characteristic of *arithmetica* practice, depended upon the exploration of material spaces.[12] The focus will be on two mathematicians, George Boole and Augustus De Morgan, and the importance of libraries for their mathematical research. Like Rotman, I will be interested in embodying the mathematician as a sign-using subject. Attention to the formal structure of Boole's algebraic logic and De Morgan's research practices, will show how the use of libraries was central to their mathematical work.

Material Environments and Virtual Spaces

The method Boole employed in his mathematics depended upon a set of practices developed at Cambridge in the first decades of the nineteenth century. Charles Babbage, a leading figure in the Analytical Society, which initiated the campaign to reform Cambridge mathematics in the 1810s, developed a "philosophy of analysis" that argued for a symbolic algebra that had developed out of, but had subsequently become completely independent of, arithmetical algebra.[13] The truth of algebraic symbols, Babbage argued, depended not on the meanings associated with them but rather on the consistency of the laws that governed how they could be manipulated.[14] Colleagues of Babbage, most especially George Peacock, would develop and teach this algebra of operations at Cambridge, where it would be taken up by a second generation of mathematicians, one of whom, Duncan Farquharson Gregory, would become Boole's most important mentor.

On the opening page of his "On a General Method of Analysis," Boole uses a quotation from Gregory to lay "down the fundamental principle of the method in these words: 'There are a number of theorems in ordinary algebra, which, although apparently proved to be true only for symbols representing numbers, admit of a much more extended application. Such theorems depend only on the laws of combination to which the symbols are subject, and are therefore true for all symbols, whatever their nature may be.'[15] It is the principal premise of Boole's paper that these rules of combination, the commutative law, the distributive law, and the index law, also govern the rules of combination for symbols of operation such as d/dx.[16] The generality Boole sought, then, came from looking for "classes of operations" that had applications across different mathematical domains. For Gregory, Boole, and De Morgan the rules that governed operations on quantities (i.e., arithmetic) were the same as those that governed operations in algebra and the use of those laws in symbolic algebra could lead to the discovery of new kinds of mathematical objects such as negative and imaginary numbers.

The operational algebraic practices Boole and De Morgan employed would also inform their work on logic. Representing logical relationships they found in books symbolically allowed the possibility of discovering laws that governed the ways those symbols combined, which in turn allowed the discovery of new kinds of logic or logical relationships. This emphasis on the general applications of laws of symbolic combination bears some analogy to the problem that writers of computer algorithms face when exploiting a small number of operations, such as "flipping, zeroing, or testing a bit," in order to make procedures that perform tasks.[17] But we need to be careful not to take the analogy too far. The twentieth-century mathematician Alan Turing, for example, famously made a parallel between a human writing in a note-book and an automatic computer.[18] Such a comparison might be useful for a mathematician but not for a historian, who must be more interested in the differences that follow from work performed in different material environments. Therefore, I am not arguing that because Boole used his mathematical practice to produce a kind of algebra, Boolean logic, that has proved useful to computer scientists, there is some kind of equivalence between those practices.[19] But I am interested that both Boole's algorithmic practice and that of the computer programmer are pursued in materially bounded worlds. Looking at how Boole and De Morgan's practice was pursued in a partic-ular virtual space afforded by their material environment will show how the materials mathematicians use shape their practice and the virtual worlds they explore.

Boole and De Morgan researched mathematics through the employ-ment of a Victorian way of "thinking with history," which the historian Carl Schorske has described as a characteristically nineteenth-century way to use the resources of the past to manage the present.[20] In so doing, Boole and De Morgan put less emphasis on mathematical rigor and more on finding histor-ical evidence that would support the truth of new kinds of mathematical rea-soning and objects. Libraries were a crucial resource for this kind of thinking with history. Boole and De Morgan used books as historical evidence of past forms of reasoning. They sought to discover laws that governed reasoning procedures; laws that might also lead them to the discovery of new kinds of logical relationships, algebras, and mathematical objects. This historical ap-proach to mathematical research had become possible after what has been called a nineteenth-century information revolution fed by developments in printing and distribution.[21] If the explosion of printed books in the second quarter of the nineteenth century contributed to the formation of a Victorian Information Society, then the nineteenth-century library was a key institution for managing it.[22]

As instruments for managing the Victorian information society, nineteenth-century libraries were not only depositories for knowledge, but also tools for maintaining social order.[23] Boole's first major library, the Lincoln and Lincolnshire Mechanics' Institute library, was part of a set of institutions disparagingly labeled the "steam intellect societies," a satirical ref-erence to Whig M.P. Henry Brougham's Society for the Diffusion of Useful Knowledge.[24] As such, it was part of a radical Whig campaign to promote both industrialization and common cause between the working classes and the industrial bourgeoisie.[25]

However, it would be wrong to assume that Brougham's campaign for the diffusion of useful knowledge and the institutions it inspired were intended to do no more than placate a belligerent working class, even if that was how the more radical members of nineteenth-century socialist movements described them.[26] Although the middle class and gentry initiated the mechanics' institutes, the creation and maintenance of services such as the library depended upon the participation of members of the moderate artisanal classes such as Boole. Furthermore, although long working hours, middle-class condescension, and subscription fees did keep much of the working class away from mechanics' institute libraries, a range of people, including some from the aspiring working and middle classes, used these libraries at a time when working people had limited access to books.[27] As we shall see, Boole's participation in the organization of the Lincoln Mechanics' Institute library deepened his knowledge of mathematics and made books available to him that went beyond the collection in the library itself. Boole's involvement with the library was crucial for his entry into the community of early Victorian research mathematicians.

The Lincoln Mechanics' Institute library included (among other things) the building, the books, a museum of collected objects, scientific instruments, the subscribers to the library, and also the information (including mathematics) that its books, members, and objects contained. As such I will call it an "assembling," which is an adaptation of the more commonly used "assemblage." An assemblage is a collection of people and things that have a temporary individual integrity, but which can, nonetheless, be broken down into constituent parts. Assemblages allow for the possibility that social groups or bodies of knowledge are more than just ideas; that they also have a material ontology. Assemblages are collectives of people and things that afford activities that individuals or isolated things cannot.[28] But if a library can be considered an assemblage, it is one that will never achieve a final form. Working libraries have a kind of life that depends upon an ongoing exchange of things and flows of people; a being-in-process that is always unfinished and is open to further addition and change in a way that facilitates social relationships and the production of new knowledge.[29] It for those reasons that I prefer to imagine the nineteenth-century library as an assembling. An assembling, as an ongoing process, appears to me to be something one can easily imagine visiting and participating in and better captures the dynamic nature of a library, which can only achieve a final assembly when it is no longer in use.[30]

The Late Regency flood of cheap books also created greater opportunity for large personal libraries, a quite remarkable example of which belonged to the first professor of mathematics at University College London, Augustus

De Morgan. De Morgan's library demands attention because he was a leading Victorian mathematician whose worked on logic would inform Boole's. We also know much more about the books that De Morgan assembled than we do about those that Boole used in his mathematics research. As a consequence, De Morgan's private library offers evidence that will provide a better understanding of the way in which Victorian mathematicians used their libraries to think with history and to explore new kinds of mathematical objects.

George Boole and the Lincoln Mechanics' Institute Library

George Boole was the son of shoemaker and so had limited opportunities for education or access to the material resources needed in order to pursue advanced mathematics.[31] His father, John Boole, was a highly literate scientific artisan who saw education as an important means of social advance and so was keen to provide his son with some formal education, beginning at a small commercial school owned by a friend. There Boole achieved a precocious grasp of Greek and Latin, perhaps as much due to the attentions of his father and family friends such as the bookseller William Brooke, as to the school. Indeed, Boole's education was possible, in large part, because of a circle of people that included tradesmen, clergy, merchants, and local gentry that participated in Lincoln's rich nineteenth-century cultural environment.[32]

Access to a broad education was one thing, but how Boole managed to find the training needed in order to participate in the Victorian mathematics research community is another. No doubt, help from mentors such as D. F. Gregory, editor of the *Cambridge Mathematical Journal*, and president of the Lincoln Mechanics' Institute, Baronet Sir Edward Ffrench Bromhead were indispensable.[33] But access to libraries was no less crucial. Both Boole and his father were deeply involved in the Lincoln Mechanics' Institute, and George Boole, especially, clearly saw the library as an important means of achieving the institute's goal of allowing greater access to learning. Boole participated in the composition of the first list of books for the library and, twelve years later, was asked to write a detailed report on the progress of the institute library, a testimony to his ongoing commitment to its maintenance and growth.[34]

George Boole's involvement in managing the Lincoln Mechanics' Institute library contributed to his ability to see further into mathematics. He helped compile and publish the institute library's printed catalog, which contains a number of valuable mathematics books including Euclid's *Elements of Geometry*, Hutton's *Mathematical Tracts*, and more recent works such as

De Morgan's *Elements of Arithmetic*. But although these works would serve as excellent introductions to mathematics for the institute library's general users, they were hardly the kind of books Boole needed to become a research mathematician. Books about advanced mathematics by French authors such as Joseph-Louis Lagrange or Silvestre-François Lacroix were expensive and could hardly be justified for purchase for the institute's library. So how did Boole gain access to them?[35] Access came through the library in two ways. First, through his work to found and grow the library, which inevitably helped increase his awareness of the advanced mathematics books that he could not justify purchasing for the library. Secondly, and directly related to the first point, his work for the institute also brought him into contact with people who could provide not only information about the British mathematics research community but also access to books. The most important of those was undoubtedly local baronet and president of the Lincoln Mechanics' Institute, Edward Bromhead.

When Bromhead became president of the Lincoln Mechanics' Institute in 1833, he was a barrister at law. During his undergraduate days at Cambridge he had been an original member of the Analytical Society reformers that had brought continental mathematics to the university, and he still retained an active interest in mathematics and science.[36] A moderate Tory, Bromhead supported radical Whig M.P. Henry Brougham's campaign for the diffusion of useful knowledge. He was also very familiar with the best place to acquire specialist mathematics books in early Victorian England, Deighton's in Cambridge, a bookshop he continued to visit, even after he had left the university, to browse its shelves for the latest publications on French mathematics.[37]

Bromhead was almost certainly in attendance for George Boole's first public lecture, which took place at the Greyfriars Chapel in Lincoln and was attended by all the major figures in the town's scientific community.[38] In April 1836, Boole and Bromhead began to correspond and Bromhead now actively encouraged what he called Boole's "scientific pursuits," not least by providing him with "valuable books."[39] Bromhead may also have played a role in arranging for Boole to meet D. F. Gregory, the editor of the first British mathematics research journal.[40] A fairly typical exchange from 1839, three years into their correspondence, has Boole apologizing for two volumes by Lagrange, which he had kept "a long time but hope that you will excuse me. They have been of great service to me and I beg to thank you again."[41] In 1841, Boole wrote to Bromhead about his (Boole's) recent investigations into "certain depts. of the Theory of Differential Equations" and his need to consult "the second volume of Lacroix's larger work on the calculus," which, "should the work be in your possession you will perhaps add to your former kindness by lending it to me for a short period." Boole was also keen to mention that "The 12 vol. of Crelle's Journal contains a memoir of Jacobi on the transformation of multiple integrals, which I am exceedingly desirous of seeing. There is a copy in the Public Library at Cambridge but I fear it will not be possible to procure it thence. Perhaps you could inform me if there is any way of procuring such books on a particular emergency without actual purchase."[42] It was Bromhead, above all, who provided Boole access to the material cognitive scaffolding he needed in order to climb the intellectual and social scales necessary to become a published research mathematician and, ultimately, gain a university position. Boole was appointed to the position of professor of mathematics at the newly opened Queens College Cork in 1849 without ever having received an undergraduate university education.

Boole's relationships with Bromhead and Gregory also changed the library. In his 1846 report on the library, Boole was critical of those parts of the library's collection that dealt with practical mathematics as well as introductory algebra and geometry. To remedy the situation, he recommended acquiring a number of books including George Peacock's *Treatise on Algebra*, a foundational text for the method of separation of symbols, which informed both Boole's and De Morgan's research. He also recommended some works in higher analysis. He recognized, of course, that such works would only be of interest to exceptional individuals and therefore should not be prioritized, but he was still willing to propose that Gregory's *Examples of the Differential and Integral Calculus* and Laplace's *La Méchanique Celeste* might gradually be acquired just in case any exceptional individuals wished to consult them.[43] Since, at its peak, after 1850, the library is reported to have had 6,000 books, it seems possible that at least some of Boole's recommendations were implemented. Access to this increasingly valuable collection remained relatively cheap, free for members or for two pennies a week for subscribers.[44]

Boole's participation in the assembling of the institute library enabled him to do research mathematics in three different ways. Social connections made at and through the library put him in contact with figures such as Bromhead and Gregory who could help him develop intellectually to the point at which he could participate in the early Victorian mathematics research community. It also provided knowledge of, and access to other sources, libraries such as the Cambridge University Public Library, booksellers such as Deighton's in Cambridge, and, no doubt, publishers at the hub of the Victorian book trade in London. Lastly, the research required to expand that collection would also have increased his awareness of not only mathematical books but also other kinds of printed materials including those in other areas of knowledge.

Boole's *Mathematical Analysis of Logic*

Boole's development as a mathematician also changed how he used libraries. In particular, Boole's involvement with the Lincoln Mechanics' Institute library provided important resources for the project he had inherited from Gregory, of generalizing the method of separation of symbols. A particularly striking example of that work was Boole's mathematical analysis of logic.

Toward the end of the Napoleonic wars, modernizing pressure began to build both inside and outside the ancient universities. If the mathematical reforms initiated by the Analytical Society were an expression of the modernizing tendency at Cambridge, then a movement led by the Provost of Oriel College Edward Copleston, the "Oriel Noetics," performed a parallel role at Oxford in roughly the same period, from 1814 to 1826. The Oriel Noetics achieved something that, as recently as the beginning of the nineteenth century, had appeared to be almost impossible: "breathe new life" into logic, transforming it from a "moribund subject" about which Aristotle and his medieval commentators appeared to be the last word, into a progressive science.[45] Central to that achievement was a collectively written article, "Logic," that appeared in the first issue of the *Encyclopedia Metropolitana* but would quickly be developed into a book, *The Elements of Logic*, published in 1827 and attributed, like the article, to Oriel College fellow and later Archbishop of Dublin Richard Whately.[46]

The Noetics campaign to revive logic attracted the attention of mathematician Augustus De Morgan, who quickly made a genuine advance, the idea of a "universe of discourse."[47] De Morgan's studies of Aristotle's Greek had led him to notice difficulties with certain basic linguistic forms, most notably negatives such as in the sentence "the weather is not hot."[48] To make a sentence in that form make logical sense, De Morgan argued, it had to be tied to a universe of discourse that contained all the possibilities the topic of such a sentence, in this case the weather, could be. Only if descriptions of weather conditions were limited to a universe of discourse containing only two categories, hot or cold and nothing between, could make saying the weather is not hot clearly mean that it is cold.[49]

Boole began corresponding with De Morgan in 1842 just as he was establishing himself as an important contributor to Gregory's *Cambridge Mathematical Journal*. But Boole did not begin thinking about logic until 1847, perhaps after a visit to De Morgan in January of that year, but more likely because of the interest he took in a dispute that erupted between De Morgan and Edinburgh professor of metaphysics and logic Sir William Hamilton in the spring.[50] However it happened, Boole began studying De Morgan's published work on logic that began with a paper, "On the Structure of the Syllogism," in which the universe of discourse was first introduced.[51] The result was two intuitions: first, that he could represent De Morgan's universe of discourse symbolically by the number 1. Second, that he could associate a category or class of things, the basic building blocks of Aristotelian logic, with a mathematical operation, the selection of a group of things sharing a common characteristic from a universe of objects. As he put it in *The Mathematical Analysis of Logic*: "Assuming the notion of a class, we are able, from any conceivable collection of objects, to separate by a mental act, those which belong to the given class."[52] So, for example, if we designate x the class of all blue things, the x represents the selection of all the blue things from a universe of objects under consideration. After the operation x has been performed, the class of things left behind, that is all those things not blue, can be represented algebraically as $(1-x)$. Boole could now combine these symbols algebraically. For example, if I combine x and $(1-x)$ I will have the universe of objects under consideration, that is, $x+(1-x)=1$, which exploits another operation, that of addition.[53] For $x+y$ to have meaning in Boole's logic, x and y had to be mutually exclusive, which means that an object in the universe of things under consideration cannot share the property that defines x and the property that defines y; in the case of $x+(1-x)$, the classes being added are those things that are blue and those that are not blue, which by definition cannot overlap.

Another kind of combination, which Boole had to take into account, were selections of groups of things that are not mutually exclusive, such as the class of things that are both blue and round (a blue ball, for example). For these kinds of combinations, he used what in an algebra involving quantities would be called multiplication. Thus, if a first group (blue things) is x and a second group (round things) is y, the class of blue round things is designated xy. I can make this group either by first selecting blue things from my universe of objects and then selecting all the round things from the group of blue things, or I can do it the other way around, with the same result. Algebraically, that means $xy=yx$, which in arithmetical algebra is known as the commutative law. A similar claim can be made for what in arithmetical algebra is known as the associative law, which is that $x(y+z)=xy+xz$ (where z is the class of cubes, say).[54] Thus far, then, Boole's algebraic logic appears to obey the same laws of combination as arithmetical algebra.

However, in the case of the third law of combination, the index law, which requires that $x^b x^c = x^{b+c}$, Boole's algebraic logic departed from arithmetical algebra. Consider, for example, the selection of all the blue things from a universe of objects, that would give x. Now let us select all the blue things in x, that would be everything in x, which is x. In fact, no matter how many times I perform the selection of all the blue things, I will always have the same number of things. Thus $x^n = x$ (or, more succinctly, $x^2 = x$). This departure from the index law meant that the symbols in Boole's mathematical logic combined in ways that were different from the way in which symbols representing quantities combined. Because Boole's mathematical logic had different laws of combination from those that governed arithmetic, Boole could claim to have discovered a new kind of algebra.

Boole considered his mathematical logical analysis a discovery and not a construction because his laws of combination yielded the syllogism, the basis of Aristotelian logic.[55] Up until Boole's work on logic, the justification for symbolic algebra had been the laws of combination for arithmetic, which were demonstrably true. This had been the justification for Boole's 1844 award-winning paper and for the method of the separation of symbols, which he had employed in it. That mathematical practice had encouraged "the notion that Mathematics are essentially, as well as actually, the Science of Magnitude." However, Boole argued, such a conclusion was "by no means necessary because "it may be doubted whether our experience is sufficient to render such an induction legitimate." Instead, Boole now concluded, symbolic algebra had been considered to be essentially quantitative because its rules of combination had always been determined through the investigation of quantity and so was "the result of the circumstances by which those forms were determined, and is not to be construed into a universal condition of Analysis."[56] In *The Mathematical Analysis of Logic*, Boole had shown that a form of reasoning that employed language and which had always been considered different from reasoning with quantities actually was a form of mathematics. What that meant for Boole was that he had discovered the "mathematics of the human intellect" or the laws of thought.[57]

Boole's investigation into the constitution of the human mind proceeded not through introspection but from "inductions" drawn from the study of the history of mathematics and logic in books and papers, especially canonical works such as Aristotle's *On Interpretation* as well as more recent publications such as Whately's *Elements of Logic*.[58] None of these books was available directly from the Lincoln Mechanics' Institute library but had to be procured from individuals, booksellers, and other libraries to which Boole had gained access through his participation in the institute library.[59] Those materials allowed Boole to make an extraordinary claim: that logic, which had hitherto been seen as system of reasoning quite separate from mathematics, was in fact algebraic. This significant gain for the domain of mathematics was justified though the books and other written and printed materials Boole had assembled, a justification, I want to suggest, which depended in part upon their physical existence. One way in which that material basis of Boole's algebraic logic found expression in its formalism, I would argue, is the insistence that any category of objects is defined by the selection of those things from a finite universe. The finitude of that operation, characteristic of algorithmic *logistica*, is beautifully captured by Boole's mathematical expression for the categories of things left after such an operation has been carried out $(1 - x)$.

Nonetheless, the material practice of assembling, which, I am arguing, was central to Boole's researches on logic, still need to be more fully described. To better appreciate the importance of those material practices, we have to turn to another library, that of Augustus De Morgan, and his relationship to the books it contained. One of the reasons De Morgan had been particularly concerned not to see Boole's *Mathematical Analysis of Logic* before he had finished his own work was because he knew that Boole shared his understanding of logic as a progressive science; that he and Boole both sought to advance logic by thinking with history. Investigating how De Morgan used his library will therefore allow a better understanding of what thinking with history actually meant for mathematicians such as De Morgan and Boole. In particular, I am interested in how an assembling contributed to Boole's and De Morgan's creativity, a creativity materially registered, we might imagine, in the Lincolnshire County Mechanics' Institute, as well as other libraries, by the addition of George Boole's *Mathematical Analysis of Logic*.[60]

Augustus De Morgan's Private Library

In 1847, when Boole was hard at work writing his *Mathematical Analysis of Logic*, his epistolary friend Augustus De Morgan wrote to him to say, "I would rather not see your investigations [into logical reasoning] till my own are quite finished.... When my sheets are printed, I will ask you for your publication [*The Mathematical Analysis*]: till then please do not send it. I expect that we are more likely to have something in common than Sir W. H. and myself."[61] De Morgan's worry was motivated in part by his recent experience of becoming embroiled in a plagiarism controversy with Scottish logician, Sir William Hamilton, but it was also driven by a deeper concern. In a draft of a letter to Boole written after both books had been published, De Morgan wrote, "Some of our ideas run so near together, that the proof of the physical impossibility of either of us seeing the other's work would be desirable to all those third parties who hold that, where plagiarism is possible $1 = a$ wherever $a > 0$."[62] As we have seen, the similarities in Boole's and De Morgan's logics had something to do with Boole's use of De Morgan's "universe of discourse." But another thing they shared was an approach to mathematical research that depended upon a similar practice of thinking through books in libraries, in particular of using books as historical sources from which the future trajectory of that logic, which for both Boole and De Morgan was a progressive science, might take.

But if their approaches to logic had some things in common, their material circumstances were very different. Unlike Boole, De Morgan had received a university education and was already a professor of mathematics at University College London (formerly London University) when he began working on logic. His economic independence and his geographical location in London, the center of the British book trade, allowed De Morgan to assemble a superb personal library that has been described as "one of the finest accumulations of books on the history of mathematics," in Victorian Britain.[63] Boole's particular assembling of books is now difficult to reconstruct but there is direct evidence of the books that De Morgan collected. To imagine De Morgan's library, therefore, will give particularity to the way in which a Victorian mathematician could think with history though the books in a library.

From a very early age, De Morgan lived in books, perhaps because of his very bad eyesight, which had made more physical pleasures such as riding or boating too difficult. He began collecting books for his library as soon as he left Cambridge and began living in London, easily the best place for obtaining both new and old books in nineteenth-century Britain. The library reached its greatest extent after he moved to Chalcot Villas in Adelaide road but even toward the end of his life, after resigning from UCL and moving to a new smaller house in Primrose Hill, it still contained more than 3,000 books. The degree to which De Morgan was attached to his library can hardly be overstated. In her memoir of him, De Morgan's wife, Sophia, describes the determination with which De Morgan, at the end of his life and despite having suffered a stroke, still insisted on personally shelving all the books in the room that was to become his last library. When Sophia worried that he could not get them all in, De Morgan replied, "with his old spirit, 'They *shall* go in, and I will put them all in myself;' and so he did."[64]

Historian Joan Richards has described the room in which De Morgan's library was housed in an earlier London home on Camden Street as De Morgan's "castle within a castle," and from Sophia's account of it there is little doubt that it was De Morgan's domain:

Not having the means to indulge in the luxuries enjoyed by richer and more affluent writers or experimentalists, he could not furnish his library with all the writing appliances and handsome bindings that ornament rich men's studies, and his old table and desk, and other cheap contrivances looked shabby enough. Any one who went to his room would be struck at first by the homeliness of the whole, and the quantity of old and unbound books and packets of papers. But when it was seen how the books were arranged and the papers labelled and put into their proper places according to subjects, the adaptation of means to ends became as apparent as in the clearness and precision with which he laid down his first principles, and showed what has to be done before making a beginning on his work. His contrivances in the way of inkstand, penholder, and blotting-block, had none of them a new or unused look, but all showed that every contingency had been carefully provided for.[65]

The arrangement, labels, and organization in the library are testament not only to De Morgan's character but also to the way in which libraries, if they are to be effective tools for managing information, must organize their user just as they are organized. As Sophia would later comment, "He [De Morgan] had the faculty of arrangement in an unusual degree, but it showed itself more in classification than in tidiness."[66]

The books were not only organized, they were also read and annotated. After De Morgan died, his collection was much sought after. After the still relatively young London Mathematical Society failed to acquire De Morgan's library, Lord Overstone would buy it for what would become the University of London Library.[67] The books are still available in University of London Senate House and so are the numerous marginal notes and illustrations, serious and otherwise, that they contain. Indeed, many of the "pictures from periodicals, notably *Punch* and other collecteana" that Sophia said that late nineteenth-century visitors to the library would find inside them are still there.[68]

De Morgan was a firm believer in historical progress, but like any good historian did not think that progress could be straightforward. For De Morgan, science, like history, developed unpredictably and the researcher therefore had to accept that they would have to confront confusion, even apparent contradiction, as they worked though the materials of the past in order to make sense of them. To fully appreciate this point, it is worth looking more closely at the way in which De Morgan investigated the history of mathematics and especially the primary materials of that investigation, the books in his library.

For De Morgan, "the history of science is almost entirely the history of books and manuscripts," and, as a result he considered mathematical bibliography as central to any historical research.[69] De Morgan's historical pieces are often discussions of one book, and one of the reasons they are valuable is because he was interested not only in the content but also in the physical form and material environment of the book, "the dates, places, titles, colophons, authorship, paper, print, &c. &c. of books, there several degrees of rarity, the libraries in which they are to be found, and all the physical history, as we may call it, of a volume."[70] Since mathematical books were historical objects as far as De Morgan was concerned, they had to be experienced directly. In De Morgan's remarkable book *Arithmetical Books*, a bibliographic study of books on arithmetic, the majority of which came from his personal library, he writes that "a useful description of a book must include not only its contents but also everything about its material features, its size, construction, form of printing, defects, waterlines, etc., so that the reader [of *Arithmetical Books*] ... [can determine] whether the book he has in his hand be the one described or not."[71] De Morgan intended his *Arithmetical Books* as a tool that could be used to properly identify the material specimens, the books, on which the production of a new "*scientific* history of arithmetic," must depend. As Joan Richards has written about De Morgan's library, "The room that contained De Morgan's books was as much a natural history museum as it was a library. His books were like fossils to him, individual artifacts left over from the long process of progressive development."[72]

Many of those fossils were carefully selected and hunted out for their rarity or singular importance. For example, in a letter to Sir John Herschel, De Morgan asks if Herschel knew "anything of, Bouillaud's or Bullialdi's *Astronomia Philolaica*[.] There is a copy in the British Museum which wants the Prolegomena, which is the very part *I* want. The matter has reference to Vieta's *Harmonicon Celeste*, which has been supposed to be lost, and which I have a faint hope might be recovered."[73] For De Morgan, like many Cambridge historians of this period, progress depended upon the recovery of truths that had to be rescued from the destructive cycles of history. That recovery depended upon finding, identifying, and assembling books in libraries.[74]

Yet, exactly because the trajectory of progress was not easy to discern from the examination of past materials, it was not entirely clear which books were the most important for a good historical analysis. As historical objects, De Morgan could learn as much, perhaps more, from lesser-known books as from canonical works by famous mathematicians such as François Viète. As De Morgan wrote in his *Arithmetical Books*, "No book that I *have* seen during the compilation [of *Arithmetical Books*] has been held too bad to appear; no book that I *have not* seen too good to be left out."[75] Even those De Morgan called "paradoxers," those authors who had written about a belief, or system of belief that was "apart from general opinion, either in subject matter, method, or conclusion" should not be forgotten. In his learned, but frankly idiosyncratic way, De Morgan insisted that Copernicus was a "paradoxer" and his heliocentric astronomical system a *paradox*.[76] Including Copernicus among the paradoxers illustrated the difficulty of knowing whether a paradoxer is mistaken and trying to take their field in the wrong direction or whether, like Copernicus, they were really on to something.[77]

Can 1 + 2 + 4 + 8 + 16 + ... equal −1?

What was a problem for De Morgan's history of mathematics also encouraged a creative approach to research. De Morgan's teacher at Cambridge, George Peacock, had used historical analysis to show that arithmetic had "suggested" symbolic algebra because the rules that governed arithmetic also governed symbolic algebra even if the symbols in symbolic algebra could have quite different meanings.[78] But, as we have already noticed, Boole's work on logic had shown that some domains of symbolic algebra might utilize different rules for combining symbols, such as those suggested by Aristotelian logic, that Peacock had not anticipated. Like Boole, De Morgan's historical investigations also suggested that the future was open. A contemporaneous "paradoxer" was not necessarily mistaken; like Copernicus, they might be pointing in an important new direction. Study of the past had shown De Morgan that it was very difficult to know what direction mathematics might take in the future. What might now appear as a paradoxical result should not be dismissed too quickly.

In his book *The Differential and Integral Calculus*, originally published in parts for the Library of Useful Knowledge and as a whole in 1842, De Morgan insisted

that the *student* ... should be taught to examine the boundary [of mathematical knowledge], as well as how to cultivate its interior. I have never scrupled, in the latter part of the work, to use methods which I will not call doubtful, because they are presented as unfinished, and because the doubt is that of an expectant learner. Experience has often shown that the defective conclusion has been rendered intelligible and rigorous by preserving thought, but who can give it to conclusions which are never allowed to come before him? The effect of exclusive attention to those parts of mathematics which offer no scope for the discussion of doubtful points is a distaste for modes of proceeding which are absolutely necessary to the extension of analysis [algebra].[79]

One of the "doubtful points" De Morgan might well have had mind, was a discussion of infinite series that appears later in the same book. There De Morgan asserts that, claims $1 + 2 + 4 + 8 + 16 + 32 + \ldots = -1$.[80] To justify the claim, De Morgan applied an algorithmic argument using the laws of symbolic combination warranted by arithmetic:

Let $S = 1 + 2 + 4 + 8 + 16 + 32 + \ldots$
Then, $S = 1 + 2(1 + 2 + 4 + 8 + 16 + 32 + \ldots)$
then $S = 1 + 2S$ and therefore $S = -1$

There are, of course, some serious problems with this argument that will be obvious to anyone familiar with the mathematics of divergent series. De Morgan was, of course, quite aware of those problems and before introducing the reader to the result, he is careful to warn that "All arithmetic is algebra, but all algebra is not arithmetic" (223). The warning is very much to the point. As De Morgan points out, our experience of arithmetic leads us to assume that a sum like $1 + 2 + 8 + 16$, must have a positive result. To find that it can lead, therefore, to -1, undermines our expectations, it is, in De Morgan's sense, a paradoxical result. More than that, for De Morgan -1 is not an arithmetical result because for him only the natural numbers are in the arithmetical domain: "therefore [when] we write $-1 = 1 + 2 + 4 + 8 + 16 + $ &c. *ad infinitum*, the student must not think we intend any arithmetical equality, or other arithmetical resemblance or analogy of any sort or kind whatsoever, between -1 and $1 + 2 + $ &c." Whatever the relationship between $1 + 2 + 4 + 8 + 16 + $ &c. *ad infinitum* and -1 might be, it cannot be arithmetical: "It has no connexion, for instance, with $1 + 2 + 4 + 8$, though the latter expression contains some of its terms; nor are we to be considered as making any approximation to its value by stopping anywhere; such an idea being reserved for arithmetical series." For De Morgan, an infinite series must not be assumed to be an arithmetical object; indeed, his *logistical* result strongly suggested that, when considered as "a whole," it was a symbolic algebraic object. A divergent infinite series was, for De Morgan, a possible new kind of mathematical object about which too little was known. Some of De Morgan's contemporaries, most famously the French mathematician Augustin-Louis Cauchy, had claimed that a divergent series (those whose sums blew up to infinity) did not have a sum.[81] Given the uncertain future that his library had opened up, De Morgan thought it too early to say to make such a strong claim. And with hindsight, perhaps he was right.[82]

Conclusions

Boole and De Morgan practiced what some historians have described as "algorithmic mathematics," a mathematics that developed procedures not only to solve algebraic equations or model reasoning processes but also to make new kinds of algebras and mathematical objects. As a kind of *logistica*, it brings attention to the material infrastructure that supported it, represented here by assemblings of books and papers in libraries. Both Boole and De Morgan developed highly abstract symbolic mathematics, but they sought legitimacy for their algebras in the virtual space that mathematics inhabited, the paper world of nineteenth-century mathematics. Their research practices involved thinking with history through books in libraries, a kind of algorithmic thinking justified by a historical perspective. Boole's work on logic, for example, depended on the investigation of mathematical and logical texts contained in books and libraries. The new algebra that he generated was justified by rules of reasoning found in classic sources such as Aristotle, and the way Aristotle's reasoning had begun to be developed by the Oxford Noetics and by De Morgan. Boole's algorithmic logic, which was a resource for early computer search engines, was justified by a historical trajectory discovered in the books he had assembled.

Boole's assembling is difficult to reconstruct because of the lack of evidence, but De Morgan's library allows for a more precise account of the kind of virtual reality the material world of the early nineteenth-century British libraries afforded. For De Morgan, the principal objects, the books assembled in his private library, were like fossils and the principal evidence for the natural history of mathematics he sought to discover. All these libraries were, necessarily incomplete, but De Morgan, especially, reveled in that incompleteness. This was not only because there were past treasures to be unearthed and added but also because as more fossils were discovered, examined, and placed in an overarching historical progress of reason, patterns would emerge that would allow the production of new mathematics, and thus new books that would carry the library into the future. De Morgan's divergent infinite series behaved paradoxically when an algorithmic process suggested that $1 + 2 + 4 + 8 + 16 \ldots = -1$. But the open future De Morgan's historical perspective had opened up to him made him reluctant to think that a paradoxical result automatically meant that the algorithmic process had been improperly applied. It might just as easily suggest a new direction for research. The relationship between -1 and an infinite geometric series, although not yet fully understood, suggested that infinite series might be new kinds of mathematical objects.

Although Boole's and De Morgan's natural historical mathematics was significantly different from twenty-first-century experimental mathematics, their insistence on "inductions" emphasizes the way in which it shares the material, nomadic creativity that informs more recent use of computers in mathematics research. Appreciating this aspect of Victorian algorithmic mathematics makes it harder to dismiss nineteenth-century British culture as philstine. utlitarian and anti idealist in the way Gray supposes. Instead, as I have tried to show here, attention to the materials mathematicians use to pursue their work suggests that mathematics does not occupy a single ideal world. There may be certain kinds of objects, the natural number three, for example, common to all mathematical worlds, but if mathematicians make their ideal worlds out of the social and material world they live in, then those ideal worlds will be significantly different in different times and places. Boole's discovery of mathematical logic, or De Morgan's view of divergent infinite series, can be described entirely in terms of ideas, but by taking seriously the work they had to do to assemble those ideas, the physical practices needed in order to collect and assemble books, as well as the intentions that motivated them, we can perhaps better appreciate the way in which they were making a specifically Victorian mathematics. That mathematics had its shortcomings, as Gray has made all too clear, but it also had virtues. The Victorian mathematical world was algorithmic. It was as an experimental science, a historical example of what Brian Rotman has called the nomadic, creative science of *logistica*.[83] That does not mean it anticipated twenty-first-century mathematics of mechanical proofs and computerized practices. But it does suggest that the new tools mathematicians are using to make their mathematical worlds will change the objects that world contains.

7

"For Computing Is Our Duty"

Algorithmic Workers, Servants, and Women at the Harvard Observatory

Andrew Fiss

This chapter investigates the many ways that different people came to the job title "computer" at the 1870s Harvard Observatory, as it set the precedent for the employment of human computers at other observatories, communications companies, engineering firms, and early aeronautical enterprises. Famously, Williamina Fleming's 1893 speech "A Field for 'Woman's Work' in Astronomy" drew attention to the observatory's practice of employing teams of women for what was called "computing."[1] Yet their employment just twenty years earlier was not so straightforward. For centuries, male servants had held positions as computers, supporting their master's work through added mathematical duties. Then in the early nineteenth century, networks of correspondence and novel scientific institutions made possible the employment of both men and women who calculated offsite, from their homes. Assigned problems remotely, they consulted personal libraries and sent back solutions.[2] Throughout, women regularly computed for male relatives on site, particularly when they lived in the observatory residence. At the 1870s Harvard Observatory, educated women from beyond observatory families did become computers, yet importantly they joined a team of servants, observatory daughters, and male assistants. The diversity of the computing team was nearly forgotten, however. By the 1890s, the administration of the Harvard Observatory had simplified its history, through their enthusiasm to encourage the employment of women as computers at other scientific institutions.[3] In recovering the 1870s workplace, I argue that broad histories of computers and algorithms should consider the professionalizing of what was called "computing" work, especially how it blurred boundaries between professional-domestic spaces and servant-familial roles.

This chapter analyzes an unusual source: the Harvard Observatory's 1879 parody of W. S. Gilbert and Arthur Sullivan's *H.M.S. Pinafore: or the Lass That Loved a Sailor*. A popular opera then and now, *H.M.S. Pinafore* takes its name

Andrew Fiss, *"For Computing Is Our Duty"* In: *Algorithmic Modernity*. Edited by: Morgan G. Ames and Massimo Mazzotti, Oxford University Press. © Oxford University Press 2023. DOI: 10.1093/oso/9780197502426.003.0008 from a fictional ship, the *H.M.S. Pinafore*. The Captain's daughter, Josephine, loves a sailor in the crew (Ralph), yet he is not appropriate to her "station," or class. In Act I, Josephine is divided between her love for Ralph and an offer to marry into a higher class, to Sir Joseph, First Lord of the Admiralty. By the end of the act, she and Ralph plan to elope, yet in Act II, the sailor Dick Deadeye informs the Captain and stops them. As is usual for Gilbert and Sullivan operas, a surprise secret transforms the choice. Little Buttercup, a mysterious woman variously joining the scenes, reveals that she had been a nurse to two babies of different classes but that she had confused them. It turns out that Ralph had the background of the Captain, and the Captain had the background of a common sailor. Ignoring the age confusion, Josephine happily marries upward, to Ralph.[4]

Harvard's *Observatory Pinafore*, as the mathematical parody was called, encapsulated gender and class confusion, as well, making the joke about those who held the title "computer." In Gilbert and Sullivan's distinctive style, the original *H.M.S. Pinafore* poked fun at the British class system, particularly as represented in naval careers and marital ambitions, and the *Observatory Pinafore* used the frame to focus on the professional hierarchies of the Harvard Observatory.[5] Including not only astronomers, professors, and assistants but also women computers, the *Observatory Pinafore* dramatized the many paths into the computing work of the observatory. Responding to recent articles in rhetoric and technical communication, I analyze the parody to investigate links to workplace cultures involving algorithms and computers.[6] Yet I also emphasize the importance of fiction for representing human computers in their roles betwixt and between.[7] As an imagined, exaggerated account of workplace stratification, the *Observatory Pinafore* can demonstrate the hybridity of computing work.

Such scientific parody provides a unique perspective on complex spaces and identities, particularly as it foregrounds performance and the many meanings of humor. As scholars of Gilbert and Sullivan have shown, humor works on multiple levels, and Laura Kasson Fiss and I have discussed the humorous dimensions of a nearly contemporaneous Gilbert and Sullivan parody: *The Mathematikado*, created and performed by Vassar College students in 1886 (parodying *The Mikado*).[8] As in other mathematical parodies, the *Observatory Pinafore* did include scientific terminology, in this case, from astronomy and the specific instrumentation of the Harvard Observatory. It also provided a commentary on the idiosyncratic personalities of the observatory's workers and especially their understandings of the professional work of astronomy. The staff of computers proved central to the *Observatory Pinafore*, as the action of Act I occurred in the Computing Room, and a Chorus of Computers followed the principal players through much of Act II.[9] Moreover, since the Computers acted as the parodic counterpart of the Sailors of *H.M.S. Pinafore*, the whole of the *Observatory Pinafore* could be read as a silly meditation on the process of socializing particular scientific workers, teaching them how to do their jobs and why.

Because of the scientific parody's place in the long history of the Harvard Observatory, women's computing emerges as a major point of concern. Though written in 1879, the *Observatory Pinafore* had its premiere performance in 1929, by which time computing had been unquestionably feminized. In the 1929 performance, the Chorus of Computers was a women's chorus, a humorous inverse of the hyper-masculine Sailors Chorus of *H.M.S. Pinafore*. In 1879, however, the Computing Room was a mixed space, reflecting the ways that Observatory Director Edward Pickering had hired a few women to work alongside teams of men. The Chorus of Computers, as written in the late 1870s, reflected such a workplace arrangement and featured both men's and women's voices. Just as the chorus harmonized diverse voice parts, the *Observatory Pinafore* as a whole assumed blurred boundaries between assistants, computers, students, and professors, and the show featured songs and spoken lines that made fun of the negotiation of stratified professional responsibilities, including women's work and men's work in the shared space. As this chapter argues, the *Observatory Pinafore* allows us to see the many paths into computing work, as well as the negotiated roles of women, servants, and assistants in the dual domestic/professional space of the Harvard Observatory.

This chapter proceeds through two main sections. The first focuses on the ways that the work of Harvard Observatory Computers bridged the domestic and professional spaces of the observatory. Quickly reviewing the history of servants' and families' computing, it presents the routine calculations performed by 1870s employees in their professionalizing roles. Considering the *Observatory Pinafore* as a dramatization of workplace stratification, I end with an analysis of the main character of Joseph/ine—jointly female and male, computer and observer, domestic and professional. The second section, then, focuses on the ways that stories of women from outside observatory families came to dominate narratives of computing work. Through analyzing the historical work of Rhoda Saunders—and her representation in the *Observatory Pinafore*—it argues that women's computing nonetheless combined roles of servants and emerging professionals. Overall, inspired by the *Observatory Pinafore*, I argue that computing work encompassed many paths in its initial professionalization and that the emergence of "computer" as a workplace category blurred boundaries between home and work and duties of servants and new professionals.

The Domestic/Professional Space of the Harvard Observatory

In the 1870s, the Harvard Observatory employed a diverse set of human computers. Contemporaneous histories of astronomy discussed how earlier astronomers and mathematicians had relied on the work of male servants, specially trained for their added duties. David Brewster's 1855 *Memoirs of the Life, Writings, and Discoveries of Sir Isaac Newton*, for instance, discussed the ways that Newton's contemporary, John Flamsteed, did not have the funds to hire another astronomer "for reducing his observations," even though he was the Astronomer Royal. According to Brewster, Flamsteed had to "pay the expenses of a servant capable of making the calculations," just as he had to construct his instruments "with his own hands."[10] Emphasizing the effects of the situation on Flamsteed while never naming the servant, Brewster's book emphasized the ways that servant-computers overworked their masters. Not only did Flamsteed have to pay a servant for his calculations, he had to train him, correct his errors, and then train another and another, as his computers periodically ran away. As Brewster and his contemporaries emphasized, other systems needed to prevail, at least for the sake of the masters.[11] Starting in the eighteenth century, institutions "employed" female family members of astronomers, leading to a situation where wives, daughters, sisters, servants, and some assistants all did the work of computing.[12] As represented in the *Observatory Pinafore*, the diverse backgrounds of the computers exposed the way that the 1870s observatory acted as a dual domestic/professional space.

So, what work did the members of the observatory staff and family perform? When thirty-year-old Edward Pickering became director of the Harvard Observatory in 1877, he had to determine the research program of the institution. Circa 1870, he had taught physics at the new Massachusetts Institute of Technology, and he designed a laboratory space and individualized experiments to aid his students' understanding of the material. Already interested in the nature of light, he devoted the Harvard Observatory to the study of photometry, measuring the brightness of individual stars. There was no standardized scale, however, so Pickering chose a British one and chose Polaris as the basis for all subsequent comparisons. In his first months at the observatory, he designed a new instrument (the "meridian photometer"), a dual telescope that would allow an observer to see Polaris and a target star at once. That observer, usually Pickering, would use the numbered dials on the side of the instrument to create the appearance that Polaris and the target star were of the same brightness. The "circle reader," always a male member of the staff, would read the dial's numbers and record them in a lined notebook.

In the morning, he would hand over his books to a computer, variously a woman, a servant, or a male assistant, who would calibrate their numbers according to the British scale. Through such a method, from 1877 to 1880, Pickering and his staff ultimately catalogued the brightness of every star visible from Cambridge. As Pickering worried about the errors that could result from overworking his employees, he regularly reassigned most of them between the night shift of reading and the day shift of computing. However, even when the observatory started employing women, Pickering was adamant that they could not be exposed to the cold and damp of observing work.[13]

Pickering, in fact, inherited the policy against women's observation, just as he had inherited a few women computers, whose presence linked to the domestic side of the observatory and whose assigned duties underscored the distinction of computing work. The Harvard Observatory was not just a scientific institution, after all, but also a home for the director and his family. Lizzie Sparks Pickering, Edward's wife, did not have an interest in computing work, unlike other women of her acquaintance, and her family connections meant she instead helped with the fundraising and bookkeeping aspects of the institution. Female relatives of former directors had different talents, however, and they continued to request computing. Selina Bond, nearly a generation older than the Pickerings, had grown up in the observatory when her father acted as its first director. Her brother had become the second director, and, under both, she learned and then applied complex mathematical formulas to correct the quantitative components of astronomical observations. Eventually assigned to assist Professor William Rogers, Bond specialized in fixing the exact positions for Harvard's zone of stars to help with the stellar mapping project called the Astronomische Gesellschaft. Receiving a male observer's tables for the times when a star passed through his eyepiece, she would spend the day applying mathematical formulas that would correct the star's apparent position for a variety of atmospheric effects (even the refraction of clear skies) and for effects of the Earth's motion (its orbit, its rotation, its wobble on its axis).[14] Anna Winlock, nearly a generation younger than the Pickerings, had started computing work shortly before they arrived. Her father, the third observatory director, had died suddenly in 1875, and she needed to support her mother and younger siblings. Learning from Bond, she too performed the daytime calculations that corrected men's nighttime observations of star positions.[15] Their workspace, the Computing Room, was a specialized library, with rows of books that contained the tables and formulas they needed to perform their calculations.[16] As Kevin Lambert explains in this volume, the amassing of a mathematical library was itself a recognizable achievement.[17] For Bond and Winlock, who had grown up in the residence connected to the observatory, it was also their once and future home, and they remained members of the observatory "family" despite new directors. Their presence in the Computing Room epitomized the dual purposes of the Harvard Observatory as a whole: both home and workplace, both familial and professional.

The Harvard Observatory was a home not only for the directors' families but also for servants and assistants, as can be seen in the exaggerated *Observatory Pinafore*. The *Observatory Pinafore* takes Gilbert and Sullivan's plot and maps it onto the career ambitions of an observatory employee. Modeled on an actual assistant, one named Joseph McCormack, the observatory's "Josephine" has to choose between remaining at the Harvard Observatory or working at a fictional private observatory in Providence, Rhode Island. The *Observatory Pinafore*, modeled on *H.M.S. Pinafore: or the Lass That Loved a Sailor*, clarifies that the choice of observatory is the same choice that Gilbert and Sullivan's "Lass" Josephine faced: between a poorly furnished home and a cosmopolitan one. Modifying Gilbert and Sullivan's jokes that aristocratic homes were full of "antiques" bought new at department stores, the observatory's "Josephine" sings about how a private observatory would feature "no dust or damp ... wonderful machines ... all the apartments lit.... No inconvenience and no failure."[18] Amid complaints about poor pay, faulty machines, noise, and cramped living conditions, the lyrics made fun of the conditions of the Harvard Observatory through indicating what they were not. Since its budget relied on private donors, individual subscriptions, and the director's salary for everyone's expenses, the place was not at all wealthy, more similar to the home of a sailor than of a British aristocrat.[19]

In such a setting, training remained piecemeal and intensely specialized to the institution, as in earlier centuries. In the *Observatory Pinafore*, it soon becomes clear that the observatory's Josephine would like to stay at the Harvard Observatory but would like a better job, one that would allow her occasional access to the distinctive instruments. Such a prospect was beyond her "station," according to the parody, not because she was of the wrong class but because she was not educated enough. At the private observatory in Providence, she could achieve better work conditions because the facilities were generally better; her job would not change. A secret manages to transform the prospects of the observatory Josephine, as it did in Gilbert and Sullivan's opera. It turns out that she had been educated in astronomy after all; in fact, she had taken a small class about how to use the observatory's specific instruments led by a member of the senior staff, one who already had the honorific "professor." Overhearing the information, the director of the Providence observatory decides that Josephine's training is too specialized for his needs, and the director of the Harvard Observatory decides it is just right for his.[20] Though fictional, the directors' stance toward education did reflect observatory practices of training female relatives and male servants for very specific tasks, developing skills that could not travel well.

Not even observatory directors had advanced degrees at the time. In the *Observatory Pinafore*, Josephine's aristocratic suitor is the director and owner of the fictional private observatory in Providence: Dr. Leonard Waldo. He is a "Harvard L.L.D.," as he repeatedly reminds the workers, making him more educated than anyone at the Harvard Observatory, as even the "professors" and director do not have doctoral degrees.[21] He therefore takes it upon himself to educate the staff, "to give instruction to young astronomers in a pleasing way."[22] Making fun of astronomers' appearance, however, his song jokes that obedience, diligence, and voluntary imprisonment (i.e., in observatories) may lead to problems in speech, as well as in the knees, neck, back, face, and eyes. The *Observatory Pinafore* here invites comparison of the physique of the attractively strong Sailors Chorus with that of the aching astronomers. Like Gilbert and Sullivan's "A British tar," Waldo's song similarly conflates theatrical appearance with professional work. After all, when formalized education was not the norm, it was paramount to look the part.

When male assistants were employed, they had to negotiate the domestic legacy of computing with their new, formal degree paths. The *Observatory Pinafore* exaggerated the gender implications of the workplace negotiation through pronoun confusion. The observatory "Josephine," after all, drew together Gilbert and Sullivan's character with the historical figure of Joseph McCormack. As scholar Carolyn Williams observes, Gilbert and Sullivan's Josephine is a humorous exaggeration of the "Lass That Loved a Sailor" from earlier nautical melodrama. She hears the romantic proposals of Ralph, her "Jolly Jack Tar," but she also cannot stop thinking about a potential engagement to an aristocratic suitor.[23] In the original version of the *Observatory Pinafore*, Josephine's name does in fact become Joseph, naming the specific assistant of the time.[24] However, because the *Observatory Pinafore* preserves so much of the dialogue, songs, and plot, Joseph/ine's pronouns shift suddenly and inexplicably between feminine and masculine, giving the character notable gender fluidity. Moreover, as the "Jolly Jack Tar" becomes a male member of the senior staff ("Professor" William Rogers), the romantic plot appears at times to be about love between two men, brought together through their common commitment to knowledge, to science, and to work.[25] Later emendations of the *Observatory Pinafore* avoided possible interpretations involving same-sex desire, as they asserted that Joseph/ine must be a woman, and they consistently used feminine pronouns for her.[26] But in the 1879 version, Joseph/ine's pronouns blur together, as did the domestic and professional aspects of computing work.

Just as the *Observatory Pinafore*'s character of Joseph/ine is not exactly a woman, so too is s/he not exactly a computer. Singing with the Computers Chorus at times, s/he also acts beyond them, consulting directly with Professor Rogers and forming elaborately complex plans for professional advancement. Furthermore, Joseph/ine's love for and with Professor Rogers demonstrates the ways that the other computers should act, and the plot ultimately rewards such combined love for co-workers and for education. An exemplary employee, Joseph/ine is also not exactly a computer: with a job title of "circle reader" instead.[27] Though later versions mapped the position onto computing duties, just as later versions asserted that the character of Joseph/ine was definitely a woman, the original version of the *Observatory Pinafore* encouraged the conflation of the job of computing (historically domestic) with the job of astronomical reading (increasingly professional).

Moreover, the computing of the Harvard Observatory reflected not only the institution's dual domestic/professional roles but also the multiple senses of the word "domestic." According to scholar Carolyn Williams, *H.M.S. Pinafore* already combined two senses of the word: the national (as in the sailors' role in "domestic" affairs) and the home (as in notions of "domestic" bliss).[28] As seen above, the Harvard Observatory did act as both a home and a workplace. Furthermore, through projects like the Astronomische Gesellschaft, it was a site for establishing national time and space, and it increasingly set US workplace norms, as well.

"Unlike their brother astronomers": Women Computers as Servants

According to the observatory's histories and to the *Observatory Pinafore* alike, the employment of Rhoda Saunders marked a drastic break in tradition, a moment when a woman joined the computing staff without any previous connection to observatory families. Some women had previously computed for other observatories and almanacs from their homes, provided they had access to specialized libraries for astronomy. Before Maria Mitchell became professor of astronomy at Vassar College, for instance, she corrected observations of Venus's position for the US Nautical Almanac. As her father was an amateur astronomer, Mitchell could use the books in their Nantucket home for her correspondence work, as she used the telescope there for her own observations, including her world-renowned discovery of a comet in 1847.[29]

Though much has been lost about Saunders's life, she had some connection to Harvard's president; it is said that he recommended her for the position. Also, it is likely she already lived locally, as certain sources indicate that she and Winlock attended the Cambridge High School together. She probably did not have more than a high school education; neither had Winlock or Bond. They did not need higher education to perform their work, assuming they received specialized training, as earlier servants had. Under the interim directorship of Arthur Searle, Saunders joined the computing team shortly after Winlock did, without much fanfare.[30] With few surviving documents, Saunders's place was instead preserved through parody and institutional lore. Through presenting her as a character, the *Observatory Pinafore* recovers some aspects of her employment and generally exposes the servant roles that women computers performed.

In fact, as the literature about women's work has made clear, professions' feminization relies on a legacy of servitude. Sociologist Judy Wajcman, science historian Ruth Schwartz Cowan, and design historian Adrian Forty have noted the emergence of American technologies marketed at housewives in the first half of the twentieth century, and they have linked the trend to the ways that domestic servants became markedly less common at the time. However, as they argue, the new marketing campaigns also needed to erase such associations, as women's tasks in the home needed to seem like love and not (servants') work.[31] Sociologist Sally Hacker concluded that similar dynamics operated in many workplaces, focusing on the cases of telecommunications, publishing, and agribusiness. American management practices, according to Hacker, have been built on the exploitation of women, not only their treatment as servants but also their socialization within and beyond the home. Moreover, for Hacker, the feminization of a position is a likely sign of its impending automation, a managerially efficient move since women workers often rationalize their termination as a judgment on their worth rather than a bureaucratic attempt to employ different kinds of workers.[32] The long history of women computers, following such accounts, could be bookended by servants' work and ultimate mechanization, as a lengthy though deliberate process of planned obsolescence. The *Observatory Pinafore*, especially its writing (1879) versus later performance (1929), uncovers a moment when women's scientific work developed out of servants' duties.

In 1879, the character named after Saunders becomes the quintessential woman computer, representing both computers and women. Though all other characters have specific analogues from Gilbert and Sullivan's original work, Saunders does not. As a named member of the Computers Chorus, she worries that Joseph/ine might "lose her [his] station."[33] Then, together with the other members of the observatory staff, she greets Dr. Waldo and sings his "educational" song about the work of astronomy. As the representative computer, she acts nearly identically to the other named members of the chorus: Joseph/ine, though "circle reader," sings the same ensemble songs, as does the "assistant observer" Winslow Upton. Unlike them, Rhoda Saunders combines pieces of many female characters from *H.M.S. Pinafore*: singing the ingénue's introductory song and serving as the leader of the women's chorus in the Act I Finale. In Gilbert and Sullivan's opera, the so-called Chorus of Sisters, Cousins, and Aunts acts opposite the Sailors Chorus, making the plot a kind of battle of the sexes. Combining various roles in the *Observatory Pinafore*, Rhoda Saunders too becomes the markedly female figure who generally fights the men around her.

The character of Rhoda Saunders furthermore reflects on the ways that women computers historically had come from observatory families. The *Observatory Pinafore* begins with scenes of computing, gently segueing into the appearance of Professor Arthur Searle, who "sells" his image of the professional work of astronomy. Professor William Rogers enters, worrying that Joseph/ine will lose his/her job at the Harvard Observatory, and finally Edward Pickering appears as the "captain" of his astronomical "crew." Joseph/ine enters only to leave again before Saunders sings, "Sorry her lot." Despite the name of the song, Saunders is not entirely unhappy, instead meditating on the history of her work:

Happy the hour when sets the sun,
Sweet is the night to earth's poor daughters,
Who sweetly may sleep when labor is done
Unlike their brother astronomers.[34]

Noting the sex segregation of computing versus observing, Saunders's song also emphasizes the family relations of observatory work. Identifying as one of "earth's poor daughters," as in the equivalent line from *H.M.S. Pinafore*, she also recognizes her co-laborers as "brother astronomers." At the Harvard Observatory and beyond, such relationships had been required for most women's computing, supporting fathers and brothers in their scientific work.

Compared to men's observing, women's computing had markedly low status, akin to servants' responsibilities. The same introductory song shifts mood, and Saunders expresses the sadness that stems from her work:

Sorry her lot who adds not well,
Dull is the mind that checks but vainly,

> Sad are the sighs that own the spell
> Symbolized by frowns that speak too plainly.
> Heavy the sorrow that bows the head
> When fingers are tender and the ink is red.[35]

Saunders's song quotes most of *H.M.S. Pinafore*'s love lament, except it substitutes adding for loving and tender fingers for dead hope. Worrying about professional prospects instead of romantic ones, Saunders therefore represents computing work as monotonous, tedious, exhausting, and generally "sad." Just as Dr. Waldo's song pokes fun at the physical effects of nighttime observing on young men, her song portrays the mental, emotional, and physical harm that women experience through computing. Computing work affected women's personal appearance as servants' labors would.

Furthermore, though male assistants computed too, women computers occupied lower positions in the Harvard Observatory's workplace culture. In the *Observatory Pinafore*, after Professor Rogers worries about being separated from Joseph/ine, he focuses on his work, finding "a fearful mistake" in the computing. The "figures" look like Saunders's, but the arrival of Pickering distracts the group from investigating. After the song "Sorry her lot," Professor Rogers confronts Saunders about her figures, which he sees reflected in her unnaturally red fingers. The color comes from the red ink used in correcting observations, she explains, and scrubbing her hands just makes the problem worse. The male assistant Winslow Upton tells her that he only corrects "others' mistakes" with red ink, and he uses blue ink for the calculations in his own book. Though Rogers disagrees with the practice, Saunders is left lamenting her fate:

> RGS: That's it! You correct your own mistakes in blue, so that they don't show so much. I think it's mean.[36]

Not able to bend the rules, because of her position, she also could not possess her own book of observations. After all, because of workplace policies that assumed that women could not be exposed to the cold of nighttime observations, Saunders could never have such a thing as her "own mistakes," that is, her own calculations of apparent positions. Her position was more like an observatory servant's than an astronomical assistant's.

The portrayal of Rhoda Saunders as insecure and irritable reflected the (perhaps misguided) attempts at humor from the author of the *Observatory Pinafore*, generally assumed to be the actual assistant named Winslow Upton. As mentioned earlier, there is a character named "Winslow Upton" in the

Observatory Pinafore. A member of the Computers Chorus, the character also acts as the analogue of Gilbert and Sullivan's Dick Deadeye. Though the undisputed villain of *H.M.S. Pinafore*, Dick is not particularly malicious in Gilbert and Sullivan's opera. Mainly, he overhears conversations between Josephine and Ralph and later reports them to the Captain. Though his intervention prevents the two lovers from leaving the ship and eloping, his villainy is not through explicit malice or danger.[37] Winslow Upton's caricature similarly tries to speak truth to power. He talks openly about the observatory's working conditions: its low pay, poor facilities, and irritating combination of domestic and professional spaces. Mainly, however, he pokes fun at Rhoda Saunders: her mannerisms, her appearance, and her computing. Even in the Act I Finale, when he prevents Professor Rogers and Joseph/ine from absconding with the meridian photometer, he tells Pickering because Saunders is also involved.[38] Within the broader context, Saunders's plight appears almost humorous.[39] Constantly thwarted by a male colleague, she cannot achieve any of her goals. Striving for better work and the beautiful hands she will never achieve, she fades into the background or perhaps leaves the observatory entirely in Act II. As Saunders is the only woman and the only computer in the script, the *Observatory Pinafore* therefore portrays women computers as servant-like, not only for their duties but also for their habit of running away.

As computing became associated with women's work in the long history of the Harvard Observatory, its connections with domestic roles (family or servant) were not well preserved, even in the *Observatory Pinafore*. Currently, there are at least two copies of the script saved in the Harvard Observatory library, representing the flurry of activity around the play in 1879 and again in 1929, and Harvard's High Energy Astrophysics Group hosted a digital edition on the personal website of a research scientist employed at the Chandra X-Ray Center.[40] The early copy, from 1879, attributed the parody to Winslow Upton but also appeared in the hand of Williamina Fleming, who was at that time a recent hire into the domestic (maid) staff of the observatory.[41] After Upton's subsequent career in federal agencies and Brown University, and after Fleming's distinguished service to the Harvard Observatory through her computing work, the first performance of the *Observatory Pinafore* occurred in 1929. The later copy, from 1929, reflected the ways that the Harvard Observatory's hiring practices reified computing as women's work. The script changed the character of Joseph to Josephine, with consistently feminine pronouns, and it featured a Computers Chorus containing only women. With leading parts for the first two graduate students of the observatory, Adelaide Ames and Cecilia Payne, the performance did emphasize the ways that women had managed to expand their opportunities beyond and through computing.[42] But the characterization of Rhoda Saunders, though played by Ames, did not change much, and the other performance choices made the observatory seem even more sex-segregated than before.

In fact, the 1879–1929 expansions of the computing team did emphasize women's work. Just four years after the *Observatory Pinafore*, the computing staff had doubled. Williamina Fleming, once the Pickerings' maid, had become a distinguished computer.[43] The others had followed Rhoda Saunders's path. Nettie Farrar, for instance, was not from an observatory family, but nonetheless was hired to process the observation notes from the meridian photometer. She, more than anyone else, tabulated the dial settings, averaged them, and then used them to compute relative values for the brightness of individual stars.[44] Though Farrar left to get married at the end of 1886, the computing team continued to grow. Louisa Winlock soon joined her sister Anna, and so did many recent graduates of new women's colleges.[45] Williamina Fleming, as the capable head of the team, encouraged magazine articles and newspaper stories about the spectacle of approximately a dozen women working at the center of a recognizably scientific institution. In 1893, she wrote (but did not deliver) the speech to the World's Columbian Exposition about computing as "A Field for 'Woman's Work' in Astronomy."[46] As the speeches and articles made clear, the computing team was a model not only for astronomy but for women's work as a whole.

Building on the subsequent stream of publicity, the 1929 performance of the *Observatory Pinafore* reimagined a history in which computing always had been women's work. The play strikingly opens with a scene of diligent workers, and, instead of the maritime labor of Gilbert and Sullivan's *H.M.S. Pinafore*, the *Observatory Pinafore* begins with the performance of computing. In 1929, a group of seven women, including the well-known Adelaide Ames and Cecilia Payne, therefore took the stage first, singing:

We work from morn 'till night,
For computing is our duty;
We're faithful and polite,
And our record book's a beauty;[47]

Though the 1879 script included both men and women singing the above lines, the 1929 performance focused on women's computing and emphasized the ways that the performers imagined past workers. Computers, they supposed, had been exclusively women, ones concerned with manners, fidelity, and aesthetics, even when applied to their everyday tools. Even the 1929 choice of costumes communicated feminine propriety, as the reviews mentioned. Though dancing, singing, and acting (passably), the women's computer chorus was "garbed in the high-necked shirt waist and flowing skirt of a bygone day," outfits that made them appear firm, formal, and feminine.[48] In sum, the *Observatory Pinafore* allowed the 1920s staff to imagine 1870s computers, though in doing so, they simplified their own history.

In particular, the 1929 performance emphasized how women's computing uncreatively built on men's mathematical work. Even in the introductory song, Ames and Payne, as well as Irma Caldwell, Sylvia Mussels, Helen Sawyer, Mildred Shapley, and Henrietta Swope, sang:

With Crelle and Gauss, Chauvenet and Peirce,
We labor hard all day;
We add, subtract, multiply and divide
And we never have time to play.[49]

Similar to the beginning of *H.M.S. Pinafore*, the Computers Chorus boasted about their work. According to the song, their acts of "labor[ing] hard" truly came from the books in their astronomy library: August Leopold Crelle's *Journal für die reine und angewandte Mathematik* (1826–), Carl Friedrich Gauss's *Theory of the Motion of the Celestial Bodies Moving in Conic Sections around the Sun* (the 1857 American translation of the 1809 text), William Chauvenet's two-volume *Manual of Spherical and Practical Astronomy* (in its first edition, 1863), and Benjamin Peirce's *Elementary Treatise on Plane and Spherical Trigonometry, with Their Applications to Navigation, Surveying, Heights & Distances, and Spherical Astronomy* (also in its first edition, 1840).[50] Referring to the books through authors' names, the 1929 women's chorus claimed to rely on well-known, male mathematicians in their labors. Through such performance choices, the Harvard staff of the 1920s imagined their female predecessors not only as overly proper but also as disappointingly unoriginal.

According to the subsequent (and currently unfinished) authorship debate about the *Observatory Pinafore*, the 1879 version could have stemmed from the humorous representation of observatory computers as servants, not exclusively as women. Poorly treated, poorly paid, and underappreciated, late-century servants nonetheless could be caricatured as "faithful," "polite," unoriginal, and dutiful. Furthermore, as we saw earlier, 1850s histories of astronomy included mention of servants who acted as computers in the past. For these reasons, Pickering's successor, the astronomer Harlow Shapley, began to doubt that Winslow Upton had written the *Observatory Pinafore* after all. Following the successful 1929 performance and a well-received encore in 1930, the observatory staff began to research the parody's origins. When they found the manuscript in 1921, Winslow Upton at first seemed the unquestionable author. Not only did his authorship seem to fit institutional lore, but oral histories confirmed that he certainly had been musical. In fact, he had been so enamored with music that he ordered an organ to be installed in his small, dormitory-style room.[51] In 1930, however, the observatory staff began to compare the manuscript's handwriting to his, and they did not find a match. In fact, it did not match the handwriting of any of the assistants or professors. Though they could not be sure, it seemed to be in the hand of the one person who explicitly combined computer and servant roles in 1879: Williamina Fleming.[52] The parody did fit Fleming's reputation for a sharp wit and genial camaraderie, and her British background perhaps made her especially interested in Gilbert and Sullivan operas. Furthermore, though many members of the 1870s staff inspired the characters of the *Observatory Pinafore*, Fleming never did appear, except through a short reference to "our Scotch maid."[53] If the author, perhaps she could poke fun at the observatory staff because, at the time, she was slightly outside their ranks. Fleming would have been especially attuned to the silly ways that human computers acted like servants of the director-master, as she did every day.

Still, no matter the author of the *Observatory Pinafore*, the later interpretations have simplified the 1870s representations of observatory life. Though historians of the observatory from Bessie Jones and Lyle Boyd to David Alan Grier to Dava Sobel have viewed the Computers Chorus as exclusively women and reinforced a notion of the late-century Computing Room as a single-sex space, it was not the case.[54] At least in the 1879 version, male voices joined the female ones, and the Computing Room welcomed male assistants and professors, who had to use the library of astronomy books, as well. Furthermore, the character of Joseph/ine, though based on an ingénue, was not unquestionably female. Joseph/ine performed the observing tasks assigned to men, and Joseph/ine's romance with Professor William Rogers reflected educational emulation just as much as adoration. Moreover, Rhoda Saunders, as the only assuredly female character, was more complex than the conventions of faithful parody, combining multiple Gilbert and Sullivan characters into one. Her representation as striving and sad was perhaps a cruel joke or, as Grier argues, maybe it was the reflection of a complicated human being.[55] The 1929 performance simplified her, as well, asserting she had to be saved through marrying Dr. Waldo at the end of the play.[56] It is easy to think of Saunders as the first in a long line of women computers hired from outside observatory families. The later performances and histories, after all, have seen her position through such a legacy. But, in 1879, her presence could have been anomalous just as well as prescient. Nothing, at the time, assured that computing would be so thoroughly feminized. Rather, as the Harvard Observatory was a dual domestic/professional space, its computers linked to servant roles.

Conclusion: From Computers to Algorithms

Therefore, the computers of the Harvard Observatory were more diverse than many histories have represented. The presence of Anna Winlock, Louisa Winlock, and Selina Bond reflected the ways that the observatory acted as both a home for astronomical families and a scientific workplace. Saunders's position, while distinctively new for the time, still did connect to servants' work. Finally, Williamina Fleming's duties as maid and then increasingly computer followed well-established trends in observatory employment, just as she worked to reshape what she called "woman's work." The Computers Room exposed how the observatory mixed domestic and professional spaces and ultimately solidified links between servants' and women's computing, even amid the observatory's publicizing of a new vision for computing work.

Since the Harvard Observatory ultimately set the standard for hiring computers, particularly after Williamina Fleming's extensive publicizing, its history has implications beyond the single case. Computing encouraged combined roles, in part because in the 1870s the computer was an emerging professional. Therefore, when vast teams of remarkable people received the job title of "computer" in the twentieth century, their professional identity carried the legacy of certain workplace tensions: scientific institutions also acting as homes and computing being servants' work as well as professionals'. In histories of science and computers, computing work has been characterized as routine, repetitive, mind-numbing calculation. Yet it also involved complicated negotiations of the divide between home and work and between servants' and professionals' duties in changing workplaces. In the *Observatory Pinafore*, the rigidity of the British class system was a model for exploring workplace stratification, and the characters called "computers" reified the tensions inherent in the creation of a new professional identity.

Yet computing work was not all serious. Someone in 1879 found time between his or her duties to create the *Observatory Pinafore*, and the staff of the Harvard Observatory decided to take the time to stage performances in 1929 and 1930. While the amateur writer and actors explored workplace expectations, even proclaiming "computing is our duty," they completed the play outside job responsibilities. Moreover, though the *Observatory Pinafore* did have serious implications, its use of humor complicates analysis. After all, anything can be explained away as just a joke. As recent work in technical communication has shown, humor is powerful precisely because of its doubling.[57] The jokes in the *Observatory Pinafore* allowed for multiple meanings to exist simultaneously—some more hidden than others.

Computing work, it seems, does contain a legacy of such covert humor, which humanities scholars can further analyze. As news reports have uncovered, jokes have been hidden in code, back to Apollo 11 and likely earlier.[58] Such instances do offer opportunities for scholars to recover meaning, particularly the ways that codes have been constructed for human readers as well as machines, according to insightful commentary from critic James E. Dobson and poet Rena J. Mosteirin.[59] Beyond code, expectations of humor have been written into the history of algorithms and computers, even as that work was being defined. Though often explained away as frivolity, the *Observatory Pinafore*'s performance history and its recurrence in scholarship demonstrate that it warrants further attention. Its representations of gender, class, and computing not only provide a fuller picture of the Harvard Observatory but also encourage further investigation of current and past practices of STEM workplaces. Joking has been a key way for algorithmic workers to imagine their past as well as construct a workplace of the future.

8
Seeds of Control

Sugar Beets, Control Algorithms, and New Deal Data Politics

Theodora Dryer

Over the past two decades, applications of machine learning and computer vision algorithms have proliferated in studies of agricultural production and management. A large segment of this research focuses on the sugar beet industry, especially the problem of predicting "good" or viable sugar beet (*Beta vulgaris*) seeds. This problem of viable seed prediction is a centuries-long math problem related to Western control of agricultural economies.[1] In these contexts, predictions of viable seeds are used to sprout sugar beets and a majority of those sugar beets are turned into sucrose, a powerful currency in modern agrarian capitalism.[2] Throughout the twentieth century, as sugar beet production expanded, it also became a central context and laboratory for the advancing powers of algorithmic oversight.

Today, computer engineers have turned to computer vision and other digital technologies for quality control inspection of sugar beets from seed counts to harvest assessments. In these virtual realms, visual digital projections of sugar beets are harvested from computer vision landscapes (e.g., predictive maps of weeds in sugar beet fields) that are calculated down to their composite structures and the chemical output of sucrose. Every dimension of sugar beet production and analysis has been written in terms of computing probabilities through algorithmic functions (Figure 8.1).

In their root functions, digital sugar beet systems are designed to yield controllable sucrose profits through the control of mathematical probabilities. Examples include: a standard expectation of discarding 35–40 percent raw sugar beet seeds; an 88 percent accuracy of detecting weeds in a sugar beet field; a 95 percent healthy yield. Computer vision techniques used for assessing and managing sugar beet production function as part of a larger program to extract probabilistic data from agricultural contexts.[3] This describes a new iteration of an older political agenda to transform agriculture into an algorithmic enterprise.

Theodora Dryer, *Seeds of Control* In: *Algorithmic Modernity*. Edited by: Morgan G. Ames and Massimo Mazzotti, Oxford University Press. © Oxford University Press 2023. DOI: 10.1093/oso/9780197502426.003.0009

Figure 8.1 A visualization from a Bonirob V3 robot operating on a sugar beet field, showing sugar beets (green) and likely weeds (red).

Source: P. Lottes, M. Hoeferlin, S. Sander, M. Müter, P. Schulze, and L. C. Stachniss, "An Effective Classification System for Separating Sugar Beets and Weeds for Precision Farming Applications, n *2016 IEEE International Conference on Robotics and Automation* (2016): 5157–5163, doi: 10.1109/ICRA.2016.7487720.

In this chapter, I put forward a significant precursor to these computer vision systems in the 1920s and 1930s United States when industrial agriculture first became a data-driven enterprise and new statistical control designs were put forward to manage it.[4] One design in particular, the Statistical Quality Control Chart (SQCC), came to prominence as a new white managerial class trope as it proliferated in telecommunications and agriculture. It was used in agricultural contexts in part to stabilize a US sugar beet economy. This corresponded with a largescale movement to transfigure sugar beet fields into laboratories for mathematical testing. For example, by the late 1930s, big sucrose companies, including the Michigan Sugar Company terraformed the US state of Michigan, known as the sugar beet state, into largescale farmland gridded into controllable area plots. These terraformed grids undergird algorithmic analysis.

By situating the design and creation of data architectures, including Statistical Quality Control Charts, in historical context, it becomes possible to map out how they came to political power and the ongoing significance of their applications and impacts. This helps determine how small-scale data architectures relate to larger historical forces including colonialism, shifting agricultural powers, and mass industrialism. At its roots, the history of algorithmic computing is a history of how information and data are conceptualized, ordered, and put to use in economic and political decision-making, and how these data-backed decisions in turn impact the environment and human life. In order to understand computing in context, it is imperative to trace computing systems at work outside of computing laboratories housed in universities or government buildings, in what I term "computing landscapes." Computing landscapes constitute the geographical areas of land or environments and corresponding cultural and political landscapes that are mediated through computational oversight.[5] These are the terraformed environments from which data is extracted. Computing landscapes feed digital systems, machine learning, and computer vision technologies.

Algorithms and other data architectures are inseparable from the culturally specific information bodies and the corresponding political and economic worlds that they manage. This history of control algorithms is a dramatic case of this rooted in the New Deal United States, in a moment of radical instability and restructuring of state policy over agricultural resources, labor, and land. The computing landscapes designed under this politics of control include agricultural systems, factory enterprises, and what colonial and industrial powers first imagined in the nineteenth century as the United States' "sugar beet belt".

Computing Control

Control is the central computing concept at work in this story. Throughout the twentieth century, control came to power in the domains of sugar beet and agricultural production as well as in industrial manufacturing. Control remains a core function of digital algorithmic society; it is a cultural value and technical apparatus at work in computer vision systems, in artificial intelligence, and digital algorithmic planning more broadly. As such, "control" is an important site of inquiry in tracing the connective tissues between early and late twentieth-century computing regimes and today.

While control has not always been a digital computing concept, there are certain shared characteristics to control computing across twentieth-century contexts. First, in computing, it has predominantly been understood as a statistical probability concept, measure, and practice. As I will discuss in this chapter, control is often synonymous with the unstable principle and practice of statistical randomization. But beyond its technical meanings, "control" holds powerful affective and political meanings. The technical and political dimensions of control are bound together down to the root scales of data and analysis. Focusing in on a particular regime of data-driven advocates in the 1930s United States, I show how new visions of a data-driven agricultural economy link the probabilistic and algorithmic formulations of control to much larger political and economic processes, in the name of a control society.

In the interwar United States and related trans-Atlantic contexts farm administrators sought to control what they deemed to be unstable agricultural production. While the assertion of instability was itself a political project, there were real underlying calamities contributing to the uncertainty of farm management in this era. First, people were recovering from the destruction of World War I. The fracturing of empires into nation-states resulting from the war caused instability in international trade systems related to agriculture: economic depressions, extreme poverty and displacement, and ongoing border disputes disrupted preexisting modes of agricultural production and the lives that depended on it.

Throughout the 1920s and 1930s, these rapid transformations in agricultural trade globally, and reflected in economic and political instability in the US context, put agriculture at the center of policy change. And these challenges were further backdropped by climatic drought and unpredictable weather, what would come to known in the US context as the "Dust Bowl" era. In the midst of these uncertainties, dreams of stable agricultural production were not equally mitigated between farmer and farm administrator. Ultimately, the multifarious desires to achieve control over agricultural production tie directly into the anxious desires of governing bodies to control people, land, and resources.

Franklin Delano Roosevelt's New Deal creed to "control and coordinate American farming," constituted a different political desire from the millions of US farmers living in impoverished conditions during this time. And farm worker labor would soon be calculated in terms of minimum and maximum objectives under the implementation of new industrial machines and mathematical oversight. US imperial and state ambitions of a control society fostered the rise of industrial agricultural and corporate agribusiness.

By 1920, what historian Deborah Fitzgerald names "the industrial ideal in agriculture" became a main feature of the US economy.[6] This is marked by the integration of industrial technologies, and Progressive Era labor policies and managerial practices into agricultural production. A significant characteristic of this integration, as Fitzgerald and others have rightfully noted, was the preoccupation with statistical bookkeeping.

Within this broader context, I argue that in the same moment when agriculture had become industrial it also became a data-driven enterprise. Within this, newly designed data architectures for counting, managing, and organizing information came to the forefront of political oversight. In the context of extreme poverty, climatic instability, and political uncertainty, a regime of mathematical planners thrust forward a new system of control logics as part of establishing a data-driven control state.

In the New Deal era, agricultural industrialists, government planners, scientists, and resource managers sought to control agricultural and industrial data through statistical experimental designs. For them, to control data meant to control the processes that the data corresponded to, namely: industrial manufacturing and agricultural production. Thus, the broader interwar moment is a significant origin story for the use of mathematical data designs and control algorithms over agricultural processing, over food production and consumption, and over the lives of farmers and farm workers.

In what follows, I introduce "control" in terms of its nineteenth-century political and affective meanings largely pertaining to technological management under corporate integration. A significant shift in control logics occurred after 1920, when "control" first became a mathematical-statistical concept. I focus in on a data architecture designed at Bell Laboratories: the Statistical Quality Control Chart (SQCC). In this context, engineers worked to mechanize "control" in order to stabilize and thereby control industrial manufacturing processes that were likewise transfigured into data-driven processes. Their newly designed statistical control chart circulated as a material emblem and algorithm for a new white managerial creed: Statistical Quality Control (SQC).

The majority of this chapter then traces the implementation of SQC logics in New Deal agriculture. I first discuss the larger imagination of a data-driven sugar beet industry corresponding to the actual reconfiguration of sugar beet fields and their corresponding managerial and labor structures.

This turn toward data-driven management and quality control logics corresponded with federal efforts to establish a control state through the mass consolidation of resources and bureaucratic reorganization. This was a moment when new algorithmic data architectures became central mechanisms in the oversight of industrial manufacturing, agricultural breeding, and administrative processes. Furthermore, this was a moment when the modern capitalist notion of "control" became algorithmic.

Unearthing these seeds of control in the New Deal context helps clarify the enduring entanglements between the environment, natural resource policy, data, and algorithm.

Control on the Factory Floor

The widespread proliferation of statistical quality control logics in interwar agricultural production is a story that begins with the American factory, the heart of US corporate capitalism and technology. Founded in Ohio in 1869, the Western Electric Company became a widely known center for American electrical engineering and technological innovation. Throughout the late nineteenth and early twentieth centuries, Western Electric drove the westward expansion of electric technologies, communication patents, and factory production.[7] The technologies and technological infrastructures developed by the company relate to a much larger social and class reorganization through the rise of factory labor and industrial managerial oversight.

In 1925, Western Electric Company partnered with AT+T and became Bell Telephone Laboratories. The Bell System was by then the center for developing managerial and organization science in a rapidly ballooning US telecommunications industry. From manufacturing telephone transmitters to expanding the electric grid and asserting a new managerial culture, Bell Laboratories was the epicenter of new corporate capitalist labor structures.

It was in this context that a group of Bell engineers designed a model for mechanizing control and industrial labor on the factory floor that they named Statistical Quality Control.

Throughout the twentieth century, the statistical mathematics community has attributed quality control logics to Bell Laboratories engineer Walter Shewhart. Shewhart first started as a "quality inspector" at Western Electric in 1920, overseeing the quality production of telephone receivers.[8] Business historian Paul J. Miranti has contextualized his hire as part of the Bell System's efforts to expand their "quality assurance" capabilities.[9] This forming corporate ideal and white managerial business creed of quality assurance equated to a promise of control over machine production and labor management. The Bell System thereby hired technical employees like Shewhart to research new inspection capabilities for stabilizing and controlling a rapidly growing telecommunications industry. It was hoped that new technical oversight would help streamline capital production through the "modernization" of industrial manufacturing processes and managerial reorganization. Quality control logics were thereby established as part of growing corporate integration for the telecommunications industry.

In 1924, Bell Telephone Laboratories technicians Walter Shewhart, Victoria Mial, and Marion Carter designed an analysis chart and model to oversee statistical data management in industrial manufacturing. The Statistical Quality Control Chart was originally entitled "Inspection Engineering Analysis Sheet." They designed this sheet to track specific industrial processes over periods of time. The "quality" analyzed in each chart related to a particular technology, piece of technology, or technological function that were each reformulated as statistical processes.

The Bell engineers asserted that there was a lack of control and therefore a lack of profitability within quality control management. As they were redesigning 'quality control' to be an inherently statistical problem, they determined that this problem could only be solved through probabilistic oversight. They therefore reconfigured the problem of quality control as a problem of statistical sampling in industrial inspection.

First, they argued that statistical sampling was needed because it was impossible to count each product and each product component one by one. For example, given the large number of telephone transmitters in a product line, it would be impossible to test every item in a reasonable amount of time. This was especially true in "destructive tests" that literally destroyed machine parts. Second, they argued that statistical sampling would help reduce inspector computational labor. At this time, Bell Telephone Laboratories manufactured 150,000 transmitters every year. It was not economical to employ inspectors to test every one of the 150,000 items, while also managing the bookkeeping. The imperatives of mass capital production were therefore baked into the mathematical design.

In taking this closer look at the Statistical Quality Control Chart (SQCC) prototype it is evident that extant Progressive Era control logics were converging with new statistical methods circulating as part of the international mathematical statistics movement. Prior to the SQCC design, US industrial manufacturing was not considered to be a mathematical or a statistical program. Rather, industrial manufacturing was more of a qualitative managerial program of training human "inspectors" to oversee technological production. These inspection regimes adhered to cultural values of control and efficiency through counting and accounting for time.

Managerial control logics fueled the Progressive Era's rapid industrialization and corporate capitalist expansion, spurring efforts to stabilize managerial control over the time logistics of expanding technological infrastructures such as the railroad and postal systems. Time and labor measurements were organized into managerial instruments such as book-keeping charts, timetables, and graphs to assert managerial authority over a limited domain of calculation.

As depicted in Alfred Chandler's 1977 *Visible Hand*, the late nineteenth-century managerial revolution gave rise to "a large number of full-time managers to *coordinate, control, and evaluate* the activities of a number of widely scattered operating units."[10] Nineteenth-century technology systems from telegraphs to the railways were linked by a new class of time managers. Control was a dominant social value in this arena of Progressive Era industrial society.[11] It was a social belief that managerial oversight and modern scientific methods could stabilize predictive expertise over market processes and thereby control capital production. In this sense, economic and technological integration were already delimited by managerial instrumentations that centered control calculations.

Likewise, with the specific managerial creed of quality control on the factory floor, inspectors assessed daily manufacturing processes by mapping out the quality of machine products over time—usually recorded on an annual and semiannual basis—to assess their profitability. Quality control was a means of achieving higher quality at lower cost by streamlining production, isolating human and machine errors, and reducing human inspector and computational labor.[12] It was a tendency toward constricting resources needed in the production processes, for profit.

As a newer iteration of this older Progressive Era formulation of control, The Statistical Quality Control Chart was designed to generate and structure industrial data for a specific mode of computational work. "Control" was thus designed in 1924 as a bounded computing mechanism that could assert mathematical limits for uncontrollability and statistical error within any manufacturing process.

Statistical control logics relegated industrial control and oversight to data-driven decision-making. The Bell Engineers designed the SQCC to isolate industrial processes over a certain period of time and quantify control within them. It comprised a formatted table of calculations pertaining to a new mode of informatics: "industrial data" manufactured on the factory floor.[13] The SQCC reformulated Progressive Era control logics in terms of probabilities that reduced managerial oversight to a limited number of factors, bounded by a computational process.

Industrial Data

The 1922 Statistical Quality Control Chart (SQCC) design merged nineteenth-century managerial practices of labor management with more newly circulated mathematical trends in probabilistic analysis. This led to a distinct conception of statistical information created on the factory floor: industrial data.

By the early 1920s, Shewhart had become very interested in the Anglophone mathematical statistics movement. He read published material and housed a copy of Karl Pearson's *Grammar of Science* in his Bell Telephone Laboratories office.[14] Additionally, he was aware of the growing interest in mathematical statistics in US agriculture. As he reviewed this literature, he came to believe that the problems seen in industrial manufacturing were inherently probabilistic and that new methods circulating the international stage could help streamline inspection work by achieving accurate predictions from a *small* set of randomized quality inspection tests.

Invoking the larger mathematical statistics movement, Shewhart maintained that it was possible to make accurate predictions using a minimal amount of observed sampling data.[15] Shewhart worked on the problem of estimating product quality using *small* sample sets in order to minimize the inspection labor needed to oversee product quality.

The Bell Laboratory engineers advanced statistical quality control as a research philosophy that aimed to translate quality inspection into a probabilistic computational process. The basic idea was that the statistical approach would allow them to frame otherwise "out of control" processes within controllable probabilistic limits. Each year, the Bell System produced upwards of 150,000 transmitters, an expensive production process that the engineers understood to be totally "out of control." They determined that total "control" was intrinsically impossible at each stage of production: from achieving homogenity in the raw granular carbon material from which transmitters were made to the analysis of their production. Under commercial conditions, this uncontrollability seeped into computational observation and estimation.

Shewhart, Mial, and Carter thereby designed the SQCC to mechanize error management and establish mathematical control metrics over what they called "out-of-control" processes. Mial and Carter conducted the work of designing the initial chart from a large-scale data analysis. After collecting tens of thousands of frequency observations, they asked: "Do these data present any evidence of a lack of control?" In answering this question, they calculated *control* as an acceptable limit to error theory, defined by the "four statistics": average, standard deviation, skewness, and kurtosis.[16] They determined that if frequency observations for one product, assessed within a finite time frame, stayed within these statistical limits, then that particular machine process could be considered to be in a state of control. In making the SQCC prototype, these women collected experimental results, conducted statistical estimation calculations, and designed graphical representations of the quality observations. In the initial publication of the chart, however, Shewhart commented that "the preparation of such a chart requires but a small amount of labor on the part of a computer."[17] And he maintained that the chart could tell "the manufacturer at a glance whether or not the product has been controlled."[18]

In the case of carbon microphones—a key component in the telephone—the physical manufacturing process was to be "controlled within very narrow limits [elsewise] wide variations [would be] produced in the molecular properties of the carbon."[19] In the production process of carbon microphones, there were many components that needed to be "highly controlled": vibrations and sound disturbances, machine chamber conditions, and the precise mixtures of gases and temperatures. The engineers noted that the high-volume needs of capital production undermined controllability in the manufacturing process, as it valued mass quantity over precision.

In translating the problem of carbon homogeneity into a statistical quality control project, the engineers collected randomized data from empirical electrical resistance test measurements. Any notable changes in the resistance measurements would indicate heterogeneity in the carbon samples (Figure 8.2). The resistance measurements were written as probabilities due to "the statistical nature of the phenomenon under question."[20] The SQC lens asserted that manufacturing was an intrinsically probabalistic process down to the molecular constitution of machine parts.[21]

Despite the heavy computing labor needed, by bounding the processes for observational error, within probabilistic limits, the SQCC asserted a mathematical vision of control over factory production. In this sense, managerial authority over the production process blurred between the human quality control inspector and the mathematical terms of the quality control chart.

The quality control inspector was now situated to think about engineering and industry production as both a physical and computational process. Underpinning this reformulation of industrial processes was the conceptualization of "industrial data" as a distinct form of highly generative probabilistic material. The design of the SQCC thus emerged from a longer history of control thinking in US industry and telecommunications, which was in turn part of a larger impetus to reinterpret manufacturing as a data-driven enterprise managed by quantitative methods. The SQCC ultimately mechanized *control* as

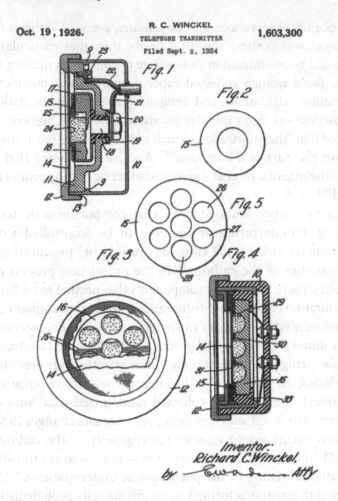

Oct. 19, 1926.

R. C. WINCKEL

TELEPHONE TRANSMITTER

1,603,300

Filed Sept. 2, 1924

Inventor:
Richard C. Winckel,
by ———— *Atty*

Figure 8.2 Patent 160330 for telephone transmitter, Richard C. Winckel, Western Electric, Filed: September 2, 1924, Published Oct. 19, 1926. The patent describes heterogeneity in the carbon samples in terms of probabilities.
Source: Richard C. Winckel. "Telephone Transmitter," US Patent US1603300A (1924).

a set of mathematical-statistical limits and computing steps drawn over industrial information.

In this same time period when Bell engineers were transfiguring the industrial managerial concept of control into a bounded computing mechanism, agricultural administrators were transforming large-scale sugar beet fields into a 428,000 square mile data-driven computing landscape. Distinct meanings of industrial data and agricultural data blurred under the same algorithmic process of quality control.

Sugar Seeds: Sugar Beet Belts and New Deal Farm Labor

In the period between 1890 and 1930, the US sugar beet industry transformed large-scale farm areas into gridded landscapes built around factory centers (see illustration of a sugar beet at Figure 8.3). While efforts to develop a sugar refinery system began in the early nineteenth century, the US sugar economy did not take off until after 1890, after nearly 150 years of failed attempts by European settlers to grow sugar. A contributing factor to these failures was the inability of US settlers to transport French and German sugar-beet breeding models onto the new soil and climate. Standardized European practices, ranging from the spacing and depth of planting sugar beet seeds to the time calculated between harvests, did not yield the same predictable results in the wildly arid climate.

Despite the ongoing climatic and control uncertainties, sugar beet production ballooned in the period 1917–1935 due to dramatic industrial transformations and forced agricultural production. These included "increased applications of electricity that made possible the use of instruments and devices for facilitating precise control of chemical processes [and] improved modes of coordinating mechanical operations ... size and shape of plot

FIG. 2.—White Improved Vilmorin Sugar Beet.

Figure 8.3 "White Improved Vilmorin Sugar Beet," Wiley, Farmer's Bulletin 52, United States Department of Agriculture. Source: *Annual Report of the Secretary of the State Board of Agriculture of the State of Michigan* (United States: Robert Smith Print. Company, state printers and binders, 1899).

in relation to field experiments with sugar beets."[22] But the relative success in forcefully creating sugar beet fields did not equate to a stable economic enterprise nor a sustainable environmental system; to the contrary.

In this context of rapid agrarian industrialism, agricultural administrators cast sugar beet breeding, like industrial manufacturing, as being out of statistical control, and described a general "lack of controllability" over US sugar production. Despite the 150 years of preceding failure, and purported lack of control, settler colonial visions of a sugar empire endured and were literally modeled after the European sugar beet, especially Napoleon's sugar empire. It was from this anxiety of imperial failure that statistical control logics were advanced.

I describe US sugar beet breeding in three rough historical stages of quantitative control programs. As discussed above, the first period, 1800–1890 constitutes an overall failure to establish and control a sugar beet economy. While sugar beet factories popped up, it wasn't until 1870 that the first commercial factory was established in California. In the period 1890–1920, "beet dreams" persisted when Anglo settlers continued to forcefully build a sugar enterprise through sheer trial and error, as there was no workable model that could be transplanted due to climatic conditions. Testifying to the failure to control in this moment, in 1899, agriculturalist H. W. Wiley wrote, "The experience of more than ten years in California has shown that the climatic data, regarded as of prime importance in beet culture in Europe, [is not] rigidly applicable to this country."[23] This resulted in imagining a statistically controlled "sugar beet belt" as an area of climatically optimal sugar beet breeding depicted as a large ribbon across the continuous United States landscape, without note of Indigenous land or variations of agricultural ownership (Figure 8.4).

The outline of the sugar beet belt was shaped by trial-and-error experimentation with climatic and weather data in comparison with European industries. This work was largely conducted by statisticians through a growing network of agricultural experiment stations, captured in the following description:[24]

For growing most crops, the weather is even more important than the soil. The conditions of climate best suited to growing the sugar beet differ from that of many crops, and the weather that would seriously impair the production in other crops, may be well suited to the crop of beets with a large content of sugar. In Germany, it has been found that a certain average temperature for the several months from May to November, and a certain average rainfall during these several months, are

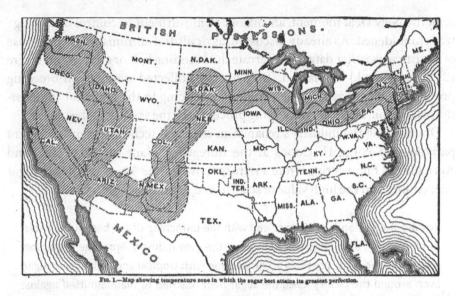

Fig. 1.—Map showing temperature zone in which the sugar beet attains its greatest perfection.

Figure 8.4 "The Theoretical Beet-Sugar Belt of the United States." Source: H. W. Wiley, "The Sugar Beet: Culture, Seed Development, Manufacture, and Statistics," *Farmers' Bulletin* 52 (1899): 5. Accessed from University of North Texas Libraries, UNT Digital Library, https://digital.library.unt.edu; crediting UNT Libraries Government Documents Department. https://digital.library.unt.edu/ark:/67531/metadc85519/.

best adapted to the growing of the crop. Such a "sugar beet belt" sweeps through the lower peninsula of Michigan.

The sugar beet belt therefore was a map of the ideal climatic conditions for sugar beet breeding following a century of failed attempts to directly model after European breeding programs. It is a map conceptualized and drawn after statistical weather information.[25] By the late nineteenth century, this computing landscape fortified and became the US sugar beet breeding ground—it was an experimental landscape of large proportions.

In the third stage of sugar beet breeding, 1920–1940, there was a clear tension between "crisis and control" in quantitative sugar beet management. During this time, data analysts asserted a lack of controllability over the sugar beet belt at large, and new models and methods for statistical oversight and data analysis were employed to fix it. Given its longer history of failure, the US sugar beet industry was seen as a program that had been made possible only through technological and scientific advancement. The sugar industry, therefore, had become industrial as well as statistical from seed to soil to refinery.

In the New Deal moment, an anxious committment to control technology was emboldened. As already discussed, agricultural administrators made use of statistical weather data to demarcate the breeding belt or area of land where sugar beets could be planted. They further collected statistics on breeding processes along dozens of axes, from seed spacing to the chemical composition of the soil, and from harvesting processes to the science of the refinery.

Dramatizing the industry's reliance on data and technology, in 1949 plant pathologist George H. Coons at the Bureau of Plant Industry, Soils and Agricultural Engineering wrote a retrospective on the "crisis" of the US sugar beet empire and the high value of science and technology in its survival:

> Nor did the demands on science end with the launching of the beet-sugar enterprise, for crisis after crisis has confronted the new industry. From the beginning, sugar from the sugar beet was in competition with tropical and subtropical sugar. Even around the factory itself, the sugar beet has had to maintain itself against other crop plants competing for the farm acreage.... plant breeders have constantly increased the productivity of the sugar beet; agronomists have discovered efficient methods of growing the crop; chemical engineers have improved the processes of sugar manufacture; and against epidemic diseases resistant varieties have been bred. Only through these contributions of science has it been possible for the sugar beet to survive.[26]

This retrospective captures how in the New Deal period, the many dimensions and processes associated with sugar beet breeding evoked interest from rising regimes of industrialists and data-driven oversight. For industrial agriculturalists, the sugar beet belt became an ideal laboratory for expanding data-driven logics of efficiency and control. In the interwar period, US sugar beet breeding became an industrialized and data-driven enterprise and this significantly impacted farm workers, agricultural production, and the environment.

Under the rising conditions of industrial agriculture, statistical administrators viewed sugar beet breeding to be an industry of calculation: of crop production and resource consumption, prices and tariffs, and the quantification of human and machine labor. And sugar factories, rather than state administrators, were the centers of oversight in these transfigurations. These factories were situated as governing centers on sugar beet fields, one of the most powerful being the Michigan Sugar Company (Figure 8.5).

As depicted in the visually mapped sugar beet belt, the Midwestern and Great Lakes regions of the United States was the center of US sugar production and the reformulation of sugar beet labor oversight.[27] By 1920, sugar

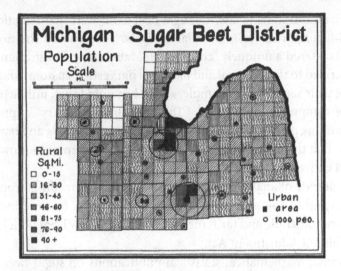

Figure 8.5 "The Michigan Sugar Beet Industry."

Source: F. A. Stilgenbauer, "The Michigan Sugar Beet Industry," *Economic Geography* 3, no. 4 (1927): 486. Permission of Taylor & Francis Ltd., http://www.tandfonline.com, on behalf of Clark University.

beet production was assessed in terms of hand labor calculations—the number of workers and time it would take to pick sugar beets by hand. Into the New Deal moment, calculations of machine time in sugar beet harvesting were being compared to hand labor calculations and for beet workers, this resulted in more strenuous work hours for lesser or no pay. New Deal legislation had further contributed to a drastic shift in labor conditions for sugar beet workers along racial lines.

In 1933 Minnesota, for example, 80 percent of field workers were white; by 1937, New Deal legislation emboldened white farm owners to "turn anew to migrant workers," and almost three-fourths of the Valley sugar beet laborers were Mexican workers.[28] In a neighboring state, the Michigan Sugar Company, situated within the largest sugar-beet producing state, staffed its fields by seasonal labor contracts with Mexican workers who had migrated from other Midwestern states.[29] The living conditions for the migrant workers were abysmal; they were grouped in thin wood buildings set up on the sugar beet fields, and were denied employment and healthcare stability.

In Detroit, Michigan, the Burroughs Adding Machine company, which produced the very same calculating devices used by the new rising ranks of statistical administrators in agriculture, was also staffed by Mexican migrant workers, many of whom additionally worked in their sugar beet fields.

For agricultural statisticians, the sugar beet fields were mathematical spaces and ideal laboratories for inferential analysis. They shared the notion that sugar beets offered a uniquely "controllable" laboratory for agrarian analysis, largely ascribed to their annual and biannual harvesting customs and the even spacing of their seeds.[30] For example, an analyst noted that, unlike the uncertainties of computing sugar cane, "The beet sugar industry ... presents no such accounting problem. The beets are planted in the spring and harvested in the fall; so that the expenditures incurred by the companies in growing beets are almost entirely confined to one fiscal year."[31]

This multitude of analytic points within sugar beet breeding, from spacing and harvesting to soil analysis and labor calculations, became increasingly tied to economic trade and tariff measures through New Deal policies such as the Agricultural Adjustment Act.

In response, mathematical statistical publications on sugar beet breeding ballooned. The US and international applied mathematics community desired to further their research in applied agricultural statistics. These included topics such as "the analysis of variance illustrated in its application to a complex agricultural experiment with sugar beet." These sugar beet applications were reflected in pedagogical training, as captured in this example: "the statistician may be asked to draw a sample of sugar-beet farms in this country, in order to provide, at minimum cost, certain information about sugar-beet growing which is needed for policymaking."[32]

Indeed, statistical procedure directly linked to state and economic policy. The sugar beet laboratory became site and context for a new pedagogical and policy regime at the US Department of Agriculture that further substantiated data analysis—and statistical quality control—as central to agrarian oversight made possible by the rapid impact of New Deal policy.

Agricultural Data: The USDA Graduate School

The concurrence between mechanizing data on the factory floor and the formation of data-driven sugar beet fields relates to a third movement to reorganize state and federal agricultural research according to SQC and new formulations of data-driven policy. This constitutes a triple concurrence of events that makes evident the extent to which control computing in the New Deal moment transfigured the US political and environmental landscape. Between 1920 and 1940, the US Department of Agriculture (USDA) acquired new authority as an intermediary between federal oversight and state policy. Efforts to assert the USDA as a leading institution in social policy were heightened again during the New Deal period (1933–1940), as President Franklin Delano Roosevelt's administration held majority congressional power in implementing new agricultural policy. The federal government went to work to try and *coordinate and control* American farming.

The USDA headquarters in Beltsville, MD, had become the central site for implementing the organizational coordination and control of US agriculture. In 1921, President Warren G. Harding had appointed Iowa-born Henry C. Wallace as secretary of agriculture. Wallace was an ideal candidate for the position having "been in contact with actual farming" his whole life.[33] Periodicals emphasized his experienced roots as a "dirt farmer" and suggested that he overcame this status through his agricultural college education, where he "witnessed the discovery of important scientific principles and their practical application to agriculture."[34] Despite his public climb to scientific and administrative power, Wallace remained vigilant in his goal of addressing class discrimination in American farming as emblemized in his posthumous 1925 book, *In Debt and Duty to the Farmer.*[35]

For Wallace, American farming was built on poverty and he believed this was an administrative problem that could be solved through advancing scientific education. His approach to social welfare resonated with institutionalist economics, a class-conscious movement emergent in 1918 that followed the work of Thorstein Veblen.[36]

Four years prior to his death, Wallace founded the USDA Graduate School to advance the education of department workers. Henry C. Wallace and his son Henry A. Wallace, who would later be appointed secretary by FDR in 1933, furthered an interest in income distribution and a socialist view of farming labor that would significantly shape pedagogical initiatives at the school. The New Deal science worker was to be trained in statistical methods and socialist politics against what was described as the powerful mythologies of corporate capitalism. Here there was a blurring of capitalist logics (profitable efficiencies in quality control) with a drive toward "first principles" that held radically different political and cultural meanings.

Immediately after taking his 1921 appointment, Henry C. Wallace announced the Graduate School in an official statement: "The Department of Agriculture proposes to establish this fall an unofficial system of advanced instruction in those scientific and technical subjects related to the work of the Department, in which adequate instruction is not otherwise available in Washington." The purpose of the unofficial graduate school, according to Wallace, was "for greater usefulness through better training and increased knowledge." Two kinds of courses were offered the first were, "lecture and drill courses on ... certain fundamental subjects in which the personnel of two or more Bureaus may be interested" and the second were "intensive graduate training in special topics."[37] The school offered extended education for USDA employees to better their research work and advance their participation in civic society.

Located on the Department of Agriculture's 13,000-acre farm in Beltsville, MD, the idea for the school stemmed from a study group conducted at the Bureau of Standards in 1908. While it never established its own building, it held classes in the Agricultural Department, in the Interior Department, and in the Smithsonian Institution.[38] Under the watchful eye of the public and congressional pundits, the school was careful not to use government resources. Classes were held only in the early mornings and the late evenings to avoid use of "government time," and only the director received direct monetary support from the government.

A Statistical Control State

On March 4, 1933, the Bell System's Walter Shewhart gave a lecture at the Graduate School on "The Specification of Accuracy and Precision." The lecture was delivered in the same moment that FDR delivered his inaugural lines, "The Only Thing We Have to Fear Is Fear Itself."[39] The new president advocated that day for engagement "on a national scale in a redistribution ... to provide a better use of the land for those best fitted for the land ... helped by national planning for and supervision of all forms of transportation and of communications and other utilities which have a definitely public character."[40] This vision of America depended on a unified political order and the formation of an administrative bureaucracy that could control these projects at the state and institutional levels.

Emboldened by New Deal policy, the USDA Graduate School became a thriving educational center and a permanent feature of the Beltsville farm, though still not endorsed or funded by government bodies, which were predominantly under the control of subscribers to the associative politics of Roosevelt's predecessor, Herbert Hoover. Throughout the 1930s, the school continued to receive widespread critique from local DC politicians and national interests that wanted separation between "university" and "government."

This political backlash heightened after the USDA began to implement New Deal legislation in 1933, the year FDR instated Henry A. Wallace as secretary of agriculture. Like his father, Wallace believed that the USDA had an important role in implementing New Deal programs on the state and regional levels. It was the USDA's role to educate the scientific community and public on what they referred to as first principled research and to implement this policy through establishing a new bureaucratic order that could sustain it.

Efforts to establish control at the state level and in scientific practice collided in the pedagogical initiatives of the USDA Graduate School that were developed for both USDA employees and the public. This included courses for USDA employees, a USDA Graduate School printing press, and a number of public lectures and conferences. A long-standing tradition of the school was to host visiting speakers from statistical experimental stations around the world. Edwards Deming directly facilitated exchanges with statisticians at Bell Telephone Laboratories and the Rothamsted Station. As early as 1927, Deming had become interested in the work of Shewhart and Bell Laboratories.

By 1933, Deming facilitated Bell Telephone Laboratories and the Department of Agriculture to co-sponsor Shewhart's commissioned talks. These talks were organized under the umbrella topic of "The Statistical Method from the Viewpoint of Quality Control," which would later become the title of Shewhart's influential 1939 book, due largely to the support of the Graduate School.[41] Throughout 1933, Shewhart gave four talks at the Graduate School on *statistical control* and the *limits of variability*.

Deming's reflections on SQC make clear the collapse between industrial manufacturing and agriculture under the formation of data-driven research. He said, "we in agriculture are faced with the same problems" as in manufacturing.[42] However, he argued that agriculture had more at stake and was in greater need of control due to the temporal delays in sugar beet breeding. He wrote:

> When machines are turning out piece parts by the thousands or even millions month, the industrial statistician does not have to wait long to see [their] predictions tested out. In agriculture, years are often required—a crop must be sowed and harvested again and again. . . . With time in our favor it is easy to become careless about fundamentals.[43]

Deming maintained that what agriculture should gain from industry was the concept of a statistical quality control state, asserting that without this oversight, "the statistician's calculations by themselves are an illusion if not a delusion."[44] While the process for achieving statistical control was indeed difficult, Deming and Shewhart promoted the need for the control chart and attention to the physical mechanism of the experiment or production process. Here agricultural testing and industry production blurred as two iterations of the same method of computation. In Deming's words: "The state of statistical control is therefore the goal of all experimentation."[45]

In this vision, achieving a statistical control state directly linked to capital market control in both industry and agriculture. This sentiment is captured in another enthusiastic assessment from Deming:

> When control (that is randomness) exists, the setting of limits within which certain percentages of the next 100 or 1000 measurements will lie is a purely statistical job, and the scientist's work is finished when [s/he] turns over the data, so obtained, to the statistician. In this ideal sate of control, attainable in industry but hardly ever in the laboratory, such limits can actually be set, and as a matter of record are working in the economic advantage of both buyer and seller of raw materials and manufactured products.[46]

In the ideal state of control advanced by SQC logics, which was a computational process as well as an economic ideal, the researcher could define the limits of uncontrollability and use this knowledge to their advantage. The idea that data could be transferred from scientist to statistician underscores the new conceptualization of statistical data as a commodified material in a modernizing state.

Under the New Deal, the USDA became a dominant institution in seeding scientific data production and collection. Increased references to "data" as "industrial data" and "engineering data," in the case of Bell Laboratories, and "agricultural data" mark a redefinition of statistics as a science of estimating and testing aggregated data in the name of economic production. Therefore, data generated and circulated in New Deal agriculture held a very particular meaning for its administrators and practitioners: it was valued in terms of its controllability.

Control Thinking for a "Better Society"

FDR's two waves of legislation in 1933 and 1935 were geared to controlling agriculture through the formation of new bureaucratic bodies such as the Tennessee Valley Authority. This was largely a response to Herbert Hoover's "failing" associative state and to the collapse of laissez-faire market capitalism more generally, which USDA administrators blamed for the current lack of economic control in agriculture.[47]

The USDA benefited from two waves of legislation, beginning in 1935 with FDR's Resettlement Act that emboldened the department's bureaucratic oversight on farming. Following the second wave of New Deal legislation in 1936, the Graduate School hosted a series of lectures on the organizational culture and initiatives of the Department of Agriculture, to help integrate New Deal policies into daily operations for administrative staff and scientific workers.[48] These lectures were organized into an official course for science workers to teach "the purpose of the work they are doing [and] how it contributes to public welfare."[49] Secretary Wallace was especially interested in treating the political "miseducation" of science workers who, he said, unthinkingly subscribed to corporate capitalist or orthodox dogma.

Throughout 1936, statistical control methods and data analysis, described as "research consultation," were promoted as a necessary mode of "clear thinking" in improving scientific research and regulatory work. USDA administrators advanced the notion that clear scientific thinking would provide the "foundation stones" to further Wallace's vision. Emblematic of this, USDA Bureau of Animal Industry's John R. Mohler gave a speech on the new future of USDA oversight that broke the politics of data analysis into three points:

1. The basic fact that research is the best means by which man *controls* and improves his environment
2. The knowledge that responsibility for conducting research rests on a relatively small group of persons having suitable ability and training
3. Research by federal and state agencies is a wise investment authorized by Congress.[50]

Here Mohler described data analysis as key to "thinking clearly" in all research work. Addressing the importance of "consultation" across disparate fields and departments, Mohler argued that cross-pollination across disciplinary divisions through this analysis would strengthen the USDA's scientific research directives. Part of the connective tissue, he contended, was knowing how to control statistical data across various scientific fields. He emphasized, "Mathematics is now playing a very important part in many lines of research work."[51]

In a colorful example of "research consultation," Mohler described a case where a biologist by the name of Dr. Dorset "consulted other specialists freely" due to his recognition that all fields in agriculture could enrich his knowledge. Dr. Dorset had wanted to know how many times he needed to repeat his sugar beet field experiment so that the results were reliable "beyond a reasonable doubt." He obtained the answer from a well-trained statistician in the Bureau of Agricultural Economics, M. Alexander Sturges:

A succession of seven positive results in a row in the same experiment gives, as the chance of error, only 1 in 128. By conducting the test three more times, getting 10 positive results in a row, the probability of error is only 1 in 1,024. In the type of work under consideration the probability of being right at least 999 times in 1,000 was deemed adequate.

Dr. Dorset's problem generated in scientific research became a problem of statistical control. This example captures the popularizing idea of a statistical control state, that it was adequate to be right 999 out of 1,000 times and the researcher could therefore be confident in their assessment. Mohler concluded his ethnographic study of scientific workers: "Knowing well that 'all progress is wrought in the mystic realm of thought,' we look to scientific workers to blaze the trail."[52]

The turn to first principles in agrarian planning under the New Deal was deeply politicized. Mohler used a popularizing terminology "superstitions" to refer to the practices in laissez-faire market capitalism and corporate governance that were thought of as currently rotting the minds of Americans, preventing them from thinking clearly. He argued that these "unscientific forces" were the enemies of progressive agrarianism, and that "the curious ceremonies" surrounding these enterprises produced "superficial" and unscientific knowledge.[53]

In 1938 and 1939 the USDA Graduate School hosted a series of widely publicized lectures on "American Democracy" just as the school was falling under widespread public critique for its use of government funds and perceived authoritarian socialist politics.[54] A public discourse of fear of *too much control* fell over the school and its research programs. In February 1938 the *Washington Times* announced the democracy talks, depicting Secretary of Agriculture Wallace as the "Farm Fuehrer."[55]

Visiting British statistician Hyman Levy spoke on the relationship between data analysis and democracy. Levy was at the time both a member of the British Labor Party and the Soviet Communist Party. His talk resonated with the importance of unifying science toward progressive ends, stressing the point that statistical work was not an apolitical enterprise. That same year his book on modern science was published, developing a thesis that offered a Marxist analysis of scientific production akin to that produced by Boris Hessen a few years earlier.

Levy maintained that "democracy" was strictly a product of industrial force, using Isaac Newton's work as an example of science that was produced by commercial and other social factors. In explicit consideration of the political nature of "first principles," Levy critiqued the Vienna circle as an apolitical campaign. He said, "There is a school of philosophy of which many of you must know, the logical positivists, who consider that the essential problem is to discover how to ask a question. They are not concerned with the actual question so much as the way in which the question should be asked." Levy also admitted that "Today in Europe it is rather dangerous to ask questions; it is much safer to discuss how a question should be asked."[56]

Despite the political backlash seen in the school's lack of support from Congress and in critical public opinion, SQC education was in full fruition by the late 1930s. In 1938, Graduate School director A. F. Woods published a letter addressed to the "statistically minded research workers in Washington" announcing the return lectures by Walter A. Shewhart of New York's Bell Telephone Laboratories.[57] By this time, Shewhart was president of a joint committee of the American Society for Testing Materials, the American Society of Mechanical Engineers, the American Statistical Association, and the American Institute of Statistics. Through the formation of these committees, SQC became a guiding approach to economic planning. It was gaining momentum as a modernizing ideal in capital production and scientific research. Woods wrote, "The officers and staff of the Graduate School are highly desirous that he contents of this letter be given as wide publicity as possible."[58]

By 1938, the Graduate School offered 125 courses, up from 70 in the previous years, and half of these were method-based courses: 17 accounting and mathematics courses, 17 in economics and law, and 16 in statistics.[59] The SQCC was increasingly thought of as the technical expression of method-based thinking in scientific research and public welfare. Deming edited dozens of talks and books on achieving statistical control in agrarian research. Out of eleven texts officially published by the USDA Graduate School, half of them were a promotion of quality control.[60] The 1939 Graduate School publication *Statistical Method from the Viewpoint of Quality Control* became the textbook on SQC logics. Within this text, the emergent SQC corporate doctrine was written as a historically revisionist view of technological progress centered on the "evolution of control logics" from the Neolithic period through the 1924 SQCC design. The textbook printing constitutes the beginning of widespread adoption of SQC in various domains of US managerial science.

Immediately after World War II, SQC expanded into global contexts. This includes a significant proliferation of SQC in Japan following the 1946 US occupation and seizure of the Japanese telecommunications industry.

Data Politics

The 1924 SQCC design constitutes an effort to mechanize *control* as a mathematical limit in machine processing and computational work. Prior to this moment, control held technical and affective but not algorithmic meanings in industrial processes. The SQCC gave a distinctive probabilistic procedure to control that adhered to progressive ideologies of corporate capitalism. It also relegated human decision-making to computational oversight. The chart was a visual representation of computing steps that the inspector "could view at a glance," thus promising a reduction in human computational and managerial labor in quality control oversight. The SQCC design reflects a larger cultural commitment to control management following from Progressive Era industrialism. SQC logics mechanized control as both a managerial labor process and a mathematical limit.

The politics of control computing shown in this chapter raise important questions about the role of data extraction and data analysis in political and economic processes. In this story, there is a complicated blurring of socialist, capitalist, and colonial ambitions at the scale of data and analysis as they formed in agricultural planning. This was especially clear in New Deal resource and social policy. During this time, scientific research initiatives in agriculture merged with industrial planning logics and were both rewritten as "essentially statistical." Thereby, radically distinct social enterprises were reconfigured as part of the same epistemic and procedural programs.

At the root of these transformations were new conceptualizations of industrial and agricultural data as probabilistic information. This probabilistic information held locally specific meanings collected from frequency observations of machine parts on an assembly line or from soil and harvesting samples. Given the large and growing levels of production in the telephone and sugar beet industries, estimation methods became the primary means of assessing labor and outcomes. These assessments relied on *small* sets of randomized testing. Ultimately, control logics secured a procedural and political dependency on statistical randomization and algorithmic thinking across these industries.

In 1947, Charles Sarle, administrator of the Bureau of Agricultural Economics, wrote on how corporate agriculture and statistical oversight further galvanized during WWII:

> As a nation's economy becomes more diversified and complex, demand for agricultural statistics increases. Demand is not only for broader coverage of agriculture, but also for other facts relating to the ever-changing agricultural process, for

statistics at more frequent intervals, and for greater accuracy. This demand is accelerated by the strain put upon a nation's economy at war. When national economies are subjected to world-wide depression and governments embark on production control and price-support programs, the demand for more and better agricultural statistics increases almost overnight.[61]

In this passage, Sarle speaks to the role of crisis in catalyzing the purported need for control and the generation of information systems that occurred in the New Deal era. As indicated here, World War II further converged military and agricultural production, especially as sucrose was used to fuel wartime production.[62] Postwar analysts noted that World War II galvanized a corporate agricultural economy that hinged on control logics: "Before the war, the tractor was calling attention to the importance of controlled spacing of seedling by precision planting. Processed seed, introduced in 1942, was almost universally adopted by 1947."[63]

In the New Deal moment, efforts to coordinate and control American farming tethered the design of statistical mathematical designs to new conceptions of probabilistic industrial and agricultural information. This ensconced a number of control principles in state and social management: statistical randomization, the law of large numbers, and inferential testing. Controlled or randomized data experiments, implemented across populations and landscapes, are a defining feature of twentieth-century quantitative governance, which is today naturalized in the applications of computer vision systems in agriculture.[64]

This history of control computing reveals stratificiations of algorithmic design and data processing, agricultural and industrial practices, institutional and political powers. Unearthing these seeds of control makes visible the entanglements between data, policy, and algorithm. It also makes visible the powerful historical precedents set in the New Deal moment toward the use of computational oversight in agricultural production and industrial manufacturing, which undergirds the entrenched political myth that algorithms are in control of data.

9

Inference Rituals

Algorithms and the History of Statistics

Christopher J. Phillips

In 1994 the distinguished psychologist Jacob Cohen penned what he noted was a distinctly unoriginal complaint about statistical testing titled "The Earth Is Round (p<.05)." The joke had bite not solely because of the implication that such an obvious claim required an article to prove it. Rather, Cohen believed statistical methods had fundamentally distorted the entire research process through the "ritual" of "mechanical" inferences made on the basis of p-values and other "sacred" statistical criteria.[1] Social scientists, blinded by methodological fetishism in their own work, might think it appropriate to use statistics even to prove the earth was round. Cohen blamed continued deference to R. A. Fisher's popular 1935 textbook *The Design of Experiments*, a book in which Fisher had proposed a series of statistical tests to enable "learning by experience" using formalized methods of "inductive inference." Fisher claimed these statistical techniques might be deployed widely to design experiments that distinguish "real" findings from results likely attributable to chance.[2] Cohen's article—taken from the address he gave after receiving a lifetime achievement award—warned that algorithmic use of statistical tests "has not only failed to support the advance of psychology as a science but also has seriously impeded it."[3]

The travesty for Cohen was that similar complaints had been made, and largely ignored, for decades. The British biologist Lancelot Hogben identified what he called a "contemporary crisis" in statistics in his 1957 book *Statistical Theory*.[4] Parts of that book were later reprinted alongside similarly critical articles in a 1970 reader on the problems of statistical inference in sociology and psychology called *The Significance Test Controversy*.[5] As early as 1960, psychologist William Rozeboom had condemned the "dogma of inferential procedure which, for psychologists at least, has attained the status of a religious conviction."[6] Six years later, David Bakan noted that the complaints about statistical inference were already becoming commonplace—"everybody" knew about them, he claimed—even

Christopher J. Phillips, *Inference Rituals* In: *Algorithmic Modernity*. Edited by: Morgan G. Ames and Massimo Mazzotti,
Oxford University Press. © Oxford University Press 2023. DOI: 10.1093/oso/9780197502426.003.0010

as journal editors continued to make significance testing a requirement for publication. Despite the complaints, Bakan admitted that researchers continued to seek a "basis for automatic inference," using statistical analysis to convert their "intuitions and hypotheses into procedures" that reliably yielded conclusions.[7]

Psychology and sociology were hardly alone in their dependence upon significance testing. Nutrition, clinical medicine, education, business, and many other fields extensively incorporated statistical procedures as a normal process of doing research. Allen Edwards's *Statistical Methods for the Behavioral Sciences* (1954) warned researchers that raw data don't reveal conclusions—observations must be processed and analyzed. For this, "statistical methods play an important role. It is primarily as a consequence of the application of statistical methods to data that the behavioral scientist decides what conclusions are warranted and what the alternatives are to the decisions made."[8] Another contemporaneous textbook, *Statistical Methods in Educational and Psychological Research*, noted that while "classical" techniques of statistics were primarily for sampling procedures, "modern" techniques of "testing hypotheses" constitute "the area of statistical inference which is of major concern to the research worker." The end result of such hypothesis testing is the determination of whether a particular hypothesis can be rejected.[9] Frederick Ekeblad's *Statistical Methods in Business* (1962) went further, noting that "The statistical method is seen as a unified body of thought concerned with the basic human problem of uncertainty and the corollary problems of risk-taking and decision-making. The emphasis throughout the text is on *a* method of thought rather than a collection of methods, or a collection of mechanical tricks of the trade."[10] Statistics constituted a singular, reliable, and seemingly uncontroversial tool many different disciplines could rely upon to make inductive inferences.

By the 1960s, statistical testing had become an "inferential algorithm," in Rozeboom's words.[11] Under any sensible definition of algorithm—specified computational protocols, precise rules for action, mechanical procedures for problem solving—the way statistics was used in practice by researchers drew on forms of algorithmic thinking.[12] This was the case well before software packages for processing statistical data were introduced for electronic computers. Rather, the steps of statistical inference were themselves algorithmic, the practice by which a researcher moved from experimental design to interpretation and conclusion. Religious terminology like "dogma," "ritual," and "sacred" pointed to the ways statistical inference tests were indeed like traditional interpretations of ritualistic behavior: proponents believed statistics transformed profane and mundane experimental findings into sacred theories and hypotheses, providing the crucial transformation of data into meaning.[13]

Professional statisticians themselves entirely rejected the use of their tools in ritualistic or mechanical ways. Articles in the *Journal of the American Statistical Association* hardly ever used the word "algorithm," and when they did, the reference was to *other* kinds of mathematical calculations. Algorithms had long referred to arithmetic, and more recently logical, procedures but not to statistics.[14] Statistics required conscious, *careful* choice of study design and continual adjustment of models; nothing important was done mechanically. Moreover, academic statistics at mid-century featured at least three distinct camps that starkly disagreed on the right way to make inferences from experimental data: Fisher's method of testing hypotheses with specified levels of significance; Jerzy Neyman and Egon Pearson's model of statistical testing as facilitating a decision between rival hypotheses; and neo-Bayesian interpretations of explicitly subjectivist personal probabilities. Most social scientists' use of statistical algorithms avoided discussion of any such "theoretical" issues and in practice simply deployed a hybrid theory combining aspects of all three camps.[15] Some statisticians saw this mechanically applied hybrid model as a threat. In 1959, statistician Theodor Sterling warned that if journal editors continued to base publishing decisions on statistically significant inference tests, over time it would result in many chance results being passed off as real phenomena, vitiating the credibility afforded to the use of statistics in experimental design.[16]

So how and why did statistics become algorithmic despite statisticians? The "how" of the spread of statistical analysis into the social sciences has been ably traced over the last three decades by Gerd Gigerenzer, Kurt Danziger, Lorraine Daston, Theodore Porter, and others. They've shown how Fisher's work on experimental design entered the American scene largely through George Snedecor, whose accessible textbook on statistical methods introduced researchers in biology, medicine, and agriculture to the latest in statistical thought.[17] An "inference revolution" followed, with hypothesis testing replacing descriptive statistics.[18] A "flood" of new textbooks on statistical practices appeared throughout the 1940s and 1950s, and just a few years after mid-century, the vast majority of articles in psychology research journals used significance tests to justify their conclusions.[19] Similar transformations took place in other social sciences.

Reasons for the rapid spread of statistics are numerous. Danziger has traced how psychologists such as Edward Thorndike promoted a "religion of numbers," providing their nascent field social status and public legitimation in the early twentieth century.[20] Educational administrators' demands for efficient mechanisms of social control (and employment opportunities for psychologists in education) provided powerful incentives to deploy statistical tools. Gigerenzer noted the advantage that "inferential statistics allowed for an application of probability theory that promised objectivity in inductive inference without threatening determinism."[21] Hypothesis testing also seemed to provide editors a clear way of distinguishing significant from meaningless scientific research or—as a retiring editor of the *Journal of Experimental Psychology* explained in 1962—a precise way of measuring the confidence we might have in results.[22] Hogben wrote in *Statistical Theory* that the previous acceptance of statistics in public health might have played a role in the transition from the use of probable error as a rough measure of significance in the nineteenth century to the "commonplace" use of inference tests across the behavioral sciences in the twentieth. Nevertheless, Hogben conceded, "that reliance on test procedures so conceived suddenly becomes so universally *de rigueur* among experimental biologists in the twenties and thirties of our own century is an enigma which admits of no wholly satisfactory solution."[23]

Though these accounts explain the presence of inferential statistics in the behavioral and social sciences by the 1950s, they don't explain why statistical procedures became *algorithmic*. Cohen, Rozeboom, and Bakan were not decrying the use of statistics and numbers per se, but an unthinking, mechanical, and ritualistic version of them. In the remainder of this essay, I'll sketch out an answer by locating statistical algorithms in the context of both Cold War models of decision-making and broader transitions in how practitioners of the human sciences dealt with the "problem" of causality. These two factors came together in the 1950s, leading to a moment in which—as Gerd Gigerenzer and David Murray characterize it—"The dream of the scientist who arrives at new knowledge by a completely mechanized process seemed to have become real."[24] By emphasizing the way new ideas of Cold War rationality enabled descriptive statistics to be turned into decision-making algorithms, I'm moving away from explanations that primarily emphasize the hunt for objectivity—the claim that weak fields adopted quantified methods as a way of claiming objective authority, or because they provided perceptions without subjective perceivers.[25] Such explanations have the problem of timelessness, applying as much to the quantification of life insurance in the seventeenth century as to sociology in the nineteenth century. Moreover, the twentieth-century human sciences treated subjectivity not as a problem to be eliminated so much as a thorny methodological issue best approached mechanically. In short, I will trace the work it took to make statistical testing—despite the pleas of statisticians themselves—algorithmic.

From Observation to Experiment and Description to Decision

Statistical methods have long been used to make inferences about probabilistic effects, calculations concerning policy, and measurements of uncertainty. Pierre-Charles-Alexandre Louis's evaluations of bloodletting, James Lind's study of scurvy, William Farr's political arithmetic, John Graunt's Bills of Mortality, Adrien-Marie Legendre's development of least squares, Carl Friedrich Gauss's formalization of the error curve, Pierre-Simon, marquis de Laplace's synthesis of what came to be known as the "central limit theorem," Daniel Bernoulli's study of smallpox inoculation, and other celebrated cases of quantified reasoning certainly treated statistical methods and mathematical models as tools for analyzing and describing complex phenomena. None of them, however, treated statistics as mechanical or ritualistic.[26] Tellingly, nearly all these cases involved individuals with extensive mathematical backgrounds; statistical techniques could not be used "out of the box" by untrained parties.

Some statistically inclined researchers did think of their work as establishing a *method* prior to the twentieth century, albeit not in the sense that psychologists would later claim. Louis's use of comparative averages in Parisian clinics in the 1820s and 1830s was known as the "numerical method." Likewise, schematic writers like M. A. Quetelet and Émile Durkheim intended statistics to provide a method for social analysis. Calling their methodological intervention "social physics" or "sociology," respectively, these writers explicitly considered their work as providing methodological models (*les méthodes*). Quetelet's methods centered on using observations of regular effects to ascertain social laws.[27] Durkheim claimed to be able to use data to identify "social" facts that pointed to "impersonal causes." For him, "collective tendencies have their own life; they are forces as real as cosmic forces."[28] Statistical methodology, prior to the twentieth century, was not in reference to the statistical techniques themselves, which were essentially limited to counting and averaging, but to *how* one constructs theories *using* numerical data.

Modern statistical concepts were not invented until after the 1880s, largely through Karl Pearson's colleagues and students in London.[29] Increasingly, methodological questions involved the *techniques themselves*—including the now-familiar concepts of correlation, regression, distribution, and standard deviation—rather than the construction of theoretical knowledge from empirical data. Those with particular applications of interest, whether inherited traits or quality control, worked to make the standard tools and concepts more relevant to their fields.[30] From the 1880s to the 1920s, research questions shifted from the derivation of social theory using basic numerical descriptors to the development of concepts and tools that made statistical analysis more relevant and powerful for answering questions of wider interest.

Statistical enthusiasts in the early part of the century consequently portrayed statistics as a collection of methods for "scientific analysis," as G. Udny Yule's influential 1911 textbook put it.[31] This shift, in Yule's terms, meant the "theory of statistics" now entailed "the exposition of statistical methods" of many different varieties, rather than the use of basic methods to extract regularities from existing data.[32] Similarly, Raymond Pearl's 1923 statistical primer for medical professionals claimed, "The statistical method is essentially a *technic*, which finds its justification in its usefulness in helping to solve the problems of the basic sciences, physics, chemistry, biology, etc." It was a "working tool," though one "probably of wider utility" than any other.[33] Henry Garrett's 1926 textbook for psychologists and educational researchers provided a helpful chart detailing when and how to use each test. By the late 1920s, even statistical novices didn't need to think carefully about procedural questions. Textbooks enabled them to search for the desired study design in their area of interest, look up the appropriate tool, and plug in values to arrive at a solution.[34] For more complicated measures, like that of the correlation co-efficient, Garrett even provided a step-by-step method of its calculation (with each sequential step clearly labeled).[35] By the end of the 1920s, statistics had moved from a disparate group of descriptive techniques to a set of procedures. But no one had yet formulated statistics as an algorithm for deciding between theories.

That next step was the result of two intersecting developments. First, the transformation of the frame of statistical analysis from observational description to experimental apparatus, and second, the promotion of statistics as the discipline uniquely capable of enabling decision-making under conditions of limited knowledge.

Statistics as Experimental Apparatus

Experiments facilitated the mechanical use of statistics. When Quetelet was describing his statistical method, it was explicitly a way of interpreting his "observations." He reserved the notion of "experiment" for interventions that might *produce* data to be observed—recording the results of people squeezing dynamometers or similar active "interrogations" of nature.

Statistical analysis itself was a part of observation, not experimentation.[36] Indeed, though long intertwined in meaning, observation and experiment had become distinct and opposed procedures by the nineteenth century: Observation was passive, whereas experiment was active.[37] That's not to say statistical researchers were passive observers before the twentieth century, of course. Frédéric Le Play, among many others, was actively developing statistical methods like the survey. Nevertheless, this was a science of observation, not experimentation.[38]

By 1935, however, Fisher called his textbook of statistical techniques *The Design of Experiments* and drew extensively from his experience heading the Rothamsted Experimental Station in Great Britain. For Fisher, experiment and statistical procedure were "only two different aspects of the same whole," namely, "the logical requirements of the complete process of adding to natural knowledge by experimentation."[39] Fisher's epigraph quoted Robert Boyle's seventeenth-century essay "Concerning the Unsuccessfulness of Experiments," lamenting the proliferation of false experimental reports, even from normally trustworthy and honest authors. Though hardly sharing Boyle's overall experimental sensibility, Fisher seemed to be implying that successful experiments were ones that had been designed in advance with a statistical interpretation in mind. Fisher's actual process of experimental design, and the implications it had for the foundation of statistical inference, have been the subject of much philosophical and historical debate over the years.[40] At base, however, he provided a method by which a researcher would first propose a particular ("null") hypothesis, predict the data that would be expected to occur during the experiment if that hypothesis were true, and then compare the actual data with the predicted data in order to derive a statement about whether there were grounds for rejecting the original hypothesis. His book provided a series of variations on this basic framework for fertilization and productivity experiments in agriculture.

Though rarely thought of as a major locus of scientific activity, turn-of-the-century experiment stations in both Europe and the United States involved a productive mix of scientific specialties: agronomy, climatology, chemistry, biology, and, eventually, statistics. Moreover, they were important sites of public research, endowed in the United States by the Federal Government through the Hatch Act of 1887. Experimental stations were visible models of the ways academic science might prove useful.[41] Not surprisingly, the primary conduit for Fisher's work in the United States, George Snedecor, was the head of the Iowa Agricultural Experiment Station, based at the land-grant institution of Iowa State (and home of the first stand-alone statistics department in the country).

Though not all those engaged in the construction of statistics as experimental tools were directly involved in agriculture experiment stations, such concerns were never far afield. William Gosset's research questions emerged out of the joint efforts of Guinness and the Irish Department of Agriculture to improve the barley crop.[42] W. Edwards Deming, the future international guru of quality control, worked in the 1930s and 1940s as a mathematician for the US Bureau of the Census. In that capacity, he taught statistics for many future US government statisticians through the Department of Agriculture Graduate School. Before coming to Berkeley, Neyman had worked in Warsaw doing agricultural experimentation.[43]

Agriculture provided the basis for thinking about statistics experimentally because it combined repeated experiments with the need to make inferences about complicated cause-and-effect relationships. Though it isn't always clear what precisely caused some plants to thrive and others to fail, there were better and worse guesses. Experiment stations had long been used to test hypotheses about increasing yields and had provided practical advice to local farmers; as such, they were visible sites of the nature of experimental practice. Statistics' use in agriculture made it visible as an experimental science.

By mid-century, no statistics textbook writer could avoid discussion of testing experimental hypotheses. Whereas the first edition of Garrett's introductory statistical text in 1926 essentially ignored the notion of an "experiment," the sixth edition in 1966 had more than one chapter on "testing experimental hypotheses," including a discussion of setting up experiments in the field and in-depth coverage of testing methodologies.[44] What was once a field of description had become one of experiment.

New statistical techniques were also developed by the 1950s to allow observations to be treated experimentally through the notion of a "quasi-experiment." Appearing largely in epidemiological settings, quasi-experiments involved the use of statistics to analyze an existing data set, even if the data had not been collected as part of an experiment. Such research was particularly prominent in studies of disease occurrence, where it was typically impossible to design experiments that involved disease states as independent variables.[45] Statistical techniques could turn observations into experiments.

Significantly, experiments were increasingly treated as step-by-step processes across the first decades of the twentieth century. When Fisher wrote of an experiment as a means of providing convincing evidence to reject a null hypothesis, he was doing so contemporaneously with the publication of Karl Popper's *Logic of Scientific Discovery*, which emphasized the importance of falsifiability, in the context of a larger unity of science movement among logical empiricists.[46] Though neither Popper nor most statisticians would explicitly describe statistics as a form of testing hypotheses algorithmically, statistical design—especially Fisher's null-hypothesis framework—provided a role for statistical tests in the routine process of scientific truth testing. Moreover, unlike the sophisticated treatment of science by Popper, the popular conception of experimental design promoted in schools was increasingly algorithmic. As John Rudolph has demonstrated, the first decades of the century were precisely when the notion of *a* scientific *method* was taking hold. Partially influenced by Karl Pearson's 1892 *Grammar of Science* and John Dewey's 1910 *How We Think*, science was increasingly taught as a series of "steps," logically leading to reliable inductive inferences.[47] At no point did scientists ever actually agree on a particular method—or even that science could be expressed as a series of steps. That did not forestall the emergence of a simplistic notion of *the* scientific method: posit a problem, suggest a potential solution, test that solution with a well-designed experiment, and then accept or reject the hypothesis.[48] Fisher's textbook fit very well with an emerging consensus, at least as popularly understood, that matters of fact could be established directly through well-designed tests of hypotheses executed in an algorithmic fashion.

When psychologist and self-described neo-Popperian Paul Meehl delivered the address "Wanted—A Good Cookbook" in 1955, it was in precisely this vein of thinking of statistics as providing step-by-step procedural rules for making inferences. His recent book, *Clinical versus Statistical Prediction*, had demonstrated that there was little evidence clinicians were that much better than mechanical algorithms at predicting outcomes in clinical settings. Now, he wanted to suggest that rote ways of applying diagnostic criteria ("recipes") might also improve on clinicians' ability to make diagnoses on the basis of personality tests. Though the brief address focused solely on recent research concerning the application of the Minnesota Multiphasic Personality Inventory, his point was about "cookbook methods" in general, ones in which any given configuration of psychometric data is numerically connected (through a correlation coefficient or probability distribution, for example) with specific personality descriptions. The "essential point is that the transition from psychometric pattern to personality description is an automatic, mechanical, 'clerical' kind of task, proceeding by the use of explicit rules set forth in the cookbook." Though acknowledging his suggestion might be "horrifying" to certain clinicians, Meehl concluded that they would have to prove their methods better than such a cookbook to justify their status as a "costly middleman who might better be eliminated."[49] Meehl implicitly wanted to treat clinical interactions not as observations leading to diagnoses but rather as an experiment testing the ability to infer diagnoses from psychometric data. The ideal of diagnosis was automatic, rule-bound, and mechanical—algorithmic—freeing up the clinician to spend his time on psychotherapy and research. In Meehl's conception, the process of diagnosis could be inherently experimental *and* statistical.

Cold War Decisions

Meehl's language—"the human brain is an inefficient recording and computing device"; "it is a truism of behavior science that organisms can *exemplify* rules without *formulating* them"—fits precisely in the context of early Cold War rationality.[50] Though the key mathematical developments came earlier, the spread and formalization of statistics were fundamentally Cold War phenomena. Accounts of Cold War social sciences have ignored the essential role that statistical analysis and formal significance testing came to play at mid-century. Historians of statistics have tended to downplay the Cold War entirely.[51] Nevertheless, the transformation of Fisher's conception of statistics as integral to experimental design in 1935 into statistics as mechanical algorithm in 1955 is inseparable from contemporary notions of what it meant to make decisions rationally.

As Paul Erickson and his coauthors have shown in detail, forms of rationality during the early Cold War centered on algorithmic thinking. By this, contemporaries meant "rigid rules that determine unique solutions"—ideally rules that were precise and yet general enough to provide a guide to making decisions especially in the context of limited information.[52] In economics, the Cowles Commission for Research in Economics conference in 1949 published its proceedings as "Rational Decision-Making and Economic Behavior." As Hunter Heyck has noted, Herbert Simon framed his organization theory as "explicitly a theory of decisionmaking."[53] Robert Dorfman, Paul Samuelson, and Robert Solow defined the field of linear economics, and linear programming in particular, as devoted to the need to make decisions about the "best" allocation of resources.[54] John von Neumann and Oskar Morgenstern introduced their seminal *Theory of Games and Economic Behavior* with the explanation that it was a "discussion of the problem of rational behavior," defined as the maximization of some quantity.[55] In their game, a player's "strategy" was simply "a plan which specifies what choices he will make in every possible situation, for every possible actual information which he may possess at that moment in conformity with the pattern of information which the rules of the game provide for him in that case."[56] Problems of resource allocation at the Pentagon, those of linear programming at the Bureau of Labor Statistics, and those of game theory at RAND all faced the challenge of choosing among available strategies to minimize or maximize certain values.[57] Strategies were algorithms. Even Daniel Kahneman and Amos Tversky's research on heuristics and humans' seeming irrationality "still adhered to ... the ideal type of Cold War rationality: good reasoning has to follow formal algorithms that optimize results and can be applied mechanically."[58] One of the major edited volumes of their work was titled "Judgment under Uncertainty."[59]

Statisticians themselves started describing their field as providing a way of making judgments under conditions of uncertainty or limited knowledge. The Applied Mathematics Panel in the war considered "probability and statistical studies" appropriate to the problems of naval warfare and the performance of torpedoes, ordnance quality control, and bombing effectiveness.[60] The emerging inference model of Jerzy Neyman and Egon Pearson emphasized the need to make choices among alternatives rather than simply rejecting or failing to reject the null hypothesis. Their theory was one of action under limited information: the inferences you make may or may not correspond to the world we live in, but we should act as if they do because it's the best information we have. This was starkly different from earlier portrayals of statistics. Whereas Deming claimed in the 1930s that statistical significance alone could *not* provide a "rational basis for action," Neyman and Pearson emphasized the way in which a properly designed statistical experiment might provide exactly that.[61]

When Abraham Wald first described his account of "sequential analysis"—the process of inference testing one unit at a time instead of a group (sample) of units at once—he also did so explicitly on the basis of decision-making. The original application of sequential analysis was in the quality control of ordnance deliveries from military contractors. Instead of sampling a number of units, finding the percentage that were faulty, and then making a probabilistic inference as to the quality of the entire lot, sequential analysis required the testing of units, randomly drawn, one at a time: {Faulty, Good, Good, Faulty, ...}. After every test, the experimenter made a decision to either (1) stop testing and accept the lot; (2) stop testing and reject the lot; or (3) continue testing. Making this decision meant following strict guidelines—there was no discretion because the experimental design prescribed an algorithm for choosing.[62] Wald's treatment of sequential analysis drew on Neyman's and Pearson's techniques in that he framed it as a way to test hypotheses under conditions of limited information.

Wald later generalized this work under the heading of "Statistical Decision Functions," and though his untimely death meant he never developed the field himself, Wald was influential in establishing the role of mid-century statistics as an aid to rational decision-making.[63] When fellow statistician Leonard Jimmie Savage reviewed Wald's contribution, he noted that it had fundamentally shifted statistical theory. The "traditional" goal of the field had been "to draw statistical inferences, that is, to make reasonably secure statements on the basis of incomplete information." Wald's work instead centered on the "problem of statistical action rather than inference, that is, deciding on a reasonable course of action on the basis of incomplete information."[64] Savage's own 1954 synthesis of statistical theory opens with a summary of the overall situation: "Decisions made in the face of uncertainty pervade the life of every individual and organization."[65] Statistics enabled action in a mid-century context dominated by uncertainty. This belief that decisions always had to be made with imperfect information was notably different from Fisher's earlier justification for randomization as a way to outwit a clever "devil" trying to foil the experiment.[66] During the early years of the Cold War, the "devil" was less an abstraction of unknown possible influences than a very real geopolitical adversary whose motives could never fully be known. The emergence of statistical decision theory in such a moment obviously does not imply every test was designed to outwit Soviet subterfuge, but rather that statistical techniques were framed as a solution to the problem of rational decision-making given pervasive uncertainty.

The pioneers of this new way of reasoning were widely dispersed across economics, mathematics, operations research, and many other fields. This was in part because new tools like game theory did not have natural disciplinary homes, but instead relied upon interdisciplinary units (and military funding).[67] The RAND Corporation, for example, in July 1948 hosted a colloquium on game theory and the "theory of planning," which drew not only from military researchers and linear programming specialists such as William Horvath, George Dantzig, and Marshall Wood, but also from statisticians including Merrill Flood, David Blackwell, and John Tukey.[68] Tellingly, the new methods of economic theory and measurement found initial footing in precisely those mathematics departments where leaders were welcoming to statisticians, including Harold Hotelling at Columbia University, Samuel S. Wilks at Princeton University, and Neyman at Berkeley.[69] Before publishing on statistical decision theory, Wald had worked with Morgenstern at the Cowles Commission for Research in Economics on economic theory, and Hotelling at Columbia.[70] Their relatively weak disciplinary identity meant that statisticians often moved frequently between fields and interests; their work was increasingly in dialogue with other disciplines.

By the 1970s, the sum total of these intertwined research efforts was known in the statistical world—appropriately enough—as the "Decision Model."

In its most general form, as statistician Thomas Ferguson wrote, the techniques "provide a model for individual behavior in mundane problems of everyday life."[71] The mid-century statistics of Wald, as well as Savage, Neyman, Pearson, and others, provided a step-by-step way of making decisions. The "sciences of choice"—with their approach to decisions as formal models of selecting among alternatives—were new to postwar social science.[72] Statistics became algorithmic only when the field was conceived not as a descriptive technique for observation but as a method of making decisions about the outcome of experiments.

Human Sciences

Not every field took advantage of new statistical theories. Most physical and natural sciences proceeded without them. Moreover, computational biology and the data sciences were essentially nonexistent in the 1970s, as were other subfields that would flourish with high-speed computing.[73] Practitioners of the human sciences—from psychology to medicine and sociology—were enthusiastic about the possibilities offered by postwar statistics, however. Perhaps the most enthusiastic of all were psychologists, who facilitated what Gigerenzer and Murray's *Cognition as Intuitive Statistics* labeled an "inference revolution" between 1940 and 1955. Though their book was primarily concerned with changing metaphors of the mind in the twentieth century, they argued that a "hybrid" form of significance testing, combining concepts from Fisher, Neyman, and Pearson, was adopted as a "common method" unifying psychology. Though noting this "hybrid" methodology was ultimately an "illusion," neither mathematically accurate nor philosophically rigorous, it was a *fait accompli* by the 1970s.[74]

The "inference revolution" did not mark the origin of psychology's use of statistics. Rather, as one historian of the field's mid-century transformation explained, "Modern psychology has always insisted on being a statistical science, since its emergence near the turn of the century; but its use of statistics has changed."[75] Danziger has shown in great detail, in fact, how psychologists initially turned to statistics around World War I when researchers stopped relying on the measurement of individuals in favor of aggregated data. The field's research objects were transformed from credible, representative, and specific individuals, whose accounts of reasoning could stand proxy for phenomena in general, to generic experimental subjects who "happened to participate as sources of data in a particular psychological investigation."[76] Statistics enabled psychologists to stop caring who provided their data.

The transformation was both methodological and theoretical. Danziger explains, "It seemed that the hypothetical distribution of the statistical analysis could be identified with the characteristics of real psychological systems, thus permitting a bridging of the gap between data that referred to groups and theoretical constructs that referred to individuals."[77] Gigerenzer and Murray note that by including statistical measures of uncertainty, psychology could claim a stronger theoretical basis: the aim was to "reduce uncertain situations ... to certain ones by the *applications of rules*. We cannot be sure that a given psychological conclusion drawn from data is correct, but we can be sure that it was reached in the *one and only correct way*."[78] Put differently, the methodological consensus might have been more attractive given psychology's lack of theoretical consensus. Early in the century, numbers functioned for Thorndike and contemporaries as a way to distinguish psychology from philosophy and spiritualism, as well as from lay knowledge of the mind, marking the discipline as an "exact science." By mid-century, the use of numbers was well established, but now statistical methods, by facilitating the "automatization of scientific decision making," were treated as theory-neutral, and therefore as uniquely powerful tools for designing and interpreting experiments. Theories could be confirmed or refuted simply by rote application of null hypothesis testing.[79] The contemporary psychologist Rozeboom agreed, noting that statistical tests attained a "ritualistic status" because they eased methodological "insecurity" while also proving "useful" by promoting clear language of accepting or rejecting hypotheses. In practice, though, he complained, psychologists almost never actually jettisoned a theory after one experimental rejection while perversely increasing their faith in a theory after one statistically significant result was reached.[80] No serious psychologist would change his mind about a theory simply because a statistic came out on one side or the other of a specific threshold. Ironically, Rozeboom suggested, the faith in statistical testing as a ritual or algorithm meant that practitioners didn't actually have to take its details that seriously. It was essentially just a tool to be manipulated until a desired conclusion was reached.

Psychology was hardly the only mid-century science turning to the step-by-step use of inference tests as a powerful methodology. Emerging postwar research on food quality and taste provides an excellent example of similarly algorithmic uses of statistical inference.[81] When University of California–Davis enologist Maynard Amerine and mathematician Edward Roessler wanted to put the evaluation of wine quality on a sound basis, they did so by creating experimental designs that could be interpreted mechanistically through statistical tests. A judge, for example, might be asked to taste two samples and indicate which was sweeter, or taste three samples and pick out the one that was not like the others. Panels of judges might rank wines on various measures, or provide numerical scores of certain qualities (color, clarity, bouquet, etc.). In either case, statistical tests could be used to determine which judgments were meaningful. These tests for "sensory judgment" drew on the formalization of taste panels in industrial laboratories over the 1930s and 1940s, particularly those that had been used to determine the palatability of soldiers' food during the war. The problem was that the conceptual basis of the judgments—what precisely caused the sensation of bitterness or a particular aroma—was not always clear.[82] To circumnavigate the conceptual difficulty, researchers like Amerine echoed psychologists' strategy by advocating a methodological turn to statistics.

Amerine and Roessler's 1976 book, *Wines: Their Sensory Evaluation*, summarized the procedures and results that they had obtained over the previous thirty years of research. (Despite the book's date, nearly all the research and experimental design work had taken place in the 1950s and so was contemporaneous with similar postwar developments in psychology and other sciences.) They emphasized that while experimental designs could eliminate some unwanted subjective influences by painting the glasses black, or serving samples in randomized batches, or controlling the lighting and other ambient conditions, there would *always* be some uncertainty and variability due to chance and to the unavoidable differences in subjective taste. In their book, Amerine and Roessler present statistical methods precisely in the mid-century mode of decision-making under conditions of limited information. Two judges might always disagree on the measurement of a particular aroma (and one judge in two different settings might disagree with herself). Since this level of variability cannot be eliminated, they thought it should be managed statistically. For them, "statistical significance" came to mean "significant, objective quality difference."[83] Statistics could be used to determine meaningful—real—differences between inherently and unavoidably variable taste judgments. By turning taste evaluations into numerical scales, aggregating them, and analyzing them statistically, subjective differences in tastes could gain stability and reliability from having been transformed into statistical differences.

One could also find examples in other fields. Sociologists increasingly used statistical tests for analyzing samples as well as for making inferences from samples to hypothetical populations. Denton Morrison and Ramon Henkel lamented that mid-century sociological research was "characterized by a substantially ritualistic and naïve use of the tests."[84] Also in the 1950s, Henry Beecher's groundbreaking work on pain measurement noted that the statistical interpretation of experiments (he enlisted Harvard statistician Frederick

Mosteller) could provide a way of gaining an objective understanding of the conceptually slippery notion of pain perception.[85] As with enology's conceptual confusion regarding the theoretical basis of taste sensations, statistics provided a way to avoid underlying questions while continuing to label experimental results as significant or meaningful. Across a number of fields, statistical tests gained currency as mechanical modes of inferring scientific truths because they enabled subjective experiences to be treated as objective, quantifiable phenomena.

In a sense, this finding is consistent with a long-standing explanation of why disciplines embraced quantification: to address real or perceived weaknesses of credibility or authority.[86] The case of statistics entering clinical medicine in the same period, however, pushes back on the notion that numbers emerge from relatively weak disciplines or from theoretical confusion. In 1900, physicians largely ignored the tools of statistical analysis. Clinicians and laboratory researchers saw themselves as fundamentally opposed to the burgeoning field of statistics: they were interested in biomedical causation, but statistics was focused on probability distributions and measures of correlation; they were focused on exceptions and idiosyncrasies, but statistics was focused on norms and averages; they were determinists, but statisticians were probabilists. There were essentially no statistical analyses in medical journal articles and no statistical training required for medical school. Over the next seventy years, clinical medicine was transformed into a statistical discipline through the emergence of randomized clinical trials, meta-analyses, and risk factors, but without any crisis in theoretical concepts. By the 1920s and 1930s, statistically trained reformers such as Raymond Pearl, Jesse Bullowa, and A. Bradford Hill claimed they had the ability to interpret experimental results with newfound power and precision. Formal clinical trials gained credibility over the following decades, and by mid-century, medical reformers had a list of now-familiar desiderata for statistical trials: tests of therapeutic efficacy should involve comparison, use blinding to reduce bias, deploy control groups and placebos to know how much of the effect can be attributed to the treatment, and rely on formal inference tests to determine whether differences in outcomes may have been due to chance. Medical reformers were able to convince skeptical doctors that, in the absence of decisive laboratory evidence, studies of efficacy in the aggregate could produce reliable knowledge about the treatment of individuals.[87]

A massive influx of new "wonder drugs" in the 1940s and 1950s—antibiotics, antipsychotics, steroids, and diuretics—provided a platform for, and a test of, the reforms. Pharmaceutical companies and regulators alike saw the benefits of the new tools. By 1972 the Food and Drug Administration clarified that proof of safety (required after 1938) and effectiveness (required after 1962) should both be demonstrated by controlled clinical investigations. The gold standard for evaluating therapeutic interventions was now the statistically interpreted randomized clinical trial.[88]

Over the same period, the longitudinal Framingham Heart Study (initiated in 1948) and the Surgeon General's 1964 report *Smoking and Health* provided evidence that certain behaviors are so strongly correlated with bad outcomes that the government may have an interest in making public health interventions. These large-scale studies also suggested that health might be redefined through the measurement of weight, cholesterol, blood pressure, and other indicators. The tools of epidemiology increasingly affected every individual clinical encounter.

The turn to statistics in medicine was not the result of underlying theoretical confusion about disease or the replacement of conceptual muddlement with methodological clarity. Statistics were praised because they enabled medical researchers to make causal claims even as subjective bias—both the patients' and the doctors'—remained endemic. Reformers in the American Medical Association initially pushed for statistically interpreted tests of therapeutic interventions only after it was clear that neither laboratory research nor ad hoc case reports from physicians enabled them to reliably distinguish effective treatments from quack cures. The Framingham Heart Study and the Surgeon General's report on smoking were explicitly meant to expand the ability of the government's public health officials to make pronouncements about threats to health even in the absence of clear and convincing biomedical mechanisms of causation. Clinicians and federal researchers were not confusing causation with correlation; they were trying to use statistics to *expand* the kind of causal claims they could make about human health.

The case of clinical medicine suggests the use of statistics to turn subjectivities into objectivities was not fundamentally about the spread of techniques into weak sciences, though of course most human sciences were marginal from the point of view of chemists and physicists. Nevertheless, food scientists, psychologists, sociologists, and physicians didn't describe statistics as providing their fields with mathematical authority; they saw statistical methods as providing an off-the-shelf technique to make reliable causal judgments in inescapably subjective settings.

Statistical tests enabled routinized judgments to be made even in the absence of clear causal relationships. As Danziger explained for psychology, when Wilhelm Wundt witnessed regularity in the actions of his nineteenth-century experimental subjects it had immediate causal significance, but as statistical regularities across groups of individuals became the central source of data, causality was no longer emphasized.[89] By mid-century an increasing number of psychological researchers—just like those evaluating bitterness in wine, or the relative worth of rival sociological theories, or the effectiveness of a new drug—wanted to make reliable judgments even in the absence of traditional evidence of causality, and to do so they turned to using statistical tests algorithmically.

The transformation of psychologist Paul Meehl is indicative of this expanding role of statistical methods. In the 1950s he wanted a statistical "cookbook" for clinical prediction and diagnosis but there was no *causal* component to his claims. By the 1970s, he changed his tune and started writing articles against statistical methods, on the grounds that people were now confusing their predictive value with the making of causal claims. In his 1978 article "Theoretical Risks and Tabular Asterisks," he indicated that "soft" psychology fields had done little to establish the stability of theories or constructs over the previous two decades, even as they united under the methodological use of hypothesis testing. Meehl in fact *blamed* the use of hypothesis testing: "I believe that the almost universal reliance on merely refuting the null hypothesis as the standard method for corroborating substantive theories in the soft areas is a terrible mistake, is basically unsound, poor scientific strategy, and one of the worst things that ever happened in the history of psychology."[90] The problem for Meehl was not with the use of statistics per se but with the use of statistical algorithms to decide between two causal claims. Substantive theories of psychology were about the causal structure of the world, whereas statistical hypotheses were not. In agriculture, Meehl suggested, this was not a big problem. When Fisher concluded that plots of corn with potash had statistically significant differences in yield compared to plots without potash, it was not a great leap to conclude that potash made the difference. But for the mind, data that might cast doubt on a particular hypothesis almost never establish a rival hypothesis—there are "all sorts of competing theories" that also explain the statistical difference. In Meehl's account, statistics used to try to make conceptual claims in psychology failed precisely because psychologists wanted conceptual claims to be about causes. The field around him had moved far past his initial hope that psychologists would turn to statistics as a tool. Now, statistical inference was *the* tool for making judgments between causal claims, a transformation Meehl came to regard as a "terrible mistake."

Critics of the 1960s and 1970s—Meehl, Hogben, Rozeboom, Bakan, and many others—were indeed outspoken about the newfound authority ceded to algorithmic use of statistical methods, but they remained on the outside. Their emphasis on the epistemic virtue of trained judgment in the use of statistics—and thus their argument that inference tests should be used carefully and deliberately—was often drowned out by others who focused on the epistemic virtue of mechanical objectivity—and who argued that inference tests were beneficial precisely because they could be deployed algorithmically as interpretive tools.[91] That critics were so consistent and vehement in their complaints over the subsequent years suggests their failure: statistical inference tests remained ingrained in practice. An increasing number of journals and researchers in the human sciences continued using statistics as a routinized form of decision-making because the ability to circumvent thorny conceptual questions with an algorithm was considered a feature, not a bug, of statistical methods.

Statistics became algorithmic most prominently in scientific settings in which subjectivity could not be eliminated because it was the *object* of the investigation: taste, pain, sociality, mind, health. It was possible to make objective causal claims out of subjective data only if statisticians' pleas were ignored and statistical techniques were treated as mechanical methods of logical inference rather than as fallible, negotiable, and contested. Critics could and did complain that the mechanical use of statistics did not reveal biochemical or physical causal mechanisms, but that is because statistics facilitated a seemingly reliable way to make conclusions and render judgments without such explanations.

* * *

This move—the circumnavigation or redefinition of causality through statistics—might be thought of as the defining feature of the subsequent rise of data science and "big data," and in particular new forms of machine learning like artificial neural networks. The world is a messy place, data science advocates claim, and it isn't always clear how one thing leads to another. Gather enough data, however, and you might still be able to reach conclusions without having figured out the precise cause-and-effect chain.[92] It's a mistake to think of this transformation as a product of the rise of computer algorithms, or of high-speed electronic computing in general. Instead, it seems much more closely aligned with the way statistics were increasingly deployed in the human sciences from the 1940s to the 1970s. The rise of statistics as algorithmic suggests that it wasn't only technological change that led directly to big data, but rather a new way of thinking about phenomena in general, and the ability of statistics to tell us something meaningful about them. The rise of recidivism- or crime-predicting algorithms in this century, for example, is deeply rooted in how researchers in the twentieth century treated statistics and causality in the human sciences.[93]

By the time of the development of statistical software in the late 1960s, the algorithmic application of statistics had been well established.[94] Software, especially as electronic computing power and storage capability expanded, proved ever more powerful and convenient and undoubtedly enabled larger data sets to be processed. Nevertheless, as Jon Agar has described more generally, early electronic computing often took the place of existing material practices of computation.[95] When that computation was itself algorithmic, as in the case of statistical inference tests, the introduction of electronic computing was even more extensive. Indeed, not only could statistical tests be applied mechanistically after the 1960s, but researchers no longer even needed to do the computations. *Both* the number crunching *and* the philosophical foundations of statistical inference were black-boxed: given a set of data, researchers can just run the software to quickly figure out whether they have "significant" (and therefore publishable) results.

Neither computing power nor statistical software created the algorithmic use of statistics. Rather, I've tried to sketch out how statistics were already thought of in algorithmic terms because they had been treated as powerful methods of decision-making and experimental design, methods that had been applied most forcefully in fields where subjectivities seemed unavoidable. Statistics has, for too long, been treated as a field apart from other sciences. By reintegrating it into the history of experimentation, social science, and the Cold War more broadly, we can better understand Cohen's complaint of the mechanized and ritualized method of decision-making on the basis of some arbitrary yet sacred criteria. Statistics became algorithmic because mid-century social scientists increasingly saw the world's subjectivities as legible only through mechanized analysis of experimental data.

10

Decision Trees, Random Forests, and the Genealogy of the Black Box

Matthew L. Jones

> Well, it's one of those ideas that's continually re-invented. Everybody who re-invents it thinks this is their "Nobel Prize" moment.
>
> —**Jerome Friedman**

Suppose you wish to automate your firm's processing of applications for a credit card. For decades, your analysts have looked at each applicant and answered yea or nay. You have loads of data, like that shown in Figure 10.1.

Wouldn't it be nice to automate the process, to have a computer take the attributes of potential credit cardholders and classify each one as creditworthy or not?

If a model trained on a computer could correctly classify nearly all of the human-classified data, then it could potentially be used with some assurance on new data that has not yet been classified by hand. Automating such a classification process is an example of what is called "supervised learning": the process of having a computer learn how to classify data based on an existing human classification of a subset of the data.

This paper concerns the genesis and the development of one of the foremost families of algorithms for supervised learning: decision trees. A decision tree puts sets of data into classes by automatically constructing a hierarchical set of decision branches based on the human classification of some subset of the data.

Since the 1980s, decision trees, in ramifying forms, have exploded in popularity within scientific, medical, commercial, defense, and intelligence work. In 2007, two algorithms for producing trees were ranked in the top ten algorithms for data mining.[1] The state-of-the-art variety of trees, called random forests, ranks at the moment of writing with convolutional neural nets

Matthew L. Jones, *Decision Trees, Random Forests, and the Genealogy of the Black Box* In: *Algorithmic Modernity*. Edited by: Morgan G. Ames and Massimo Mazzotti, Oxford University Press. © Oxford University Press 2023. DOI: 10.1093/oso/9780197502426.003.0011

Number	Attributes				Class
	account	balance	employed	monthly expense	
1	bank	700	yes	200	accept
2	bank	300	yes	600	reject
3	none	0	yes	400	reject
4	other inst	1200	yes	600	accept
5	other inst	800	yes	600	reject
6	other inst	1600	yes	200	accept
7	bank	3000	no	300	accept
8	none	0	no	200	reject

Figure 10.1 Example "training" data for decision tree.

Source: Chris Carter and Jason Catlett, "Assessing Credit Card Applications Using Machine Learning," *IEEE Expert* 2, no. 3 (September 1987): 71–79, at 73.

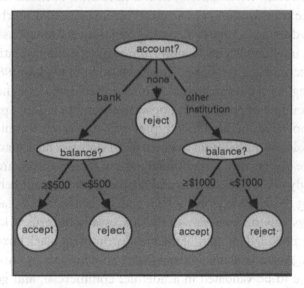

Figure 10.2 Example decision tree for credit card applications.

Source: Chris Carter and Jason Catlett, "Assessing Credit Card Applications Using Machine Learning," *IEEE Expert* 2, no. 3 (September 1987): 71–79, at 73.

as the best machine learning predictors for many tasks. We know relatively few details about how centrally they figure in quotidian decision-making in commercial and governmental sectors, except that their reach is enormous. Decision trees were central for predicting clicks for ads on Facebook well into 2014 at least.[2] Random forests figure prominently, for example, at the US National Security Agency, which may have underwritten some of their funding. Trees and forests have robust implementations in almost all platforms for the analysis of data, and now require little to no programming ability to use, even on large and complicated data sets.

Decision trees emerged with the spread of digital computers as a research tool from the 1960s onward. "The tree methodology discussed in this book," begins *Classification and Regression Trees* of 1984, "is a child of the computer age. Unlike many other statistical procedures which were moved from pencil and paper to calculators and then computers, this use of trees was unthinkable before computers."[3] Computers did not cause researchers to create, to embellish, and then to celebrate trees, but they made them thinkable and tractable. And as they came into broad use, trees helped to change the boundaries of what is acceptable in making predictions in science, business, law enforcement, and intelligence activities.

Before the recent renaissance of neural networks, decision trees had been central to the transformation of artificial intelligence and machine learning since 1980: a dramatic shift to a focus on prediction at the expense of concerns with human intelligibility and a shift from symbolic interpretation to potent but fairly inscrutable black boxes. Trees exploded in the late 1980s and early 1990s as paragons of interpretable algorithms; in the late 1990s they developed into a key example of powerful but opaque ensemble models, predictive but almost unknowable and hard to interpret. In some cases, techniques for creating ensembles imported from mathematical statistics have made them even less attractive objects within traditional statistics, even as they are embedding in increasingly ubiquitous systems making judgments on our behalf.

A growing literature explores the increasing centrality of black-boxed algorithmic systems on everyday life.[4] This essay concerns a key family of algorithms embedded in many of the most prominent black boxes of the past twenty years. We understand precious little about how black-boxed predictive systems came to be validated in academic, commercial, and governmental spheres. Neither data, nor statistics, nor indeed machine learning would necessarily have come to embrace such opaque algorithms as legitimate. Decision trees offer a privileged optic that allows us to see the computation and epistemic shifts culminating in the ubiquitous black boxes we find a grave challenge to comprehend, much less regulate or audit. We need to explain, rather than take as given, the shift in values to prediction—to an instrumentalism—central to the ethos and practice of the contemporary data sciences. Opacity needs its history—just as transparency does.[5]

What sort of history does opacity need? The history of trees does not cleanly divide into a theoretical and an applied stage; an academic and a commercial phase; a statistical and a computational stage; or even an abstract algorithmic design and a practical implementation stage. This arborous history is iterative: the implementation of algorithms on actually existing computers with various limitations drove the development and transformation of the techniques.[6] Confrontations with "real-world" data sets encouraged the iteration of both the abstract algorithm and its implementations. The wide scale adoption of the algorithm in various platforms required large amounts of data and funding, but can't be reduced to the explosion of data or of funding imperatives.

While attending to the materiality of implementation, we must equally attend to ideation of implementation. In case after case, the creators of different forms of trees deployed "applied" philosophies of science in critiquing contemporary practices, epistemic criteria, and even promotion practices in academic disciplines. Faced with increasing amounts of high-dimensional data, these authors time and again advocated a data-focused positivism that they set against the model-based, confirmatory statistics via hypothesis testing that dominated mathematical statistics after World War II. While the National Science Foundation funded academic statistics, the US Department of Defense and the US and allied intelligence agencies funded a more positivist program grounded in computational statistics through the entire second half of the twentieth century.[7] And while much of the story involves criticisms of academic statistics, it rests equally on those provinces of academe focused squarely on storing data and creating techniques appropriate to "very large data bases."[8] So decision trees are an important part of recent encomia to the "end of theory," or, less hyperbolically, to the triumph of predictive models based on large amounts of high-dimensional data over theory and causal models. The expansion of trees, along with other learning algorithms, took place amid a profound critique of artificial reason and the ambition of producing tools with less rigor than mathematical statistics but capable of dealing with more of the records of the social and natural worlds.[9]

My focus here is less surveying the prominent technological systems in which trees have been used than in the development and transformation of various algorithmic forms.[10] In the pages that follow, I trace a variety of researchers who came obliquely to trees: a data-driven, practically oriented statistician, a machine learning expert focused on larger data sets than usual at the time, social scientists unhappy with multivariate statistics, and a physicist interested mostly in computers who eventually became a tenured star of his statistics department. Each researcher was askew to the dominant practices and epistemic virtues of their fields. These plural contexts echo in the technical qualities of the algorithms these diverse people produced. The anthropologist Nick Seaver has argued that focusing too exclusively "on context keeps algorithms themselves untouched, objective stones tossed about in a roily social stream. We should not abandon concern with technical details or 'close readings' of algorithmic function." He calls for study of algorithmic systems, where " 'cultural' details *are* technical details—the tendencies of an engineering team are as significant as the tendencies of a sorting algorithm."[11]

Current implementations of decision trees carry layers of code stemming from the algorithm's multiple genealogy. The highly optimized code in the DecisionTreeClassifier of scikit-learn, currently the premier machine learning platform in the Python language, offers among its parameters a choice of measuring "the quality of a split." The two choices, using "information gain" or "Gini impurity," testify to the two key genealogies of decision trees, one emerging from a challenge to mathematical statistics, the other from a challenge to existing forms of artificial intelligence.[12] This essay will explore this emergence of decision trees from these two critiques in turn: first, the critique of academic statistics as overly mathematical; second, the critique of artificial intelligence as inadequately accepting of the difficulties of replicating human expertise. From there, the essay turns to the explosion of different forms of trees in the 1990s and 2000s and the subsequent loss of interpretability. The account here concerns tendencies, beliefs, and practices of teams of practitioners, often in liminal spaces between academia, government, and private industry, and the tendencies of the algorithms they envisioned, implemented, iterated, and celebrated.

The Near Death of Newborn Trees: Survey Data and the Limits of Statistical Practice

The Survey Research Center at the Institute for Social Research at the University of Michigan in the early 1960s offered the soil for one early sprouting of decision trees. Confronted with a large volume of high-dimensional data produced in wide-ranging social surveys, researchers there came to view the predominant statistical methods for multivariate analysis of the postwar era as too narrow, too focused on hypothesis testing for their needs.

Two convictions motivated the Michigan researchers to create a computer program they called "Automatic Interaction Detection," or AID:

> First, is the belief that the multivariate statistical techniques in common usage are often inadequate for the analysis of the rich body of data from cross-section sample survey, and second is the conviction that a large-scale digital computer can be used for more than just a high-speed adding machine.

In particular, they chaffed at the inadequacy of the empiricism of the practice of statistics in the Fisher-Neyman-Pearson synthesis.[13] The researchers explicitly sought to use the computer to foster epistemic habits distinct from those dominant in mathematical statistics:

> We have tried to break away from the habit of asking the question, "What is the effect of x on y when everything else is held constant?" This has been replaced with, "what do I need to know most in order reduce predictive error a maximum amount?"

A focus on predictive error, they argued, characterized exploratory scientific, as opposed to statistical work:

> This is the type of question that might be asked by a research scientist working in a substantive area in which theory is not yet very precise. Once he receives an answer, he may well ask, "Now that I know this, what additional information would help to reduce predictive error still further?" and so on.[14]

Faced with large amounts of little-explored data, the authors called for returning to a modest positivism about potential interactions, rather than jumping too soon to hypothesis testing. Existing causal models for understanding survey data simply limited our purview far too much. They "focused on a particular kind of data-analysis problem, characteristic of many social science research situations, in which the purpose of the analysis involves more than the reporting of descriptive statistics, but may not necessarily involve the exact testing of specific hypotheses."[15]

The authors gave an example of attempting to predict income based on age, race, and a few other factors. They described their algorithm for preparing such trees: the key is to devise a way of "splitting" groups of observations of a variable into branches of the decision tree to maximize predictive ability.

From the start, the creators were concerned with implementing the algorithm for practical purposes with existing storage and computational technology. Computing splits, they explain, is easiest if all the variables can be stored in fast memory, rather than the slower input/output mechanism: while "data may be stored on tape, in the interest of computing efficiency, all of the information for any particular analysis, including predictors, dependent variables, weights, etc. are kept in core storage." The system in use could accommodate no more than 100 input variables, and the maximum number of groups to split the observations was 63.

Critics soon eviscerated the work at Michigan, led by scholars who drew upon the rigor and competencies of postwar mathematical statistics in remaking the social sciences, the culture Christopher Phillips discusses in chapter 9 of this volume. In "Alchemy in the Behavior Sciences," an assistant professor at Chicago named Hillel Einhorn blasted the Michigan researchers' computational positivism:

> proceeding via a "dustbowl" empiricism is dangerous at worst and foolish at best.... The purely empirical approach is particularly dangerous in an age when computers and packaged programs are readily available, since there is temptation to substitute immediate empirical analysis for more analytic thought and theory building.[16]

The author singled out AID: "it is poor science to 'test' hypotheses on the same data that generated those hypotheses. If AID is used to gain insight and hypotheses from data, these hypotheses can only be tested on another set of data."[17] These lessons would soon become gospel in machine learning, in which the separation of testing and training data became an essential step taught to all beginners. Proponents of AID responded, "The dichotomy [Einhorn] raises between mindless ransacking and precise testing of a theory is a false one. In the real world no one does either. It would be absurd to take any set of data of any richness and test a single model with it."[18]

Trees produced by AID and similar systems have a singular danger: they "overfit." That is, they predicted classifications of the data they were trained on far too accurately to be applied to any other set of data; the algorithm failed utterly to distinguish noise in the data from signal. In other words, the trees came to have almost as many subdivisions as they did data points, to become too "bushy." To use them for another data set, a user would need to "prune" in some systematic way—but that technique didn't yet exist. Great at predicting their training data, decision trees were widely seen as disastrous in predicting anything else, much less giving any causal scientific knowledge.

By the late 1960s, because of overfitting, decision trees "were called 'a recipe for learning something wrong,'" says Dan Steinberg. "This was a death sentence, like a restaurant with E. Coli. Trees were finished."[19] Trees seemed moribund.

And yet, despite this near-death experience, in the 1970s and 1980s trees came alive anew in several communities, each somewhat askew to the dominant assumptions of different academic disciplines, and took on forms with distinct philosophical assumptions and algorithmic implementation. They emerged within critiques of the adequacy of disciplinary focus within academic programs. First, we'll look at their revivification within a critique of adequacy of mathematical statistics in the age of large-scale data. Second, we'll look at a distinct revivification within a critique of the ability of researchers in artificial intelligence to understand how experts reason.

Part 1: The Critique of Mathematical Statistics

Two major researchers, Leo Breiman and Jerry Friedman, independently reinvented trees in the mid-1970s, and subsequently began a long period of collaboration. In a 1984 book coauthored with two additional researchers, they explained that the key motivation for the development of trees came from the inadequacy of many statistical techniques when faced with the high-dimensional data sets increasingly available by the 1970s:

> Many of the presently available statistical techniques were designed for small data sets having standard structure with all variables of the same type; the underlying assumption was that the phenomenon is homogeneous.... This led to models where only a few parameters were necessary to trace the effects of the various factors involved.[20]

Increasingly, such homogeneity was no longer assured, and dreams of having but a few parameters seemed inadequate. "With large data sets involving many variables, more structure can be discerned and a variety of different approaches tried." The sample size of data, however, "does not necessarily imply a richness of structure." The complexity of the data is key, and may include:

> High dimensionality
> A mixture of data types
> Nonstandard data structure
> and, perhaps most challenging, nonhomogeneity; that is, different relationships hold between variables in different parts of the measurement space.[21]

No longer are such data sets rare, they argued, given "accelerating computer usage." Existing multivariate statistics involved dimensionality reduction, and came with substantial drawbacks. "To analyze and understate complex data sets, methods are needed which in some sense select salient features of the data, discard the background noise, and feed back to the analyst understand-able summaries of the information."[22] Decision trees, the authors explained, offered "a flexible nonparametric tool to the data analyst's arsenal."[23] (A nonparametric statistical tool doesn't presume a particular form of probability distribution.) These trees offered insight into the data, insight that was to be understood by human beings; while this insight differed from dominant statistical modeling, such traditional practices were hard pressed to deal with the form of data now increasingly in question. What sort of knowledge was produced, however? "An important criterion for a good classification procedure is that it not only produce accurate classifiers (within the limits of the data) but that it also *provide insight and understanding into the predictive structure of the data*."[24] Classifiers were to be predictors, but also much more.

Their 1984 volume *Classification and Regression Trees* proffers a largely technological determinist narrative: complex, big data required new algorithms and new attitudes about statistics.[25] How did Breiman and Friedman each come to adopt a version of these views and begin collaborating? How did the challenge of large complex data sets come to seem a necessary one? And how did they understand the transformation of acceptable knowledge required by larger and more complex data sets?

From Academia to Industry and Back Again

When Breiman moved from industry and defense work back into academia at UC Berkeley, he was startled. He later described it as being in *"Alice in Wonderland."*

> I knew what was going on out in industry and government in terms of uses of statistics, but what was going on in academic research seemed light years away. It was proceeding as though it were some branch of abstract mathematics.[26]

Having left a promising career in mathematical statistics at UCLA, he took on a wide range of statistical work for the Department of Defense and the then new Environmental Protection Agency as a consultant with the Technology Services Corporation. Working outside of academic statistics, he explained, he came to focus on prediction over making causal claims using models or doing rigorous hypothesis testing:

> We were working on prediction problems like next day ozone in the Los Angeles basin, carbon monoxide levels on freeways, but also things such as could we recognize the sender of handset Morse code—this was something we were doing for

the spook agencies—or could we recognize from sonar returns whether the other submarine was Russian or American?[27]

Outside of academia, Breiman underwent—or perhaps cemented—a fundamental shift in his epistemic values and mathematical practices, away from explanation to prediction. Statistics was born from making sense of data about diverse populations and systems analyzing data, and yet in the eyes of practitioners like Breiman, the discipline had gone far astray; only then, around 2000, was statistics beginning to "recover" from what he called its "'overmathematization' in the post World War II years."[28]

Like the creators of AID before him, Breiman and a few other dissident statisticians had a heroic ancestor, someone both central to the statistics profession and at a distance from it, the mathematician John Tukey. In a programmatic 1962 paper on the "Future of Data Analysis," Tukey called for a data-focused statistical practice at a significant remove from mathematical statistics.

Data analysis ... must then take on the characteristics of a science rather than those of mathematics, specifically:
(1) Data analysis must seek for scope and usefulness rather than security.
(2) Data analysis must be willing to err moderately often....
(3) Data analysis must use mathematical argument and mathematical results as bases for judgment rather than as bases for proofs or stamps of validity.[29]

Like that of Breiman, Tukey's approach was nourished outside academic statistics, particularly in defense work during and after World War II. Thanks to "war problems," Tukey explained, "it was natural to regard statistics as something that had the purpose of being used on data—maybe not directly, but at most at some remove. Now, I can't believe that other people who had practical experience failed to have this view, but they certainly—I would say—failed to advertise it."[30]

Tukey's paper doesn't just seem prophetic in retrospect: he actively worked for decades with a range of collaborators to bring the tools for data analysis into being. His access to classified systems in the defense and intelligence worlds gave him a preview of what was to come in the commercial world, in terms of data storage and processing, as well as computational power. In the meanwhile, working mostly at Bell Labs, Tukey led a research program into concrete forms of "exploratory data analysis," done by pen and pencil up into the early 1980s, and thereafter in software and graphics systems he and a remarkable set of researchers at Bell helped bring into being.

Breiman worked along the lines Tukey celebrated. Breiman's published work on behalf of various government agencies confirms the history told in his retrospective accounts. In his work for the Environmental Protection Agency during the 1970s, Breiman and his collaborators undertook a highly empirical examination of the variety of causes of ozone fluctuations in Los Angeles. "It is obvious," they wrote, "that the oxidant formation and transport process is quite complex. . . . The present study uses available data to perform an exploratory analysis of the amount of information required to explained observed data."[31] The techniques used were not tree like, but they shared a common motivation in undertaking exploratory analysis to understand the foremost predictive elements in complex data sets. "Through the extensive use of a nonlinear, nonparametric, exploratory regression technique . . ., we found that virtually all of the predictive capability was contained in three variables,

> the current O_3 level
> the current solar rational reading
> the current NO_2 level.[32]

The ozone question was central to Breiman's consulting practice in the 1970s. The data exemplified the large data set with many variables: "The learning sample for the ozone classification project contained 6 years (1972–1977) of daily measurements on over 400 meteorological variables and hourly air pollution measurements at 30 locations in the Los Angeles basin."[33] Creating a classifier involved discerning which variables strongly predicted change. "The work toward understanding which meteorological variables and interactions between them were associated with alert-level days was an integral part of the development of a classifier."[34] The sheer number of observations was less a problem than the number of variables associated in each observation, referred to as the "high-dimensionality" of data.

As a consultant at Technology Services Corporation, Breiman sought methods that leveraged large amounts of high data without precipitously throwing it out. Existing methods, he and his collaborator Meisel argued in a report to the US Air Force, required the analyst to choose, without reference to the data, a means for reducing its dimensionality:

> In the usual pattern recognition approach, the dimensionality is made reasonable by the selection of an apriori mapping from measurement space to "feature" space. That is, depending on certain physical or heuristic principles, the large amounts of

detailed information regarding any one object are aggregated and summarized in a small number of variables that comprise the feature vector.[35]

He explained that the usual algorithms "give a one gulp answer" that requires a "drastic reduction in dimensionality ... to make the sample size sufficiently dense in the space to define the problem and to make it computationally feasible." Reduction was a one-way street: "having made the reduction in dimensionality, one is stuck with it. The loss in information is irrevocable."[36]

For Breiman and his collaborator Meisel, the excitement of decision trees was that they dealt with large amounts of data piecemeal, not all at once, so they did not require any apriori reduction: "in a tree decision structure, there is possible a sequential Interaction between classification and information. As one progresses down the branches of the tree, more and more detailed information can be called for by the tree construction algorithm."[37]

If trees were a major potential answer to doing this analysis, they still suffered from the array of problems that doomed them in the 1960s. Two fundamental problems had to be solved: the questions of how to make splits in the tree and how to "prune" the tree so that it would be predictive for other sets of data, to overcome the problem of overfitting that bedeviled trees from the start.

"The whole story is in finding good splits and in knowing when to stop splitting."[38] They defined the goodness of a split in terms of its decrease in impurity, though noting a number of differential plausible splitting criteria. After spending much of their effort at attempting to refine their splitting methods, they undertook a "fundamental shift in focus":

> Instead of attempting to stop the splitting at the right set of terminal nodes, continue the splitting until all terminal nodes are very small, resulting in a large tree. Selectively prune (recombine) this large tree upward, getting a decreasing sequence of subtrees. Then use cross-validation of test sample estimates to pick out that subtree having the lowest estimated misclassification rate.[39]

Einhorn, the critic of the Michigan efforts, denounced the earlier AID classifier from the point of view of science understood as a statistical practice; Breiman and his collaborators drew on statistics to improve the making of the classifier, but toward ends distinct from the focus of mathematical statistics on hypothesis testing.

Up the coast from Los Angeles, another researcher found himself developing trees with big data sets, once again distant from the central concerns of mathematical statistics.

From SLAC to Statistics

The question of how to produce efficient algorithms for the high-dimensional data produced in physics experiments motivated Jerome ("Jerry") Friedman, who would in time collaborate with Breiman. Trained as a high-energy physicist, Friedman found himself more interested in the computational problems around doing physics in the late 1960s and early 1970s, eventually getting a job at the Stanford Linear Accelerator Laboratory that afforded him unusual access to powerful computers and lots of data. As a computational physicist, he explained, he constantly used statistics, but not as mathematical statisticians did. "Physicists didn't do much hypothesis testing and things like that; it was mostly exploratory, automatically making scatter plots, histograms, various other kinds of displays mostly displayed on hardware of the time."[40]

In an interview later in life, he explained that physicists became interested in applying a key pattern recognition family of algorithms called "nearest neighbors" to their data sets. Many of the presumptions built into such algorithms began falling apart when the dimensions of the data sets began to get bigger. "I thought, well, if we are going to use this approach in applications with bigger data sets like those in high-energy physics, we'll need a fast algorithm to find nearest neighbors in data sets." Nearest neighbor algorithms, he explained, are an "n-squared operation: for each point you have to make a pass over all the other points"—so they get vastly slower as the data gets large and high-dimensional. "So I started working on fast algorithms for finding near neighbors, without too much success."[41] Soon another researcher introduced him to a tree algorithm that "involved recursively partitioning the data space in boxes."[42] Friedman soon came to realize he could use the boxes for classification. "So it occurred to me that in the nearest-neighbor algorithm you could recursively find the variable with the largest spread and split it at the median to make boxes. Why don't we find the variable that has the most discriminative power and split it at the best discriminating point?" Ultimately he realized he could do the classification without using nearest neighbors and its algorithmic inefficiencies at all. In his paper on recursive partitioning, he underscored two important aspects of his algorithm. First, it was fast: "Computationally, the procedure is quite fast both in the training and classification stages"; and second, it could handle missing data common in real-world applications. "Methods for using vectors with missing coordinates in both training and classification are presented."[43]

Thus the computational challenges associated with using a pattern-recognition algorithm led Friedman to approaches to draw on data more efficiently, and then eventually to the idea of growing classification trees by finding variables with the most discriminatory power. Two colleagues recognized the affinity of the approaches of Breiman and Friedman and brought them together. The result included the 1984 monograph discussed above and an accompanying commercial software package, CART.

The impact of statisticians' decision tree package and approach was initially limited. "We take pride in the fact that CART came ten years earlier than C4.5, but it was Quinlan and the machine learners who popularized trees. We did CART and it just sat there: statisticians said, 'What's this for? What do you do with it?'"[44] Upon return to academia, Breiman reflected intensely on where mathematical statistics had gone wrong.

Breiman and the Two Cultures of Statistics

Some years later, with his algorithm and his orientation toward statistical reasoning having grained traction outside statistics, Breiman offered a philosophical account of the shift in values he had undergone. He described a dramatic alteration toward a data-focused practice:

(a) Focus on finding a good solution—that's what consultants get paid for.
(b) Live with the data before you plunge into modeling.
(c) Search for a model that gives a good solution, either algorithmic or data.
(d) Predictive accuracy on test sets is the criterion for how good the model is.
(e) Computers are an indispensable partner.[45]

Attendant upon this change in practice, he described a radical contrast between a "Data modeling culture" used by an estimated "98% of all statisticians" and an "algorithmic modeling culture," used by "2% of statisticians" but also by "many in other fields." In the data modeling culture dominating academic statistics, model validation come through "Yes-no-using goodness-of-fit tests and residual examination." In contrast, the algorithmic culture focused on "predictive accuracy." Restricting oneself to the limited range of models of contemporary statistics was to abandon vast arrays of data, to demand more certain knowledge of causes than often possible, and to limit the creation of new tools. Algorithmic culture had too much to offer, even if it meant loosening epistemic demands.

Part 2: The Critique of Artificial Intelligence

Not only in statistics did critics revivify trees. Programming—or teaching—computers to perform acts of judgment motivated considerable work in artificial intelligence in the 1960s and 1970s, especially in projects of "expert systems." Notable successes included attempts to formalize the judgment of scientists concerning organic chemical structures, as in the case of the expert system DENDRAL.[46] By the early 1970s, nevertheless, many practitioners worried greatly about the challenge of converting human expertise into "knowledge-bases" and formal inference rules. In a move akin to Harry Collins's reinvigoration of "tacit knowledge" in the sociology of science, artificial intelligence researchers became worried about a fundamental problem they dubbed the "knowledge acquisition bottleneck."[47] However good experts may be at performing actions or making judgments on the basis of sense perceptions, they all, from art connoisseurs to physicists, struggle to convert their expertise into explicitly stated rules. J. Ross Quinlan noted that part "of the bottleneck is perhaps due to the fact that the expert is called upon to perform tasks that he does not ordinarily do, such as setting down a comprehensive roadmap of a subject."[48]

Echoing lines in some of Turing's classic articles, Donald Michie, one of the leaders of this branch of investigation, noted:

> Mastery is not acquired by reading books—it's acquired by trial-and-error and teacher supplied examples. This is how humans acquire skill. People are very reluctant to accept this. Their reluctance tells us something about the philosophical self-image that we, as thinking beings, prefer. It tells us nothing about what actually happens when a teacher or a master trains somebody. That somebody has to regenerate rules from example to make them an intimate part of his intuitive skill.[49]

Rather than attempting to simulate some aspect of the cognitive process of judgment or to follow explicit rules, new forms of pattern recognition and machine learning attempted to *predict* the expert judgments based on the behavior of experts in some task of classification. As a 1993 article explained,

> the machine learning technique takes advantage of the data and avoids the knowledge acquisition bottleneck by extracting classification rules directly from data. Rather than asking an expert for domain knowledge, a machine learning algorithm observes expert tasks and induces rule emulating expert decisions.[50]

One of the major forms of decision trees emerged just within this research program.

Quinlan was seeking to undertake computer induction using larger databases than typical in machine learning of the time: his approach "has two general characteristics: its aim is to discover structured information about some collection of entities, and the methodology employed is the analysis of many examples of the *genre* called *Instances*."[51] He explained further:

> The essence of the induction task is discovery. The data for the task is a collection of instances or descriptions of some set of entities in terms of their properties. A rule is some method of explaining an instance by establishing some relationship between these properties. The induction task is to discover a rule adequate to explain each instance in the data.[52]

Quinlan's interest was not in artificial or toy data sets. His goal was the discovery of a structured knowledge about entities using larger databases of instances, in an implementation within the memory and storage constraints of existing computers.

While visiting Stanford in 1978, Quinlan met Michie, who posed a problem for empirical machine learning. As so often in the history of artificial intelligence, the subject matter was chess, in particular chess endgames.[53] "We consider an endgame situation in which the pieces have been reduced to a black knight, a white rook and the two kings, and where it is black's turn to move. We wish to know whether black is safe for at least one more white move."[54] Quinlan used a slightly simplified version of the problem to generate a large database on instances. From these instances, he sought to induce the rules. While chess provided the examples, crucially, his induction algorithm was agnostic about subject matter:

> it is important to realise that the game of chess has no central role in this evaluation, but merely serves as the source of a non-trivial task that can readily be understood by human beings. Neither the induction algorithm nor the iterative technique being investigated have anything pertaining to chess associated with them. As far as the programs are concerned, *this is a task of inducing a complex class-prediction rule from a data base of two thousand elements described by fourteen other attributes.*[55]

Quinlan's project focused not on any particular induction algorithm, but rather on general methods for transforming existing algorithms for contending with data sets that cannot be stored entirely in memory. Quinlan's doctoral supervisor, Earl Hunt, along with an interdisciplinary team of collaborators, had sought to produce a "concept learning system," defined as "a device for creating a concept corresponding to some partition of a sample of objects which have been categorized" that works by "observing a subset of objects in the universe and being informed of whether or not the name is applicable to them."[56] Working with the larger data sets than envisioned in the 1960s, Quinlan sought to develop to work iteratively over subsets of the database, to produce an algorithm he called ID3, for "Iterative Dichotomizer 3."[57]

As discussed above, a major challenge in automatically producing classifications in trees is choosing a criterion for branching or for judging whether one candidate for branching is superior to another. At Stanford, Peter Gacs suggested that Quinlan use information gain as a superior criterion for deciding at each iteration which attribute to branch upon to produce splits.[58] Information gain had many positive qualities, including its computational cost. Any algorithm that involves an exponential increase in time is almost useless for large data sets. The algorithm's computational requirement is "proportional to the product of the size of the training set, the number of attributes and the number of non-leaf nodes in the decision tree. The same relationship appears to extend to the entire induction process, even when several iterations are performed. No exponential growth in time or space has been observed as the dimensions of the induction task increase, so the technique can be applied to large tasks."[59]

Statistical Trees, Machine Learning Trees

How did the revivification of trees within artificial intelligence differ from their revivification among applied statisticians? Reflecting upon Quinlan's work, Michie and his collaborator Cao Feng stressed its difference in epistemic ambition from the work of Breiman: "As important as a good fit to the data, is a property that can be termed 'mental fit.'" Breiman and colleagues, they noted, in the work on trees, saw "data-derived classifications as serving 'two purposes: (1) to predict the response variable corresponding to future measurement vectors as accurately as possible; (2) to understand the structural relationships between the response and the measured variables."[60] But such is inadequate. Machine learning "takes purpose (2) one step further." It can provide meaning and concepts to practitioners:

The soybean rules were sufficiently meaningful to the plant pathologist associated with the project that he eventually adopted them in place of his own previous reference set. [Machine Learning] requires that classifiers should not only classify but

should also constitute explicit concepts, that is, expressions in symbolic form mean-
ingful to humans and evaluable in the head.[61]

These epistemic demands found a solution in the output options of the commercially available code that Quinlan produced. Not only could his package, called C4.5, output trees, it could output rules legible to human beings: the program "contains a mechanism to re-express decision trees as ordered lists of if-then rules.... There are substantially fewer final rules than there are leaves, and yet the accuracy of the tree and the derived rules is similar. Rules have the added advantage of being more easily understood by people."[62]

Thus, despite many outward similar appearances, the machine learners and the data-focused statisticians diverged about the sort of knowledge that tree algorithms could produce. Quinlan and his later collaborators insisted on the knowledge of rules expressed symbolically, whereas Breiman focused heavily, if not entirely, on the predictive capacity.

And yet the statisticians and the machine learning soon came to subscribe to a similar sense of the impossibility of stating rules comprehensible to human beings. In the late 1990s and subsequently, the machine learning community shifted and came increasingly to focus on prediction and its challenges, leaving symbolic meaning and concepts aside. For all their differences, a concern for producing implementations of algorithms capable of contending with large, real-world data sets in non-exponential time underlay the work by Breiman as much as Quinlan.

The abandonment of interpretability is central to the dramatic reemergence of artificial intelligence as successful—but an artificial intelligence shorn of many of its deepest ambitions, reduced, in many cases, to the power to predict, a return to saving the phenomena.

Part 3: Trees Ramify

In the late 1980s and up into the present, a pattern has emerged in the data mining and machine learning literature: people I'll call algorithmic advocates take up a particular algorithm, offer a series of suggested improvements, often as part of doctoral study, and then celebrate refinements of those algorithms across various scientific and industrial domains.[63] Such advocates often circulate among postdoctoral positions, faculty positions, or jobs in industry and various major labs; many form consulting firms even as they remain in academia. They publish papers focused on improvements to their variants of the algorithm as well as papers coauthored with practitioners from industry, the military, and specific sciences. They make available software implementations of the various algorithms either commercially or as public domain or shareware; in many cases, academics made software publicly available but insist on a license for commercial use. The creators of the CART and C4.5 algorithms, for example, published extensively about them and commercialized their algorithms.

One of the key early major advocates for data mining, or, what was branded "Knowledge Discovery in Databases," Usama Fayyad, exemplifies many facets of this process.[64] In his 1992 dissertation at the University of Michigan, working within the tradition of empirical machine learning just discussed, Fayyad sought to improve decision trees as an example of "automatic knowledge acquisition"—induction, with a particular concern for computational feasibility: showing that this task of learning with trees "can be formed by an algorithm of polynomial complexity."[65] The dissertation focused on an exemplary "task that is not well-understood, even by experts in the area," namely reactive ion etching in semiconductor manufacturing, an area with lots of data but lacking solid models that "relate how output variables are affected by changes in the controlling variables."[66] Fayyad undertook several major research projects to demonstrate the salience of improved decision trees to scientific and industrial work. Industrial work, Fayyad and his collaborators explained, presented two challenges, which inspired further transformation of the algorithms:

Noisy Data: attribute values may be erroneous due to human recording errors, imperfect sensor repeatability, or defects in process equipment or sensors.
Limited Training Data: the training data may be small in size and conducting more experiments may be too costly.

Like many practitioners in machine learning and data mining, he moved easily from industrial to scientific applications, and collaborated with experts in various fields. Fayyad's papers typically embed his development of trees within implemented algorithmic systems, either industrial or academic, connected to human expertise and labor. An early presentation slide from his work on stellar object classification includes the astronomer who produces the training data (see figure 10.3).

In reporting on the success of the stellar classification effort, Fayyad and his astronomer collaborators noted the greater than a 90 percent success rate. Decision trees were a good choice for such collaboration because "the final classifier produced is symbolic and, therefore, not difficult for domain experts to interpret (as opposed to a neural network or a pattern-recognition–based

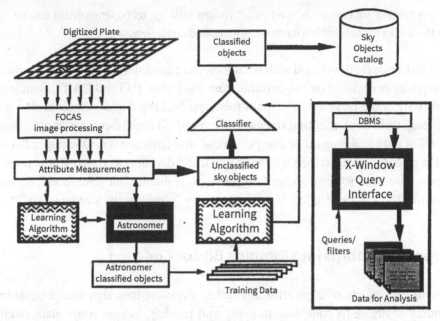

Figure 10.3 Architecture of the SKICAT cataloguing and classification process. *Source*: Usama M. Fayyad, Nicholas Weir, and S. Djorgovski. "Cataloging and Analysis of Sky Survey Image Databases: The SKICAT System," in *Proceedings of the Second International Conference on Information and Knowledge Management* (New York: ACM Press, 1993), 527–536.

approach)."[67] Such interpretable results would become rarer and rarer as the evangelical efforts of Fayyad and other spread the good and bad news around decision trees broadly over the course of the 1990s.

Very Large Databases

In graduate school, during his postdoctoral work, and then in prominent roles in industry, Fayyad became an advocate for the creation of algorithms and software packages capable of working on large data sets that could not reside in main memory, such as huge numbers of digitized stellar images. Large amounts of data broke things easily, as Fayyad noted in 1998:

A typical statistical package assumes small data sets and low dimensionality. For example, suppose you want to do some simple database segmentation by running ... a basic simple method for clustering data. Let's say you have managed to lay your hands on an implementation in some statistical library. The first operation the routine will execute is "load data." In most settings, this operation will also be the last as the process hits its memory limits and comes crashing down.[68]

While they were developed with what then were considered large data in mind, the primary decision tree algorithms from the 1980s, ID3 and CART, assumed that data could be resident in main memory. Making decision trees scale was no easy task: implementing trees as scale required careful choices about what sorts of data to disregard or computational shortcuts to take. One paper from IBM researchers described how they reworked decision tree algorithms to eschew "the need for any centralized memory-resident data structures," which permits their algorithm "efficiently" to classify "virtually any sized dataset."[69]

From Algorithms into Running Bits of Code

Thanks to efforts of algorithm advocates, decision tree algorithms became widely available in code easy to apply and modify, in ever more elaborated packages useful for new researchers and implemented on increasingly available microcomputers. At NASA, researchers released an implementation of aspects of the commercial algorithms CART and C4.5, providing "wrappers" with a consistent interface for algorithms under copyright.[70] Based on work done while students at Stanford, Ronny Kohavi and Dan Summerfeld at Silicon Graphics did much the same.[71] With no royal road to prediction, such software packages promised to aid in this process of iterative work with many variants of the different algorithms for classification and prediction. The NASA researchers explained:

> As no one decision tree building method (or, for that matter, machine learning method) is the best for all datasets, we feel that a machine learning researcher/practitioner should experiment with as many methods as possible when attempting to solve a problem.[72]

The packages offered black-box solutions, simply allowing one to run a number of the major machine learning algorithms, notably the major variants of decision trees in a couple of lines. Using MLC++, running a simple decision tree algorithm such as ID3 requires simply:

```
setenv DATAFILE iris # The dataset stem
setenv INDUCER ID3 # pick ID3
```

```
setenv ID3_UNKNOWN_EDGES no # Don't bother with unknown edges
setenv DISP_CONFUSION_MAT yes # Show confusion matrix
setenv DISPLAY_STRUCT dotty # Show the tree using dotty
Inducer
```

Such an unthinking invocation of the implemented algorithms was just the beginning. Little in what follows is automatic. For researchers, whether in academia or industry, going deeper, the copiously documented source code in these packages offered guidance for anyone looking to mash up tree-making process to draw in techniques from statistics or neural nets. The well-documented code for inducing decision trees in MLC++ is replete with comments indicating how the academic literature suggests improvements to implementations and paths not taken. The implementers spelled out the details of their often tentative design decisions, which may or may not be optimal: "We currently split instances but keep the original structure.... It may be faster in some cases to actually create a new List without the attribute."[73] The code even includes suggestions for further research along the day. "It may be an interesting research topic finding ways to prune nodes containing other categorizers (naive-bayes, for example)."[74]

A data-focused statistician surveying the field in 1997 described how human working with data is. "The problem, as I see it, is not one of replacing human ingenuity by machine intelligence, but one of assisting human ingenuity by all conceivable tools of computer science and artificial intelligence, in particular aiding with the improvisation of search tools and with keeping track of the progress of an analysis."[75] The various implementations of machine learning algorithms and their environment are efforts to bring such a vision into being. We might say this code is part of a world of the quotidian algorithmic life to be unearthed through ethnographic study not possible here: the attempt to extract something from data using a variety of algorithms in many variants, with attention to the trade-offs available and necessary to the researcher.[76]

The code of the major open-source packages of the 1990s, like that of today, speaks to the iterative quality of building predictive models, which often were highly sensitive to small changes in the underlying data—often referred to as the "instability" of the predictive model. The software enabled one to quickly try a variety of models and, increasingly, to combine the models into what became known as ensemble models. Soon this aspect of the practice of humans inspired further algorithmic development around combining multiple models.

Multiple Trees

Decision trees have many virtues: they are moderately predictive, can produce human intelligible results, and are computable in polynomial time even when applied to large data sets. Their moderate predictiveness stems from their instability: various tree algorithms are prone to grow considerably different trees with slightly different training sets or with small perturbations with the training set.

The early 1990s saw both the proliferation of alternative tree-growing methods and of best practices for dealing with these unstable algorithms. If manually growing multiple trees was a best practice, why not, practitioners began to reason, produce them in some automatic fashion? In his doctoral work and a series of papers, for example, Fayyad advocated growing multiple trees from randomly determined subsets of the training data:

> For each tree, we apply the statistical test and keep only the "good" rules. Finally, a subset of the surviving rules that covers the original training set is selected. When coupled with a method for statistical significance testing, the multiple random sampling of the training set has proven to be an effective technique for extracting a compact and reliable set of rules from the original training set.[77]

The end result of this multiple tree process is a refined and relatively small set of rules in the form of "if … then …" statements, long the desideratum in machine learning (see figure 10.4).

Combining multiple trees and other predictive models has turned out to be extraordinarily powerful, if usually opaque.[78] By the 1990s, Breiman argued, a growing literature revealed that "combining a multiple set of predictors, all constructed using the same data, can lead to dramatic decreases in test error." This predictive success came at great cost:

> At the end of the day, what we are left with is an almost inscrutable prediction function combining many different predictors. But the resulting predictor can be quite accurate…. Growing and combining 50 CART trees … gives a predictor that performs much better on these four data sets than any other classifier. But this puts us in a difficult dilemma. A single tree … has a simple and understandable structure, but combinations of many trees do not.[79]

The same held true of neural networks: ensembles of them performed better than any one predictive algorithm. Neural nets, however, were always inscrutable, whereas a fundamental virtue of trees had long been that they are interpretable.

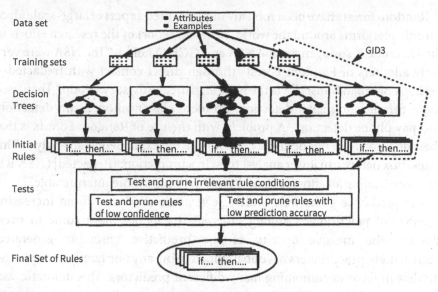

Figure 10.4 Data flow diagram for RIST.

Source: U. M. Fayyad et al., "Machine Learning of Expert System Rules: Applications to Semiconductor Manufacturing," in *Collected Notes on the Workshop for Pattern Discovery in Large Databases (NASA Ames, January 14–15, 1991)* (Moffett Field, CA: NASA Ames Research Center, 1991), 26.

A bevy of techniques with snappy names emerged to create such ensembles: bagging, boosting, arcing, etc. The creation of ensembles of trees typically involve two stages: first, a random process of selecting different subsets of training data, and then training a succession of trees; second, a voting among the grown trees, sometimes with weighting, when confronted with additional data. Some of these procedures pick out misclassified data and give greater weight to classifiers correctly classifying that data. Devotees of different methods—bagging, boosting, arcing—debated and published reports.

By far the best-known form of these approaches Breiman dubbed "random forests," in part developed in collaboration with Adele Cutler. A random forest grows thousands of different trees from different samples of the data, finds the best split for a set of random variables, and then has the trees vote for each classification. Breiman explained:

> forests are A+ predictors. But their mechanism for producing a prediction is difficult to understand. Trying to delve into the tangled web that generated a plurality vote from 100 trees is a Herculean task. So on interpretability, they rate an F.[80]

Breiman himself was both concerned with the question of interpretability and convinced that random trees could in fact impart considerable information of scientific importance.

Random forests have been robustly implemented as part of large-scale algorithmic platforms around the world. A leaked report of the research efforts of the US and UK spy agencies the NSA and GCHQ noted, "The NSA were very early adopters of Random Forests through direct contact with [redacted—perhaps Breiman himself] via the NSA Statistical Advisory Group." They used them for steganography detection, protocol classification, spam detection, and pay phone detection. "A problem with the use of Random Forests is that their decisions can not be simply and intuitively explained to an analyst. This black-box nature can lower analyst trust in a prediction. [Redacted] (NSA R1) has been leading an effort to make Random Forests more interpretable."[81]

The predictive gains were widely seen as massive, and to an increasing number of practitioners among different disciplines, have come to overshadow the massive opacity of the predictive ensemble generated. Increasingly, practitioners have abandoned using any one family of predictive models in favor of combining many different predictors. This dramatic success of ensembles amplified the ethic of prediction over interpretation. A generation ago, the inscrutability of neural nets made them deeply problematic; the renaissance of neural networks from around 2012 rests squarely on the legitimation of such ensemble models, for commerce, for spies, and for science.

Epilogue

By and large, the recent humanistic study of algorithms has focused on large-scale technological and social systems where veiled algorithms figure centrally. Indeed, we often speak colloquially of Google's implemented search platform as an "algorithm." Most of these systems are opaque, both for reasons of proprietary or state secrecy and by virtue of the kind of implemented algorithms, parameters, and data integrated into the system. In general, it is far easier to gauge their effects, good and bad, and to illustrate their biases, than to understand their becoming. Much algorithmic criticism works from a double ahistoricity. At the macro level, the particular sort of opacity of large-scale algorithmic systems built with machine learning we now find ubiquitous was not a foregone conclusion: indeed, it would have been seen as anathema within most of the artificial intelligence community thirty years ago. We need to understand how prediction ("optimizing cost-functions") became the paramount virtue, in the face of great opposition to crucial values of older forms of symbolic, statistical, and computational intelligibility. Neoliberalism or the like are inadequately granular to explain the shift in values and practices within computational statistical work. And the explosion of data in the past thirty years is at most a sufficient but clearly not a necessary cause. At the smaller scale, the making of large-scale implementation of machine learning models is a profoundly human process. Machine learning models don't just build themselves: there's no systematic, or dare we say, algorithmic process through which good models are created, tested, and then deployed. As the examples of implemented code illustrate, trial and error using algorithmic tools on large data sets is central to the everyday practices of machine learners and data miners—just as it is in the creation of any other system.

By tracking the development, fall, and explosion of decision trees we can chart the dramatic shifts in the epistemic values of computational statisticians and machine learners from the 1960s to the present. What we see, again and again, is the conscious effect to account for the explosion of high-dimensional data, not a technologically determinist account of the volume of data-altering epistemic values. And looking at the small-scale implementations of decision trees focuses our attention on the everyday work of testing models, formatting of data, and determining sample size required to constitute large-scale algorithmic systems, such as search or recommender engines.[82]

Notes

Chapter 1

1. On the role of histories of algebra in the formation of mathematical Europe, see Jens Høyrup, "The Formation of a Myth: Greek Mathematics—Our Mathematics"; and Giovanna Cifoletti, "The Creation of the History of Algebra in the Sixteenth Century," both in *L'Europe mathématique: Histoires, Mythes, Identités*, ed. Catherine Goldstein, Jeremy Gray, and Jim Ritter (Paris: Éditions de la Maison des sciences de l'homme, 1996), 103–119 and 123–142.

2. See, inter alia, Karen Hunger Parshall, "A Plurality of Algebras, 1200–1600: Algebraic Europe from Fibonacci to Clavius," *BSHM Bulletin: Journal of the British Society for the History of Mathematics* 32, no. 1 (2017): 2–16; and, on the French story, Warren van Egmond, "How Algebra Came to France," *Mathematics from Manuscript to Print 1300–1600*, ed. Cynthia Hay (Oxford: Clarendon Press, 1988), 127–144.

3. In this volume Michael Barany discusses a prominent example: Robert Recorde's 1543 *The Ground of Artes*. For a brief discussion of algorism and arithmetic, see Angela Axworthy, *Le Mathématicien renaissant et son savoir. Le statut des mathématiques selon Oronce Fine* (Paris: Classiques Garnier, 2016), 288–289, and literature cited there.

4. Anthony Grafton and Lisa Jardine, *From Humanism to the Humanities: Education and the Liberal Arts in Fifteenth- and Sixteenth-Century Europe* (Cambridge, MA: Harvard University Press, 1986); on mathematics, see Ann Moyer, "Reading Boethius on Proportion: Renaissance Editions, Epitomes, and Versions of the Arithmetic and Music," in *Proportions: Science, Musique, Peinture & Architecture*, ed. Sabine Rommevaux, Philippe Vendrix, and Vasco Zara (Paris: Honoré Champion, 2012), 51–68; and Richard J. Oosterhoff, *Making Mathematical Culture: University and Print in the Circle of Lefèvre d'Étaples* (Oxford: Oxford University Press, 2018).

5. Giovanna Cifoletti, "La question de l'algèbre: Mathématiques et rhétorique des hommes de droit dans la France du 16e siècle," *Annales: Histoire, Sciences Sociales* 50, no. 6 (November–December 1995): 1385–1416; "L'utile de l'entendement et l'utile de l'action: discussion sur la utilité des mathématiques au xvie siècle," *Revue de synthèse* 4, no. 2–4 (April–December 2001): 503–520.

6. Katherine Neal, *From Discrete to Continuous: The Broadening of Number Concepts in Early Modern England* (Dordrecht: Springer, 2002); David Rabouin, *Mathesis Universalis: L'idée de mathématique universelle d'Aristote à Descartes* (Paris: Presses universitaires de France, 2009).

7. On sixteenth-century algebra, the essential starting point is Sabine Rommevaux, Maryvonne Spiesser, and Maria Rosa Massa Esteve, eds., *Pluralité de l'algèbre à la Renaissance* (Paris: Honoré Champion, 2012).

8. See contributions in Sabine Rommevaux, Philippe Vendrix, and Vasco Zara, eds., *Proportions. Science, Musique, Peinture & Architecture* (Turnhout: Brepols Publishers, 2012).

9. See Rabouin (2009); and, earlier, Jacob Klein, *Greek Mathematical Thought and the Origin of Algebra*, trans. Eva Brann (Mineola, NY: Dover Publications, Inc., 1992 [reissue of 1968

edition]). The issue is broached in a new way, from a pedagogical perspective, in Oosterhoff (2018).

10. Debates about the certitude of mathematics, even when they question the status of Euclidean demonstration vis-à-vis Aristotelian syllogism, obviously accept that demonstration (rather than, say, algorism) is the relevant point of comparison to logic. On such debates, see Nicholas Jardine, "The Epistemology of the Sciences," in *The Cambridge History of Renaissance Philosophy*, ed. Charles Schmitt, Quentin Skinner, Eckhard Kessler, and Jill Kraye (Cambridge: Cambridge University Press, 1988), 685–711; Paolo Mancosu, "Aristotelian Logic and Euclidean Mathematics: Seventeenth-Century Developments of the *Quaestio de Certitudine Mathematicarum*," *Studies in the History and Philosophy of Science* 23, no. 2 (1992): 241–265.

11. Girolamo Cardano, *Artis magnae, sive de regulis algebraicis, liber unus* (Nuremberg: Johannes Petreius, 1545); see also the recent critical edition, *Artis magnae, sive, De regulis algebraicis liber unus*, ed. Massimo Tamborini (Milan: FrancoAngeli, 2011); and in English translation, *The Rules of Algebra*, trans. T. Richard Witmer (Mineola, NY: Dover Publications, Inc., 1992 [reissue of 1968 edition]). For a discussion of the overall work, see Jacqueline Stedall, *From Cardano's "Great Art" to Lagrange's "Reflections": Filling a Gap in the History of Algebra* (Zürich: European Mathematics Society, 2011), 3–17. On Cardano's career, see Anthony Grafton, *Cardano's Cosmos: The Worlds and Works of a Renaissance Astrologer* (Cambridge, MA: Harvard University Press, 1999).

12. Cardano distinguished different cases where we would see just one because he only accepted positive values for the root. Thus the equation $ax^2 + bx + c = 0$ would give rise to multiple cases depending on whether a, b, and c are positive or negative.

13. Cardano (1545), cap. 6, fol. 14v and cap. 11, fol. 29v = Cardano (1992), 48 and 96. These two heads are discussed in ch. 11 and 12 of Cardano's text. For $(b/3)^3 > (c/2)^2$ the latter mode is irreducible in real numbers, and Cardano restricted his rule to the alternative case. See Cardano (1545), fol. 31r = Cardano (1992), 103; and for discussion, see Stedall (2011), 10.

14. Cardano (1545), cap. 1, fol. 6v = Cardano (1992), 20.

15. All the cases where none of the monomials has a 0 value (ch. 17–23) use it, and several of the earlier cases (ch. 11–16) use variations on it.

16. Cardano (1545), cap. 12, fol. 31r = Cardano (1992), 102. Later demonstrations drew on this one: see, e.g., cap. 19, fol. 41r.

17. Cardano (1545), cap. vi, fol. 16r = Cardano (1992), 52. On reducibility and generality in Cardano's mathematics, see Stedall (2011), 11. Cardano identifies Niccolò Tartaglia in this context; on Tartaglia's later reception in by the mathematicians I discuss below, see Giovanna Cifoletti, "Mathematics and Rhetoric: Jacques Peletier, Guillaume Gosselin, and the Making of the French Algebraic Tradition" (PhD diss., Princeton University, 1992), ch. 2.

18. On commercial arithmetic, see the well-known studies by Natalie Zemon Davis: "Mathematicians in Sixteenth-Century French Academies: Some Further Evidence," *Renaissance News* 11 (1958): 3–10, and "Sixteenth-Century French Arithmetics on the Business Life," *Journal of the History of Ideas* 21, no. 1 (1960): 18–48; on the incorporation of arithmetic at the universities, see the important article by Jean-Claude Margolin, "L'Enseignement des mathématiques en France (1540–70): Charles de Bovelles, Fine, Peletier, Ramus," in *French Renaissance Studies 1540–70: Humanism and the Encyclopedia*, ed. Peter Sharratt (Edinburgh: Edinburgh University Press, 1976): 109–155.

19. Jacques Peletier, *L'Aritmetique ... departie in quatre Livres* (Poitiers: [Jean de Marnef], 1549); Pierre Forcadel, *L'Arithmeticque* (Paris: Guillaume Cavellat, 1557).

20. Cifoletti (1995), 1392–1393 and 1412; more generally, see Cifoletti (2001), esp. 514. For Gosselin's text, see the critical edition in Guillaume Gosselin, *De arte magna libri IV Traité d'algèbre suivi de Praelectio/Leçon sur la mathématique*, ed. and trans. Odile Le Guillou-Kouteynikoff (Paris: Les Belles Lettres, 2016), 406–483, at 422–425.

21. On the use of geometric theorems, see also Jacques Peletier, *L'Algèbre ... departie an deus Livres* (Lyon: Jean de Tournes, 1554), 113.

22. Cardano (1545), cap. xiv, fol. 33r = Cardano (1992), 110.

23. Forcadel (1557), livre I, fol. 74r. For discussion of Forcadel's use of rules, see François Loget, "L'algèbre en France au XVIe siècle," *Pluralité de l'algèbre à la Renaissance*, ed. Rommevaux, Spiesser, and Massa Esteve (2012): 69–101.

24. Proclus, *In Primum Euclidis Elementorum Librum Commentarii*, 203 ff.; *A Commentary on the First Book of Euclid's Elements*, trans. Glenn Morrow (Princeton, NJ: Princeton University Press, 1970), 159–162. For discussion of the scheme, see Reviel Netz, "Proclus' Division of the Mathematical Proposition into Parts: How and Why Was It Formulated?" *Classical Quarterly* 49, no. 1 (1999): 282–303.

25. See John Wallis's letter to Christian Huygens of August 12/22, 1656, in John Wallis, *Correspondence*, ed. Philip Beeley and Christoph Scriba, 4 vols. (Oxford: Oxford University Press, 2003-), I:195: "Deinde exponit unam rectam *AB*, et super hanc constructum triangulum demonstrat esse aequilaterum: Tum subsumendum relinquit, (quod nempe quilibet supplere debet,) Atque eadem methodo, super quamvis aliam rectam ita construetur triangulum, et pariter demonstrabitur. Ergo, In quavis &c. Nec quidem aliter constat vis istius aliusve fere demonstrationis, quam ex suppositione, quod nullus casus in contrarium urgeri possit cui non applicabitur exposita demonstratio." Cf. John Wallis, *Due Correction for Mr. Hobbes* (Oxford: Leonard Lichfield, 1656), 42.

26. On the reception of Diophantus in the Renaissance, see Joann Stephanie Morse, "The Reception of Diophantus' 'Arithmetic' in the Renaissance" (PhD diss., Princeton University, 1981); Ad Meskens, *Travelling Mathematics—The Fate of Diophantos' Arithmetic* (Basel: Springer, 2010).

27. See Diophantus of Alexandria, *Rerum arithmeticarum libri sex*, ed. Wilhelm Xylander (Basel: Eusebius and Nicolas Episcopius, 1575), 9. For Xylander's organization of Diophantus' problems, see *passim*. For a critical edition, see Diophantus of Alexandria, *Opera omnia ...*, ed. Paul Tannery, 2 vols. (Leipzig: B. G. Teubner, 1893–1895), but note well that Tannery's edition likewise exaggerates Diophantus' resemblance to Euclid: for discussion, see Reviel Netz, "Reasoning and Symbolism in Diophantus: Preliminary Observations," in *The History of Mathematical Proof in Ancient Traditions*, ed. Karine Chemla (Cambridge: Cambridge University Press, 2012), 327–361.

28. To what degree early modern geometric diagrams were understood as undetermined (in magnitude and position) or particular is beyond the scope of this chapter; for discussion, see Vincenzo De Risi, ed., *Mathematizing Space: The Objects of Geometry from Antiquity to the Early Modern Age* (Cham et al.: Springer/Birkhäuser, 2015).

29. Simon Stevin, *L'Arithmetique de Simon Stevin de Bruges: Contenant les computations des nombres Arithmetiques ou vulgaires: Aussi l'Algebre, avec les equations de cinc quantitez. Ensemble les quatre premiers livres d'Algebre de Diophante d'Alexandrie, maintenant premierement traduicts en François* (Leiden: Christophe Plantin, 1585).

30. On Stevin's life and career, see E. J. Dijksterhuis, *Simon Stevin: Science in the Netherlands around 1600*, trans. C. Dikshoorn (The Hague: Martinus Nijhoff, 1970).

31. Stevin (1585), 431 ff., *passim*.

32. For the effects of reading Diophantus on the arithmetical style of another early modern reader, the algebraist Rafael Bombelli, see Morse (1981), 88 ff.

33. Guillaume Gosselin, *De arte magna ... libri quattuor* (Paris: Gilles Beys, 1577), fol. 57v, 59v, 61v = Gosselin (2016), 346 ff. See also Cifoletti (1995), 1409. On Gosselin's patchy biography, see Gosselin (2016), 49–58.

34. John Wallis, *Opera Mathematica*, 3 vols. (Oxford: Sheldon Theater, 1693–9), I:118–126 and I:183–193.

35. Petrus Ramus also used the term "canon" in his *Algebra* ([Paris: Andre Wechel, 1560], fol. 13v, 15v, 16v). The very close similarity between Ramus' use and Gosselin's (both use canons to differentiate between what Cardano called "modes" of the equation, and they are the same three modes) suggests that Ramus was Gosselin's source. On Ramus' canons, see François Loget, "De l'algèbre comme art à l'algèbre pour l'enseignement: Les manuels de Pierre de La Ramée, Bernard Salignac et Lazare Schöner," *Revue de synthèse* series 6, 132, no. 4 (2011): 495–527, at 500.

36. Gosselin (1577), fol. 59r = Gosselin (2016), 351.

37. Gosselin (1577), fol. 59v = Gosselin (2016), 351.

38. Gosselin (1577), fol. 60r = Gosselin (2016), 351–353. Le Guillou-Kouteynikoff argues on the contrary for the generality of Gosselin's demonstrations on the grounds that the demonstration could be repeated for different numbers: see Gosselin (2016), 127. But note that whether the calculations can be performed "in the same way" on other numbers is the very question at stake in the difference between arithmetic and geometry.

39. On Gosselin's appeals to axioms, see Cifoletti (1995), 1409.

40. Questions of formality in geometry are complicated by the role of "formulas" in Greek expression: see Reviel Netz, *The Shaping of Deduction in Greek Mathematics* (Cambridge: Cambridge University Press, 1999), ch. 4.

41. On the related epistemic state of "assent" as a goal of geometrical demonstration, see Abram Kaplan, "Analysis and Demonstration: Wallis and Newton on Mathematical Presentation," *Notes and Records of the Royal Society* 74, no. 2 (2018): 447–468.

42. On instrumentation, see, inter alia, Alexander Marr, ed., *The Worlds of Oronce Fine: Mathematics, Instruments, and Print in Renaissance France* (Donington: Shaun Tyas, 2009); and discussion of the organic construction of curves in Niccolò Guicciardini, *Isaac Newton on Mathematical Certainty and Method* (Cambridge, MA: MIT Press, 2009). A general discussion of the "organic" approach in early modern geometry remains a desideratum.

43. Oosterhoff (2018); on the importance of math for Ramus, see Robert Goulding, *Defending Hypatia: Ramus, Savile, and the Renaissance Rediscovery of Mathematical History* (Dordrecht: Springer, 2010).

44. Pierre Hérigone, *Cursus Mathematicus, nova, brevi et clara methodo demonstratus ... Tomus Primus* (Paris: Henry Le Gras, 1634), Ad lectorem/Au lecteur (not paginated). On Hérigone, see also Maria Rosa Massa Esteve, "The Roel of Symbolic Language in the Transformation of Mathematics," *Philosophica* 82 (2012): 153–193.

45. William Oughtred, *Key to the Mathematics*, trans. Robert Wood (London: Thomas Harper for Richard Whitaker, 1647), B4i. For discussion of this quotation, see Helena Pycior, *Symbols, Impossible Numbers, and Geometric Entanglements* (Cambridge: Cambridge University Press, 1997), 114.

46. See, e.g., John Pell's three-column tables for organizing algebraic manipulations: Noel Malcolm and Jacqueline Stedall, *John Pell (1611–1685) and His Correspondence with Sir Charles Cavendish: The Mental World of an Early Modern Mathematician* (Oxford: Oxford University Press, 2004), 268 ff.

47. See the title and "Ad lectorem" to Hérigone (1634). Hérigone's text was printed in double-column in both Latin and French and the two are sometimes lightly different. On brevity and clarity, see Cifoletti (1992), *passim*.

48. Cifoletti (1995), 1389–1397. Judgment was another key concept imported into mathematical thinking from the rhetorical tradition, in particular by Ramus.

49. Jacques Peletier, *L'algebre* (Lyon: Jean de Tournes, 1554), 1; on Peletier's understanding of the relationship between Euclidean demonstration and syllogism, see Giovanna Cifoletti, "From Valla to Viète: The Rhetorical Reform of Logic and Its Use in Early Modern Algebra," *Early Science and Medicine* 11, no. 4 (2006): 390–423, at 403–406.

50. Stevin (1585), 169, 232. On Stevin and the rhetorical tradition, see also Jean-Marie Coquard, "Mathématiques et dialectique dans l'oeuvre de Simon Stevin: l'intérêt des series de problèmes," *SHS Web of Conferences* 22, no. 12 (2015).

51. See Gosselin (1577), fol. 23r–24v = Gosselin (2016), 277–281.

52. I cannot pursue this question here. But for near contemporary recognition of the problem see Wallis (1695), 53.

53. On the role of commonplaces as "proofs" in algebra, see Cifoletti (1995), 1391.

54. See Gosselin (1577), fol. 57v ff. = Gosselin (2016), 347 ff.

55. Cf. Xylander's edition, where both the Greek scholiast Maximus Planudes and Xylander himself introduce fractional examples for problems that Diophantus broaches in whole numbers. At one point Planudes explains the practice: "Idque exercitationis causa plenioris demonstremus in numeris fractis" (Diophantus [1575], 35). The juxtaposition between "we demonstrate" and "for the sake of fuller exercise" suggests that "demonstration" does not have the philosophical significance often attributed to the word. On this significance, see discussion of the "question of the certitude of mathematics" in Jardine (1988) and Mancosu (1992).

56. Gosselin (1577), fol. 59v = Gosselin (2016), 351.

57. Boethius' *De institutione arithmetica* was often central to this presentation. See Moyer (2012); Oosterhoff (2018).

58. For one example, see discussion of Peletier in Natalie Zemon Davis, "Peletier and Beza Part Company," *Studies in the Renaissance* 11 (1964): 188–222. For Peletier's philosophy more generally, see Sophie Arnaud, *La voix de la nature dans l'œuvre de Jacques Peletier du Mans (1517–1582)* (Paris: Honoré Champion, 2005).

59. Peletier (1554), 122–123.

60. See Gosselin (2016), 417–419 = fol. 5r/v.

61. Hérigone (1634), "Ad lectorem."

62. Hérigone (1634), "Ad lectorem."

63. For Arabic precedents for Viète's mathematics, see Roshdi Rashed, *The Development of Arabic Mathematics: Between Arithmetic and Algebra*, trans. A. F. W. Armstrong (Dordrecht, Boston, and London: Kluwer Academic Publishers, 1994).

64. See Klein (1992); Michael Mahoney, "The Beginnings of Algebraic Thought in the Seventeenth Century," in *Descartes: Philosophy, Mathematics and Physics*, ed. Stephen Gaukroger (Sussex: Harvester Press, 1980), 141–168; Marco Panza, "What Is New and What Is Old in Viète's *Analysis Restituta* and *Algebra Nova*, and Where Do They Come

From?" *Revue d'histoire des mathématiques* 13 (2007): 85–153; and, for discussion of these alternatives, Rabouin (2009).

65. On parallels between Viète's zetetica and Diophantus' *Arithmetic*, see Paolo Freguglia, "Viète Reader of Diophantus: An Analysis of *Zeteticorum Libri Quinque*." *Bollettino di storia delle scienze matematiche* 28, no. 1 (2008): 51–95.

66. For example, Pell's three-column method mentioned earlier keeps track of just this kind of manipulation. See Malcolm and Stedall (2004).

67. François Viète, *Opera Mathematica*, ed. Franz van Schooten (Leiden: Elsevier, 1646), 1ff.; and in English translation, see François Viète, *The Analytic Art*, trans. T. Richard Witmer (Kent, OH: Kent State University Press, 1983), 11ff.

68. See, e.g., *Zeteticorum libri quinque*, lib. II, zetetica 3–8, in Viète (1646), 51–52.

69. Grafton and Jardine (1986).

Chapter 2

1. Robert Recorde, *The grou[n]d of artes teachyng the worke and practise of arithmetike, moch necessary for all states of men* (London: R. Wolfe, 1543), STC (2nd ed.) 20797.5, accessed on Early English Books Online [EEBO]; Stephen Johnston, "Recorde, Robert (c. 1512–1558)," *Oxford Dictionary of National Biography* (Oxford: Oxford University Press, 2004), doi:10.1093/ref:odnb/23241. When quoting Recorde, I have substituted j for i and v for u where warranted and expanded contractions with bracketed letters, but have otherwise preserved Recorde's spelling. Readers without access to EEBO can follow most of this chapter's discussion of the text with a free digital version of a later edition, e.g. the 1618 edition in Google Books, https://books.google.com/books?id=i8NJomIVzlgC (accessed 2019).

2. Joy B. Easton, "The Early Editions of Robert Recorde's *Ground of Artes*," *Isis* 58, no. 4 (1967): 515–532; John Denniss and Fenny Smith, "Robert Recorde and His Remarkable Arithmetic," in *Robert Recorde: the Life and Times of a Tudor Mathematician*, ed. Gareth Roberts and Fanny Smith (Cardiff: University of Wales Press, 2012), 25–38; John Denniss, "Learning Arithmetic: Textbooks and Their Users in England 1500–1900," in *Oxford Handbook of the History of Mathematics*, ed. Eleanor Robson and Jacqueline Stedall (Oxford: Oxford University Press, 2009), 448–467.

3. Keith Thomas, "Numeracy in Early Modern England," *Transactions of the Royal Historical Society* 37 (1987): 103–132; Francis R. Johnson and Sanford V. Larkey, "Robert Recorde's Mathematical Teaching and the Anti-Aristotelian Movement," *Huntington Library Quarterly* 7 (1935): 59–87; E. G. R. Taylor, *The Mathematical Practitioners of Tudor & Stuart England* (Cambridge: Cambridge University Press, 1954), 15, 167; S. K. Heninger Jr., "Tudor Literature of the Physical Sciences," *Huntington Library Quarterly* 32, no. 2 (1969): 101–133, on 107–110.

4. See, e.g., Florian Cajori, *A History of Mathematical Notations, vol. I: Notations in Elementary Mathematics* (London: Open Court, 1928), 164; Jack Williams, *Robert Recorde: Tudor Polymath, Expositor and Practitioner of Computation*, History of Computing (London: Springer Verlag, 2011); Gordon Roberts, *Robert Recorde: Tudor Scholar and Mathematician* (Cardiff: University of Wales Press, 2016); "Equals Sign," *Wikipedia*, online, https://en.wikipedia.org/w/index.php?title=Equals_sign&oldid=767759138, accessed 2017; J. J. O'Connor and E. F. Robertson, "Recorde Summary," *MacTutor History of Mathematics*, http://www-history.mcs.st-and.ac.uk/Mathematicians/Recorde.html, accessed 2017. Without

being able to say with certainty whether Recorde knew of them, Jens Høyrup identifies Italian manuscript examples predating Recorde's print use by more than half a century. Jens Høyrup, "Hesitating Progress—The Slow Development toward Algebraic Symbolization in Abbacus- and Related Manuscripts, c. 1300 to c. 1550," in *Philosophical Aspects of Symbolic Reasoning*, ed. Albrecht Heeffer and Maarten Van Dyck (London: College Publications, 2010), 3–56, on 36–37.

5. Lisa Wilde, "'Whiche elles shuld farre excelle mans mynde': Numerical Reason in Robert Recorde's Ground of Artes (1543)," *Journal of the Northern Renaissance* 6 (2014), online, http://www.northernrenaissance.org/whiche-elles-shuld-farre-excelle-mans-mynde-numerical-reason-in-robert-recordes-ground-of-artes-1543/; Roberts and Smith, eds. (2012); John V. Tucker, "Foreword: Robert Recorde and the History of Computing," in Williams (2011), v–xi.

6. On Recorde's vernacularization, see Michael J. Barany, "Translating Euclid's Diagrams into English, 1551–1571," in *Philosophical Aspects of Symbolic Reasoning in Early Modern Mathematics*, ed. Heeffer and Van Dyck (2010), 125–163, on 128–137.

7. "Augrim," *The Middle English Dictionary*, online, http://quod.lib.umich.edu/cgi/m/mec/med-idx?type=id&id=MED3019, accessed 2017; "algorism, n." and "arithmetic, n.1," *OED Online* (Oxford University Press, 2016). Nia M. W. Powell, "The Welsh Context of Robert Recorde," in Roberts and Smith, eds. (2012), 123–144, on 130–131, 140 (n. 39).

8. Ibid.

9. "Arse" is a British rendering of "ass" specific to the latter's vulgar anatomical (as opposed to zoological) meaning. On the ass/arse pun in Early Modern English, see Will Stockton, "'I Am Made an Ass': Falstaff and the Scatology of Windsor's Polity," *Texas Studies in Literature and Language* 49, no. 4 (2007): 340–360.

10. Glending Olson, "Measuring the Immeasurable: Farting, Geometry, and Theology in the *Summoner's Tale*," *Chaucer Review* 43, no. 4 (2009): 414–427; Timothy D. O'Brien, "'Ars-Metrik': Science, Satire and Chaucer's Summoner," *Mosaic: An Interdisciplinary Critical Journal* 23, no. 4 (1990): 1–22.

11. Valerie Allen, *On Farting: Language and Laughter in the Middle Ages* (New York: Palgrave Macmillan, 2007).

12. The preface is unpaginated. This quotation is from the verso page beginning "it is thonely thyng (all most) that separateth man from beastes," image 8 in the EEBO scan of the book. On Recorde's dialogue form, see Johnson and Larkey (1935). On the contradictions and complexities of dialogue as a pedagogical genre, from Plato onward, see Daniel Boyarin, *Socrates and the Fat Rabbis* (Chicago: University of Chicago Press, 2009).

13. Henry Cowles, *A Method Only: The Evolving Meaning of Science in the United States, 1830–1910* (PhD diss., Princeton University, 2015), on 235–247.

14. See Matthew L. Jones, "How We Became Instrumentalists (Again): Data Positivism since World War II," *Historical Studies in the Natural Sciences* 48, no. 5 (2018): 673–684.

15. Donald MacKenzie, *Mechanizing Proof: Computing, Risk, and Trust* (Cambridge, MA: MIT Press, 2001).

16. Francis R. Johnson, "Latin versus English: The Sixteenth-Century Debate over Scientific Terminology," *Studies in Philology* 41, no. 2 (1944): 109–135, on 114.

17. Robert Recorde, *The ground of artes teaching the woorke and practise of arithmetike, both in whole numbres and fractions, after a more easyer and exacter sorte than any lyke hath hytherto beene set forth: with divers new additions* (London: R. Wolfe, 1558), STC (2nd ed.)

20799.5, accessed on Early English Books Online (2017). The pagination in this EEBO edition is unclear; the quoted passage is on image 29 of the EEBO record.

18. Cf. Chaucer's invocation, c. 1390, of "nowmbres of augrym," which Powell (2012, 130) identifies as among the earliest English references to Hindu-Arabic numerals.

19. See, e.g., Jay Mathews, "10 Myths (Maybe) about Learning Math," *Washington Post*, May 31, 2005; David Klein and R. James Milgram, "The Role of Long Division in the K–12 Curriculum," 2000, online (accessed 2017), http://www.csun.edu/~vcmth00m/longd ivision.pdf; Patricia A. Sellers, "The Trouble with Long Division," *Teaching Children Mathematics* 16, no. 9 (2010): 516–520.

20. Here the Master teases the Scholar when the latter asks if the method works equally well for small currencies like halfpennies, farthings, kewes (eighth-pennies, denoted "q"), and cees (sixteenth-pennies, denoted "c"): "You thynke you be at Oxforde styll, you brynge forthe so faste your q, and c" (36^r).

21. Robert Recorde, *The pathway to knowledg, containing the first principles of Geometrie, as they may moste aptly be applied unto practise, bothe for use of instrumentes Geometricall, and astronomicall and also for projection of plattes in everye kinde, and therfore muche necessary for all sortes of men* (London: R. Wolfe, 1551), STC (2nd edn.) 20812, accessed on Early English Books Online and in the Cambridge University Library. See Barany (2010).

22. On this contrast, which remained a key to interpreting mathematical discourse at least through the nineteenth century, see Paolo Mancosu, *Philosophy of Mathematics and Mathematical Practice in the Seventeenth Century* (Oxford: Oxford University Press, 1996), ch. 1–3; Michael J. Barany, "God, King, and Geometry: Revisiting the Introduction to Cauchy's *Cours d'analyse*," *Historia Mathematica* 38 (2011): 368–388, on 372–379.

23. See Joel Kaye, "Money and Administrative Calculation as Reflected in Scholastic Natural Philosophy," in *Arts of Calculation: Quantifying Thought in Early Modern Europe*, ed. David Glimp and Michelle R. Warren (New York: Palgrave Macmillan, 2004), 1–18.

24. Allen (2007).

25. O'Brien (1990), esp. 2–8.

26. Allen (2007), 2.

27. Olson (2009), 415–419; Kaye (2004).

28. Geoffrey Chaucer, *The Canterbury Tales*, ed. F. N. Robinson (1957), University of Michigan Corpus of Early English Prose and Verse, https://quod.lib.umich.edu/c/cme/CT/1:3.6?rgn=div2;view=fulltext, accessed 2019, III:2222–2223.

29. Olson (2009); O'Brien (1990); Amir Alexander, "The Imperialist Space of Elizabethan Mathematics," *Studies in the History and Philosophy of Science* 26, no. 4 (1995): 559–591.

30. O'Brien (1990), 18.

31. Ben Jonson, *The New Inne* (London: Thomas Harper, 1631), STC (2nd ed.) 14780, EEBO, Act II, Scene 5.

32. Matthew Jones, *Reckoning with Matter: Calculating Machines, Innovation, and Thinking about Thinking from Pascal to Babbage* (Chicago: University of Chicago Press, 2016).

33. See Ludwig Wittgenstein, *Remarks on the Foundations of Mathematics*, trans. G. E. M. Anscombe, ed. von Wright, Rhees, and Anscombe (Oxford: Basil Blackwell, 1956); Michael Lynch, "Extending Wittgenstein: The Pivotal Move from Epistemology to the Sociology of Science," in *Science as Practice and Culture*, ed. A. Pickering (Chicago: University of Chicago Press), 215–265. Tallies and finger-counting would play a pivotal role in late

nineteenth-century debates about the origins and meaning of modernity; Michael J. Barany, "Savage Numbers and the Evolution of Civilization in Victorian Prehistory," *British Journal for the History of Science* 47, no. 2 (2014): 239–255.

34. Nicolas Bourbaki, *Elements of Mathematic: Theory of Sets* (Paris: Hermann, 1968), 7–8.

35. Michael J. Barany, "Integration by Parts: Wordplay, Abuses of Language, and Modern Mathematical Theory on the Move," *Historical Studies in the Natural Sciences* 48, no. 3 (2018): 259–299.

36. See, e.g., among many explicitly to consider determinist fantasies of formal calculation as symptomatic of modernity's mystifications, Max Horkheimer and Theodor W. Adorno, *Dialectic of Enlightenment: Philosophical Fragments*, trans. Edmund Jephcott, ed. Gunzelin Schmid Noerr (Stanford: Stanford University Press, 2002).

Chapter 3

1. On the role of Galileo and his disciples, and in particular Bonaventura Cavalieri and Evangelista Torricelli, in developing and disseminating the method of indivisibles see Amir Alexander, *Infinitesimal: How a Dangerous Mathematical Theory Shaped the Modern World* (New York: Farrar, Straus & Giroux / Scientific American, 2014), ch. 3, "Mathematical Disorder," 80–117.

2. Bonaventura Cavalieri, *Exercitationes Geometricae Sex* (Bologna: Iacob Monti, 1647), Book 1, prop. 19, p. 35.

3. On the fight over the method of indivisibles, which extended to issues of religious and political authority, see Alexander, *Infinitesimal*, esp. chap. 4, "Destroy or Be Destroyed: The War on the Infinitely Small," 118–149, and chap. 5, "The Battle of the Mathematicians," 150–180.

4. For Cavalieri's calculation of the area enclosed in an Archimedean spiral see Bonaventura Cavalieri, *Geometria Indivisibilibus Libri VI* (Bologna: Clementis Ferroni, 1635), prop. 19, p. 238. Torricelli's calculation of the slope of an "infinite parabola" can be found in Gino Loria and Giuseppe Vassura, eds., *Opere di Evangelista Torricelli* (Faenza: G. Montanari, 1919–1944), vol. 1, part 2, 322–333. For a discussion of Cavalieri's and Torricelli's procedures see Alexander, *Infinitesimal*, 99–101, 114–117.

5. See J. E. McGuire and Martin Tamny, *Certain Philosophical Questions: Newton's Trinity Notebook* (Cambridge: Cambridge University Press, 1983).

6. McGuire and Tamny, *Certain Philosophical Questions*, 337.

7. McGuire and Tamny, *Certain Philosophical Questions*, 343.

8. Galileo Galilei, "The Assayer," in *The Discoveries and Opinions of Galileo*, ed. and trans. Stillman Drake (New York: Anchor Books, 1957), 217–280, at 237–238.

9. McGuire and Tamny, *Certain Philosophical Questions*, 344–347.

10. McGuire and Tamny, *Certain Philosophical Questions*, 347.

11. Isaac Newton, "A Method for finding theorems concerning Quaestiones de Maximis et Minimis," in *The Mathematical Papers of Isaac Newton*, ed. D. T. Whiteside (Cambridge: Cambridge University Press, 1967), I:272–273. Newton did not use the term "subnormal," but that is the modern equivalent for the magnitude he was seeking.

12. On Newton's discovery of the binomial theorem by expanding on the Pascal triangle see Niccolò Guicciardini, *Isaac Newton on Mathematical Certainty and Method* (Cambridge, MA: MIT Press, 2009) 148–152.

13. Isaac Newton, *De Analysi per aequationes infinitas*, in *The Mathematical Papers of Isaac Newton*, ed. Whiteside, II:202–245. For a comprehensive discussion of Newton's binomial theorem and the *De Analysi* see Guicciardini, *Isaac Newton*, 148–167.

14. Newton, *De Analysi*, in *Mathematical Papers*, ed. Whiteside, II:243–245.

15. See Guicciardini, *Isaac Newton*, 180–182. The text of *De Methodis Serierum et Fluxionum* can be found in Whiteside, *Mathematical Papers*, vol. 3.

16. According to Marco Panza, Newton's progression from the techniques of *De Analysi* to the Method of Fluxions in *De Methodis* involved two distinct steps: first, he redefined the curves of *De Analysi* as the trajectories of moving points, independent of any specific polynomial equation. Second, he redefined the curve trajectories and the speed of their generation as "fluents" and "fluxions" respectively, these being universal abstract quantities, whose relationships are independent of any particular curve or motion. Newton's progression is from particular relationships, which were detected in the world, to universal relationships, applicable to any possible quantities. See Marco Panza, "From Velocities to Fluxions," in *Interpreting Newton: Critical Essays*, ed. Andrew Janiak and Eric Schliesser (Cambridge: Cambridge University Press, 2012), 219–254.

17. Newton, *De Quadratura Curvarum*, in *Mathematical Papers*, ed. Whiteside, VIII:122–135.

18. Newton, *De Quadratura Curvarum*, 123.

19. See discussion in Guicciardini, *Isaac Newton*, 226–227.

20. Newton, *De Quadratura Curvarum*, 127–129; discussed in Guicciardini, *Isaac Newton*, 226–227. Note that unlike in modern notation, in which n signifies a natural number, for Newton it could also signify fractions, including negative ones.

21. Berkeley's famous quip can be found in George Berkeley, *The Analyst, or A Discourse Addressed to an Infidel Mathematician* (Dublin: S. Fuller, 1734), 57.

22. The fact that a mathematical system is founded on rigorous proofs does not, in itself, guarantee that it will be amenable to treatment by standard algorithms. Euclidean geometry, for example, which for millennia set the standard for truth and certainty founded on rigorous proof, hardly ever makes use of standard procedures, treating each object in accordance with its unique geometrical properties. The reverse, however, is certainly the case: if a standard algorithm can be proven to apply to a mathematical field (e.g., of polynomials), then this field must be regular, predictable, and amenable to mathematical proof.

23. Newton, *De Quadratura Curvarum*, 123.

24. Isaac Newton, "Preface to the Reader," in *The Principia: Mathematical Principles of Natural Philosophy*, ed. and trans. I. Bernard Cohen and Anne Whitman (Berkeley: University of California Press, 1999), 382. First published by the Royal Society of London in 1687.

25. On Henry More's influence on Newton's early thought, see J. E. McGuire and Martin Tamny commentary in *Certain Philosophical Questions*, ch. 1.

26. On the Cambridge Platonists and their contention that the world was pervaded by a "spirit of nature" whose operation could not be explained by regular laws or mechanical principles, see Danton B. Sailor, "Cudworth and Descartes," *Journal of the History of Ideas* 23, no. 1 (January–March, 1962): 133–140; Michael Boylan, "Henry More's Space and the Spirit of Nature," *Journal of the History of Philosophy* 18, no. 4 (October 1980): 395–405; and Scott Mandelbrote, "The Uses of Natural Theology in Seventeenth-Century England," *Science in Context* 20, no. 3 (2007): 451–480. On the Cambridge Platonists' influence on the young Newton, and his gradual rejection of their views on nature, see J. E. McGuire and P. M. Rattansi, "Newton and the 'Pipes of Pan,'" *Notes and Records of the Royal Society of London* 21, no. 2 (December 1966): 108–143; J. E. McGuire, "Force, Active Principles, and Newton's Invisible Realm," *Ambix* 15, no. 3 (1968): 154–208; P. M. Rattansi, "Reason in Sixteenth

and Seventeenth Century Natural Philosophy," in *Changing Perspectives in the History of Science*, ed. M. Teich and R. M. Young (London: Heinemann Educational, 1973), 148–166; and John Gascoigne, "The Universities and the Scientific Revolution: The Case of Newton and Restoration Cambridge," *History of Science* 23, no. 4 (December 1985): 391–434.

27. The early Royal Society's preference for Baconian empiricism is set out in Thomas Sprat, *History of the Royal Society of London* (London, 1667); in the experiments and writings of Robert Boyle, the Society's most illustrious member in those years; and in the works of Robert Hooke and other Society fellows. See also Steven Shapin and Simon Schaffer, *Leviathan and the Air-Pump: Hobbes, Boyle, and the Experimental Life* (Princeton, NJ: Princeton University Press, 1985); Barbara J. Shapiro, *Probability and Certainty in Seventeenth Century England* (Princeton, NJ: Princeton University Press, 1983); Steven Shapin, *A Social History of Truth: Civility and Science in Seventeenth Century England* (Chicago: University of Chicago Press, 1995); and Amir Alexander, *Infinitesimal: The Dangerous Mathematical Theory That Shaped the Modern World* (New York: Farrar, Straus & Giroux / Scientific American, 2014).

28. Newton's reservations about the spirit-imbued world of the Cambridge Platonists is already in evidence in a tract that became known as *De Gravitatione et Equipondio Fluidorum* written around 1670. See McGuire and Rattansi, "Newton and the Pipes of Pan," 124. But his most public rejection of these views came in the "General Scholium" to the second edition of the *Principia* (1713).

29. In the "General Scholium" Newton wrote: "And thus much concerning God; to discourse of whom from the appearances of things, does certainly belong to Natural Philosophy."

30. On the positive view of miracles among early members of the Royal Society see Mandelbrote, "The Uses of Natural Theology."

31. Newton, "General Scholium."

32. On the political implications of Newtonianism see Margaret C. Jacob, *The Newtonians and the English Revolution, 1689–1720* (New York: Gordon and Breach, 1990); and Steven Shapin, "Of Gods and Kings: Natural Philosophy and Politics in the Leibniz-Clarke Disputes," *Isis* 72, no. 2 (June 1981): 187–215.

Chapter 4

1. "algorithm, n." *Oxford English Dictionary Online*. Oxford University Press, December 2019, available at https://www.oed.com/view/Entry/4959 (accessed January 14, 2020); "algorithme," *Le Grand Robert de la langue française: dictionnaire alphabétique et analogique de la langue française*, 2nd ed., ed. Alain Rey (Paris: Le Robert, 2001), https://grandrobert-lerobert-com.ezp2.lib.umn.edu/robert.asp; Niccolò Tommaseo, *Dizionario della lingua italiana* (Torino: Soc. L'Unione Tipografico-Editrice, 1865), http://www.tommaseobellini.it/#/.

2. Edmond Huguet, *Dictionnaire de la langue française du seizième siècle* (Paris: Honoré Champion, 1925–1956).

3. Ibid.

4. "algorithm," OED.

5. Ibid.

6. "algoritmo," in *Grande Dizionario Italiano dell'Uso*, ed. Tullio del Mauro (Torino: UTET, 1999–2007).

7. Niccolò Guicciardini, *The Development of Newtonian Calculus in Britain, 1700–1800* (Cambridge: Cambridge University Press, 1990); Guicciardini, *Newton on Mathematical*

Certainty and Method (Cambridge, MA: MIT Press, 2009); Douglas Jesseph, *Berkeley's Philosophy of Mathematics* (Chicago: University of Chicago Press, 1993); Morris Kline, *Mathematics: The Loss of Certainty* (Oxford: Oxford University Press, 1980); Paolo Mancosu, *The Philosophy of Mathematics & Mathematical Practice in the Seventeenth Century* (Oxford: Oxford University Press, 1999); J. B. Shank, *Before Voltaire: The French Origins of "Newtonian" Mechanics, 1680–1715* (Chicago: University of Chicago Press, 2018).

8. "algorithm," OED.

9. Ibid.

10. Ibid.

11. Ibid.

12. Ibid.

13. *Commercium epistolicum D. Johannis Collins, et aliorum de analysi promota: jussu Societatis Regiæ in lucem editum* (London: Pearson, 1712). See also "An account of the book entituled [*sic*] *Commercium Epistolicum Collinii & Aliorum, De Analysi Promota*; published by order of the Royal-Society, in relation to the dispute between Mr. Leibnitz and Dr. Keill, about the right of invention of the method of fluxions, by same call'd the differential method," *Philosophical Transactions of the Royal Society of London* 29, no. 342 (December 1714).

14. On the calculus priority dispute overall, see Guicciardini, *Newton on Mathematical Certainty and Method*, chs. 16–17; A. R. Hall, *Philosophers at War: The Quarrel between Newton and Leibniz* (Cambridge: Cambridge University Press, 1980).

15. On these epistemological complexities, see especially Guicciardini, *Newton on Mathematical Certainty and Method*, chs. 16–17. Also Shank, *Before Voltaire*, chs. 5, 7, and 9, and Shank, "A French Jesuit in the Royal Society of London: Father Louis-Bertrand de Castel, S.J. and Enlightenment Mathematics, 1720–1735," *Journal of Early Modern Studies* 1 (November 2012): 151–184.

16. "algorithm," OED.

17. On this, see esp. Guicciardini, *Newton on Mathematical Certainty and Method*, chs. 16–17.

18. Niccolò Guicciardini, "Did Newton Use His Calculus in the *Principia*?" *Centaurus* 40, no. 3–4 (October 1998): 303–344.

19. Shank, *Before Voltaire*.

20. Jean Sylvain Bailly, *Histoire de l'astronomie moderne depuis la fondation de l'école d'Alexandrie, jusqu'à l'époque de M.D.CC.XXX [M.D.CC.XXXII]*, 3 vols. (Paris, 1779–1782; new edition, 1785), 471–472.

21. Ibid., 472–473.

22. Joseph-Louis Lagrange, *Mécanique analytique* (Paris: La Veuve Desaint, 1788), 159.

23. See esp. Craig Fraser, *Calculus and Analytical Mechanics in the Age of Enlightenment* (Aldershot, Hampshire: Ashgate, 1997); Sandro Caparrini and Craig Fraser, "Mechanics in the Eighteenth Century," in *Oxford Handbook of the History of Physics*, ed. Jed Buchwald and Robert Fox (Oxford: Oxford University Press, 2013), 358–405

24. Thomas Hankins, *Science and the Enlightenment* (Cambridge: Cambridge University Press, 1985); Henry Guerlac, *Lavoisier. The Crucial Year. The Background and Origin of His First Experiments on Combustion in 1772* (Ithaca, NY: Cornell University Press, 1961), and more recently Mi Gyung Kim, *Affinity, That Elusive Dream: A Genealogy of the Chemical Revolution* (Cambridge, MA: MIT Press, 2003); John L. Heilbron, *Electricity in the 17th & 18th Centuries: A Study in Early Modern Physics* (Berkeley: University of California Press, 1979).

25. "algorithm," OED.

26. "algorithme," *Dictionnaire de l'Académie Française*, 4th ed. (Paris: Chez la Veuve B. Brunet, 1762).

27. "algorithme," *Dictionnaire de l'Académie Française*, 6th ed. (Paris: Firmin Didot Frères, 1835).

28. "algorithme," *Le Grand Robert*.

29. "algoritmo," in Del Mauro ed., *Grande Dizionario Italiano dell'Uso*.

30. Jean Piaget, *Epistimologie des Sciences de l'Homme* (Paris: Gallimard 1971), 90.

31. Eric W. Weisstein, "Binet's Fibonacci Number Formula," *MathWorld*, http://mathworld.wolfram.com/BinetsFibonacciNumberFormula.html.

32. E. Roy Weintraub, *How Economics Became a Mathematical Science* (Durham, NC: Duke University Press, 2002).

33. See Keith Michael Baker, "Enlightenment and the Institution of Society: Notes for a Conceptual History," in *Civil Society. History and Possibilities*, ed. Sudip Kaviraj and Sunil Khalnani (Cambridge: Cambridge University Press, 2001), 84–104; Margaret Schabas, *The Natural Origins of Economics* (Chicago: University of Chicago Press, 2007).

34. See esp. Foucault's lectures at the Collège de France from the late 1970s, now translated into English as: *Society Must Be Defended. Lectures at the Collège de France 1975–76* (New York: Picador, 2003); *Security, Territory, Population. Lectures at the Collège de France 1977–78* (New York: Picador, 2009); *The Birth of Biopolitics. Lectures at the Collège de France 1978–1979* (New York: Picador, 2010) ; *On the Government of the Living. Lectures at the Collège de France 1979–1980* (New York: Picador, 2016).

35. Ian Hacking, *The Taming of Chance* (Cambridge: Cambridge University Press, 1990); Lorraine Daston, *Classical Probability and the Enlightenment* (Princeton, NJ: Princeton University Press, 1995).

36. Ronald Meek, *The Economics of Physiocracy* (Cambridge, MA: Harvard University Press, 1962); Georges Weulersse, *La Physiocratie à la Fin du Règne de Louis XV* (Paris: Presses Universitaires de France, 1959); Weulersse, *La Physiocratie sous les Ministres Turgot et Necker* (Paris: Presses Universitaires de France, 1950); Elizabeth Fox Genovese, *The Origins of Physiocracy: Economic Revolution and Social Order in Eighteenth-Century France* (Ithaca, NY: Cornell University Press, 1976).

37. Loic Charles, "The Visual History of the *Tableau Économique*," *European Journal of the History of Economic Thought* 10, no. 4 (December 2003): 527–550.

38. Steven L. Kaplan, *Bread, Politics, and Political Economy in the Reign of Louis XV*, 2 vols. (Dordrecht: Springer, 1976).

39. Judith Miller, *Mastering the Market: The State and the Grain Trade in Northern France, 1700–1860* (Cambridge: Cambridge University Press, 1998).

40. Steven L. Kaplan, "The Famine Plot Persuasion in Eighteenth-Century France," *Transactions of the American Philosophical Society* (December 1982).

41. For a fresh take on the scientific modernity of Adam Smith and his economic thought, see Eric Schliesser, *Adam Smith: Systematic Philosopher and Public Thinker* (Oxford: Oxford University Press, 2017).

42. Adam Smith, *An Inquiry Concerning the Nature and Causes of the Wealth of Nations* (London: Methuen & Co. Ltd., 1776), Ch. I—"On the Division of Labor."

43. Donald Winch, *Riches and Poverty: An Intellectual History of Political Economy in Britain, 1750–1834* (Cambridge: Cambridge University Press, 1996).

44. Monique Pelletier, *Les cartes des Cassini. La science au service de l'Etat et des régions* (Paris: Comité des travaux historiques et scientifiques, 2002).

45. Jean Boutier, *Les Plans de Paris*, 2nd ed. (Paris: Bibliothèque nationale de France, 2007).

46. *Mémoire sur la réformation de la police de France. Soumis au roi en 1749 par m. Guillauté, officier de la Maréchaussée de l'Ile-de-France. Illustré de 28 dessins de Gabriel de Saint-Aubin. Introduction et notes par Jean Seznec* (Paris: Editions Hermann, 1974).

47. Claude Nicolas Ledoux, *L'architecture considérée sous le rapport de l'art, des moeurs et de la legislation* (Paris: H. L. Perronneau, 1804); Anthony Vidler, *Claude-Nicolas Ledoux: Architecture and Social Reform at the End of the Ancien Régime* (Cambridge, MA: MIT Press, 1990).

48. The most comprehensive analysis of Ledoux's dream is found in Vidler, *Ledoux*.

49. Michel Foucault, *Discipline and Punish: The Birth of the Prison*, trans. Alan Sheridan (New York: Vintage Books, 1995).

50. Keith Michael Baker, *Condorcet. From Natural Philosophy to Social Mathematics* (Chicago: University of Chicago Press, 1974), esp. Part II. On the Arrow-Condorcet connection see Christian List, "Social Choice Theory," *The Stanford Encyclopedia of Philosophy* (Winter 2013 Edition), https://plato.stanford.edu/archives/win2013/entries/social-choice/.

51. Theodor Adorno and Max Horkheimer, *The Dialectic of Enlightenment*, 2nd ed. (London: Verso, 1997), 26–27.

52. Ibid., 6.

Chapter 5

1. This essay is based on "Numbers for the Innumerate: Everyday Arithmetic and Atlantic Capitalism," originally published in *Technology & Culture* (April 2017). I am grateful to the journal and Johns Hopkins University Press for the opportunity to reprint a revised and expanded version here.

2. For some of the many recent perspectives on algorithms, see, e.g., Brian Christian and Tom Griffiths, *Algorithms to Live by*; Safiya Noble, *Algorithms of Oppression*; Ruha Benjamin, *Race after Technology*; Christopher Steiner, *Automate This*; Ed Finn, *What Algorithms Want*; Frank Pasquale, *The Black Box Society*; Virginia Eubanks, *Automating Inequality*; Cathy O'Neill, *Weapons of Math Destruction*. For reviews of the emerging critical history of data, see Theodora Dryer, "The New Critical History of Surveillance and Human Data."

3. Noah Webster, *A Dictionary of the English Language*, 381.

4. Peter Barlow et al., *Encyclopaedia of Pure Mathematics*, 510.

5. Edward Cocker and J. Hawkins, *Cockers Arithmetick*, 102.

6. Adams, *The Scholar's Arithmetic*.

7. Emmor Kimber, *Arithmetic Made Easy to Children* (Philadelphia: Kimber, Conrad, & Co., 1805), 222–230.

8. Kimber, *Arithmetic Made Easy to Children*, 222–230.

9. Kimber, *Arithmetic Made Easy to Children*, 109.

10. For many examples see "Record of Indentures of Individuals Bound Out as Apprentices."

11. Smyth, *The New American Clerk's Instructor*, 51; Montefiore, *The American Trader's Compendium*, 27; Member of the Philadelphia Bar, *The Form Book*, 226.

12. *Abridgment of the Public Permanent Laws of Virginia*, 148. Littell, *The Statute Law of Kentucky*, 192, 677. Over the course of the nineteenth century, many other states adopted similar provisions. For example, an Ohio "Act Concerning Apprentices and Servants" passed in 1831 specified that male servants bound to serve five years or more must be

taught "to read and write and so much arithmetic as will include the single rule of three."
Public Statutes at Large of the State of Ohio, 2479–2480.

13. Cocker and Hawkins, *Cocker's Arithmetick*, 103.

14. William Stokes, *Stokes's Rapid Arithmetic*, 78–79.

15. Cohen, *A Calculating People*, 121.

16. Francis Place, *The Autobiography of Francis Place*, 48.

17. Page, "Advancement in the Means and Methods of Public Instruction," 127–129.

18. *Southern Literary Messenger*, 445. For another critique of methods of teaching by rote, see *Pennsylvania School Journal*. Pennsylvania State Education Association, 107–108.

19. Rothenberg, *From Market-Places to a Market Economy*, 72–76.

20. Lincoln, fragment from copy book.

21. Lincoln, *Words of Lincoln*, 40.

22. Olaudah Equiano, *The Interesting Narrative of the Life of Olaudah Equiano; or, Gustavus Vassa, the African* (The Author, 1789), 171. Wickman, "Arithmetic and Afro-Atlantic Pastoral Protest," 192–195.

23. G. W. Offley, *A Narrative of the Life and Labors of the Rev. G. W. Offley*, 11–12.

24. Bragg, *Men of Maryland*, 92. Pennington's own narrative does not mention the rule, though he described arithmetic as a favorite subject. Pennington, *The Fugitive Blacksmith*, 43.

25. Wickman, "Arithmetic and Afro-Atlantic Pastoral Protest," 194.

26. "Commensuration as a Social Process," 315.

27. On cost-benefit analysis, see Porter, *Trust in Numbers*; on commensuration within the household, Zelizer, *The Social Meaning of Money*; on commodity grading, Cronon, *Nature's Metropolis*; and on commensuration in legal contexts, Sunstein, "Incommensurability and Valuation in Law."

28. *The Merchants and Shipmaster's Assistant*, 1822, ch. II, 42.

29. *The Merchants and Shipmaster's Assistant*, ch. II, 42.

30. McCusker, "The Demise of Distance."

31. Weston, *The Complete Merchant's Clerk*.

32. *The Merchants and Shipmaster's Assistant*, ch. III, 65.

33. Nathaniel Allen, Cyphering Book. 1805. [AAS]

34. Sarah Pollock, "Practical Arithmetic." 1810. [AAS]. For more on Pollock and similar examples, see chapter 2 Rachel Knecht, "Visionary Calculations: Inventing the Mathematical Economy in Nineteenth Century America," Brown University, dissertation filed May 2018.

35. Edward Cocker, *Cocker's Arithmetic*, 37.

36. Daniel Fenning, *The British Youth's Instructor*, 3–4 and frontmatter.

37. Alfred Arnold, *Arnold's Ready Reckoner*. For the use of narrative instructions to solve more complex problems, such as calculations of compound interest, see Deringer, William. "Compound Interest Corrected: The Imaginative Mathematics of the Financial Future in Early Modern England." *Osiris* 33:1, 2018.

38. C. P. Huestis, *The Mechanics' and Laborers' Ready Reckoner*, 4.

39. John Calvert, *Calvert's Pocket Wages Table, 56½ Hours, and Compendium of the New Factory Act, 1874*.

40. *The "Fifty-One Hour" Wages Reckoner*.

41. Thomas Jefferson, *Report of the Secretary of State on the Subject of Establishing a Uniformity in the Weights, Measures and Coins of the United States*. Other manuals targeting women sometimes also included the rule. See Hannah Glasse, *The Servant's Directory or Housekeeper's Companion*.

42. Booksellers Association of Great Britain and Ireland and Publishers' Association, *Bookseller: The Organ of the Book Trade* (J. Whitaker, 1871), 690.

43. Marx, *Capital*, 13.

44. *The Accountant* (Lafferty Publications, 1887), 147. For reference, see also "Ready Reckoners" by Bruce O. B. Williams and Roger G. Johnson in "IEEE Annals of the History of Computing."

45. Warwick, "The Laboratory of Theory; or, What's Exact about the Exact Sciences?" 318.

46. Dodd, "Calculating and Registering Machines," 11, [83].

47. "A New Calculating Machine," *Eclectic Magazine*, May 1857.

48. For two different approaches to the expanding language of capital in the early and late nineteenth century, see Tamara Thornton, *Nathaniel Bowditch and the Power of Numbers*; and Eli Cook, *The Pricing of Progress*. For a parallel analysis of another nineteenth-century calculative technology, see Michael Zakim, who has described bookkeeping as "ideology," arguing that it served to legitimize a new capitalist social order. As he writes, the detailed procedures of accounting made it appear as a "an island of disinterested neutrality" amid a "tidal wave of profit seeking." Bookkeeping put everything in its proper place, "naturalizing and then taming the market"; Zakim, "Bookkeeping as Ideology." See also "The Business Clerk as Social Revolutionary; or, A Labor History of the Nonproducing Classes.". Ian Baucom goes even further than Zakim in his analysis of the story of the *Zong*—a slaving ship where men and women were tossed overboard and later used to claim an insurance payout. In Baucom's analysis, the tragedy was a "catastrophically exemplary event" in the arrival of modernity—a reflection of the elevation of the language of financial capital and the culture of speculation to "dominant aesthetic modes." Baucom, *Specters of the Atlantic*.

49. Benjamin, *Race after Technology*, 48; see also Noble, *Algorithms of Oppression*.

50. Noble, *Algorithms of Oppression*, 165.

51. Benjamin, *Race after Technology*, 185.

Chapter 6

1. George Boole, "On a General Method of Analysis," *Philosophical Transactions of the Royal Society of London* 134 (1844): 225–282.

2. Luis Laita, "The Influence of Boole's Search for a Universal Method of Analysis on the Creation of his Logic," *Annals of Science* 34 (1977): 163–176

3. So, for example, in his "On a General Method in Analysis," Boole designates the differential operator, d/dx, D so that a linear homogeneous differential equation of the form $a(d^2y/dx^2) + b(dy/dx) + c = 0$ can be written as an algebraic equation $aD^2 + bD + c = 0$. The quadratic equation can then be solved and its roots, let's call them p and q, will immediately suggest solutions to the original equation, $y = e^{px}$ and $y = e^{qx}$.

4. Uta C. Merzbach and Carl B. Boyer, *A History of Mathematics* 3rd ed. (New Jersey: John Wiley & Sons, 2011), 509. George Boole would generalize the method of treating the differential operator as if it were a symbol of quantity further in his *Treatise on Differential Equations* (Cambridge: Macmillan & Co., 1859).

5. Jeremy Gray, "Overstating Their Case? Reflections on British Pure Mathematics in the 19th Century," in *Mathematics in Victorian Britain*, ed. Raymond Flood, Adrian Rice, and Robin Wilson (Oxford: Oxford University Press, 2011), 397–414, on 414.

6. For a wonderful historical examination of Cauchy's calculus, see Judith Grabiner, *The Origins of Cauchy's Rigorous Calculus* (New York: Dover, 1981).

7. Michaela Giebelhausen and Tim Berringer, eds., *Writing the Pre-Raphaelites: Text, Context, Subtext* (New York: Routledge, 2016). Marcia Werner, *Pre-Raphaelite Painting and Nineteenth-Century Realism* (Cambridge: Cambridge University. Press, 2005. Elizabeth Prettejohn, *The Art of the Pre-Raphaelites* (Princeton, NJ: Princeton University Press, 2000)

8. Donald MacKenzie, *Mechanized Proof: Computing, Risk, and Trust* (Cambridge, MA: MIT Press, 2003), 149.

9. David Epstein, Silvio Levy, and Rafael de la Llave, "About This Journal," *Experimental Mathematics* 1 (1992): 1–3, on 1. See also David Epstein and Silvio Levy, "Experimentation and Proof in Mathematics," *Notices of the AMS* 42 (1995): 670–674.

10. Ian Hacking, *Why Is There Philosophy of Mathematics at All?* (Cambridge: Cambridge University Press, 2014), 64.

11. *Logistica* is therefore employed as an analytical category. Boole and De Morgan would perhaps have found problems with the characterization. For example, in a letter to banker, barrister, mathematician, and astronomer Sir John William Lubbock, Boole wrote that "the province of the [human astronomical] computer may certainly be separated from that of the mathematician." Boole to Lubbock, February 16, 1849, Lubbock papers in the Royal Society Archive, LUB. B. 362. As Tony Crilly has demonstrated, many "pure mathematicians" of the Victorian period such as Arthur Cayley were committed Platonists (and thus would see themselves very much in the *Arithmetica* camp). The mottos on the volumes of the *CMJ* journals were from Plato: Tony Crilly, "The Cambridge Mathematical Journal and Its descendants: The Linchpin of a Research Community in the Early and Mid-Victorian Age," *Historia Mathematica* 31 (2004): 455–497. See also Crilly's biography, *Arthur Cayley: Mathematician Laureate of the Victorian Age* (Baltimore: Johns Hopkins University Press, 2006), 87, 383. But on the other hand, Peacock, Boole, and De Morgan had a strong commitment to Locke; see M. J. Durand-Richard, "Genèse de l'algèbre symbolique en Angleterre: Une influence possible de John Locke," *Revue d'Histoire des Sciences*, 43 (1990): 129–180. Joan L. Richards, "The Art and Science of British Algebra," *Historia Mathematica* 7 (1980): 343–365, "Rigor and Clarity: Foundations of Mathematics in France and England, 1800–1840," *Science in Context* 4 (1991): 297–319, and "God, Truth, and Mathematics in Nineteenth-Century England," in *The Invention of Physical Science*, ed. Mary Jo Nye, Joan Richards, and Richard H. Stuewer (Dordrecht: Kluwer, 1992), 51–78.

12. Brian Rotman, "Counting on Non-Euclidean Fingers," in *Mathematics as Sign: Writing, Imagining, Counting* (Stanford, CA: Stanford University Press, 2000), 125–153.

13. Babbage's philosophy only circulated in manuscript form, a version of which, "Essays on the Philosophy of Analysis," is in the British Library (Add MSS 37202).

14. The literature on the so-called Analytical Revolution at Cambridge is huge. The early accounts of the now familiar story include C. J. Enros, "The Analytical Society: Mathematics at Cambridge University in the Early Nineteenth-Century" (PhD diss., University of Toronto, 1979), "Cambridge University and the Adoption of Analytics in Early Nineteenth-Century England," *Social History of Nineteenth-Century Mathematics*, ed. H. Mehrtens, H. Bos, and I Schneider (Boston/Basel/Stuttgart: Birkhauser, 1981), 135–148, and "The Analytical Society (1812–1813): Precursor to the Renewal of Cambridge Mathematics," *Historia Mathematica* 10 (1983): 24–47. J. M. Dubbey, "The Introduction of the Differential Notation to Great Britain," *Annals of Science* 19 (1963): 37–48, "Babbage, Peacock and Modern Algebra," *Historia Mathematica* 4 (1977): 295–302, and *The Mathematical Work of Charles Babbage* (New York: Cambridge University Press, 1978).

15. Boole, "On a General Method," 225. It is not clear to me exactly what Boole is quoting from here but a similar statement can be found in D. F. Gregory, "On the Real Nature of Symbolic Algebra," originally published in *Transactions of the Royal Society of Edinburgh* 14 (1840): 208–216, but reproduced in *The Mathematical Writings of Duncan Farquharson Gregory*, ed. William Walton (Cambridge: Deighton, Bell & Co.; London: Bell and Daldy, 1865), 1–13, esp. 2. See also Patricia Allaire and Robert E. Bradley, "Symbolical Algebra as a Foundation for Calculus: D. F. Gregory's Contribution," *Historia Mathematica* 29 (2002): 395–426.

16. In terms of symbols, the commutative law can be written as ab = ba, the distributive law as a(b + c) = ab + ac, and the index law as $a^b a^c = a^{b+c}$.

17. David Harel, *Algorithmics: The Spirit of Computing*, 2nd ed. (New York: Addison Wesley, 1992), 3. Flipping changes the bit's value, zeroing makes sure that the bit ends up in the 0 position, and testing does one thing if the bit is already in the 0 position and another if it does not.

18. Alan Turing, "On Computable Numbers, with an Application to the Entscheidungsproblem," *Proceedings of the London Mathematical Society* 2, no. 1 (1937): 230–265, who writes on 231 that "We may compare a man in the process of computing a real number to a machine." It is also worth noticing that, as Simon Schaffer among others has pointed out, Turing was a keen reader of Babbage. Schaffer, "Babbage's Dancer and the Impresarios of Mechanism," in *Cultural Babbage: Technology, Time and Invention*, ed. Francis Spufford and Jennifer S. Uglow (London: Faber & Faber, 1996), 52–80.

19. In fact, significant work had to be done to make a "Boolean logic" useful for computer engineers. See Paul J. Nahin, *The Logician and the Engineer: How George Boole and Claude Shannon Created the Information Age* (Princeton, NJ, and Oxford: Princeton University Press, 2013).

20. Carl E. Schorske, *Thinking with History: Explorations in the Passage to Modernism* (Princeton, NJ: Princeton University Press, 1998). On a nineteenth-century representation of Boole's progressive thinking, see Kevin Lambert, "Victorian Stained Glass as Memorial: An Image of George Boole," in *Visions of the Industrial Age, 1830–1914: Modernity and the Anxiety of Representation in Europe*, ed. Minsoo Kang and Amy Woodson Boulton (Aldershot/Burlington: Ashgate, 2008), 205–226. On the importance of the history of mathematics to De Morgan, Joan L. Richards, "Augustus De Morgan, the History of Mathematics, and the Foundations of Algebra," *Isis* 78 (1987): 7–30. Adrian Rice, "Augustus De Morgan: Historian of Science," *History of Science* 34 (1996): 201–240.

21. Aileen Fyfe, *Steam Powered Knowledge: William Chambers and the Business of Publishing* (Chicago: University of Chicago Press, 2012). James A. Secord, *Victorian Sensation: The Extraordinary Publication, Reception, and Secret Authorship of Vestiges of the Natural History of Creation* (Chicago: University of Chicago Press, 2000).

22. Alistair Black, "The Victorian Information Society: Surveillance, Bureaucracy, and Public Librarianship in 19th-Century Britain," *The Information Society* 17 (2001): 63–80.

23. Michel Foucault, *Discipline & Punish: The Birth of the Prison*, 2nd ed. (New York: Random, 1995).

24. Ian Inkster, ed., *The Steam Intellect Societies—Essays on Culture, Education and Industry circa 1820–1914* (Nottingham: Department of Adult Education, University of Nottingham, 1985). Thomas Love Peacock used the name "The steam intellect society" in his satiric novel *Crochet Castle* (1831).

25. Steven Shapin and Barry Barnes, "Science, Nature and Control: Interpreting Mechanics Institutes," *Social Studies of Science* 7 (1977): 32. John Edward Royle, "Mechanics' Institutes and the Working Classes, 1840–1860," *The Historical Journal* 14, no. 2 (1971): 305–321.

26. The classic quote is "Lord Brougham thinks to stop our mouths with Kangaroos," which a Scottish judge, Sir Archibald Alison, attributed to some "operatives of Manchester." Quoted in Shapin and Barnes, "Science, Nature and Control," 56.

27. On the early history of the Mechanics' Institutes see, Mabel Tylecote, *The Mechanics' Institutes of Lancashire and Yorkshire before 1851* (Manchester: Manchester University Press, 1957), 1–25. On the "inadequate supply of books ... for self-educated men" in this period, Tylecote, *The Mechanics' Institutes*, 31. Mudie's Circulating Library was not established until 1842, and the first free public libraries were opened in Salford and Manchester in 1850. Boyd Hilton, *A Mad Bad and Dangerous People? England 1783–1846* (Oxford: Clarendon Press, 2013), 173–174.

28. The idea of an assemblage owes something to the ideas of the French philosophers Gilles Deleuze and Félix Guattari, especially in their *A Thousand Plateaus: Capitalism and Schizophrenia*, trans. Brian Massumi (Minneapolis: University of Minneapolis Press, 1987). See also Manuel De Landa, *A New Philosophy of Society: Assemblage Theory and Social Complexity* (London: Bloomsbury, 2006). Bruno Latour, *Pandora's Hope: Essays on the Reality of Science Studies* (Cambridge, MA: Harvard University Press, 1999). Bruno Latour, *Reassembling the Social: An Introduction to Actor-Network-Theory* (Oxford: Oxford University Press, 2007)

29. For an extremely useful discussion of the ongoing process of assembling knowledge, see Kathryn J. Franklin, James A. Johnson, and Emily Miller Bonney, "Towards Incomplete Archeologies?" in *Incomplete Archaeologies: Assembling Knowledge in the Past and the Present*, ed. Emily Miller Bonney, Kathryn J. Franklin, and James A. Johnson (Oxford: Oxbow Books, 2016), ix–xvii.

30. I wish to thank Emily Bonney for suggesting the word "assembling" to me.

31. For biographical details of Boole's life see Desmond MacHale's *The Life and Work of George Boole: A Prelude to the Digital Age* (Cork: Cork University Press, 2014). The principal sources for all works on the life of George Boole include a lengthy memorial to Boole written by the Reverend Robert Harley that first appeared in the *British Quarterly Review* in July 1866 and is reproduced in George Boole, *Studies in Logic and Probability*, ed. R. Rhees (London: Watts and Co., 1952), 425–472. There are also some details in Mary Everest Boole, *Collected Works*, 4 vols., ed. E. M. Cobham (London: The C. W. Daniel Co., 1931), especially "Home Side of a Scientific Mind," 1:1–48. There are unpublished works in the Lincolnshire Central Library and the Lincoln archive: L. Elvin, "The Life of George Boole," and G. Layton, "Some Notes on the Boole Family." There is also a fragment by George Boole's sister Mary Ann Boole among the Boole papers in University College Cork.

32. Sir Francis Hill, *Victorian Lincoln* (Cambridge: Cambridge University Press, 1974), esp. 147–151. Lincoln clergymen the Revs. G. S. Dickson and E. R. Larken were particularly important for Boole's education.

33. For the problems of training in theoretical work, see Andrew Warwick, *Masters of Theory: Cambridge and the Rise of Mathematical Physics* (Chicago: University of Chicago Press, 2003), 1–48.

34. "Minute Book, Lincoln Mechanics Institute," Local History Collection of Lincolnshire County Library [5108 LCL], 70. On George Boole's participation in the Lincoln Mechanics'

Institute, see MacHale, *The Life and Work of George Boole*, ch. 3. The 1846 report has not survived but Boole appears to have served on the committee until he left for Cork in 1848. *A Report on the Library of the Mechanics' Institute* for 1848 is available in the Local History Collection of the Lincolnshire Central Library (UP 1035).

35. That he did have access to them is clear from a paper Boole sent to the *Cambridge Mathematical Journal* in 1838. It finally appeared in the number (IX) for May 1840. George Boole, "On Certain Theorems in the Calculus of Variations," *Cambridge Mathematical Journal* II (Cambridge: E. Johnson, 1841), 97–102. The paper begins: "It would perhaps have been more just to entitle this communication, 'Notes on Lagrange.' The papers from which it is selected were written towards the close of the year 1838, during the perusal of the *Mechanique Analytique* [by Lagrange]," 97. See also MacHale, *The Life and Work of George Boole*, 56–57.

36. For example, he continued to publish. Between 1836 and 1840 he submitted a number of papers on botany, in which he was keenly interested, two to the *Edinburgh New Philosophical Journal*, two to the *Philosophical Magazine*, and three to the *Magazine of Natural History*. For a short biographical sketch of Bromhead, see Cannell, *George Green*, ch. 5. The information about his later scientific work is at 262 n.18.

37. Bromhead to Babbage, October 1, 1819, British Library Add. Mss. 37, 182 f. 99. On Deighton's, see Jonathan Topham, "A Textbook Revolution," in *Books and the Sciences in History*, ed. Marina Frasca Spada and Nicholas Jardine (Cambridge/New York: Cambridge University Press, 2000), 317–337. J. R. Topham, "Two Centuries of Cambridge Publishing and Bookselling: A Brief History of Deighton, Bell, and Co. 1778–1998, with a Checklist of the Archive," *Transactions of the Cambridge Bibliographical Society* 11, no. 3 (1998): 350–403.

38. It was published as a pamphlet: George Boole, *An Address on the Genius and Discoveries of Sir Isaac Newton* (Lincoln: Gazette Office, 1835). On the talk and its publication, see MacHale, *The Life and Work of George Boole*, 38–43.

39. Bromhead to Boole, August 29, 1937, George Green Collection, University of Nottingham Library, Kings Meadow Campus manuscripts dept. (hereafter GG), 4D 6 V. The Bromhead correspondence is in possession of Patrick German, but I have not yet been able to gain access. Typed transcripts of the correspondence between Boole and Bromhead letters made by A. M. Rollett and his son J. M. Rollett are held at the Kings Meadow Campus archive in the University of Nottinghamshire and it is those that I have consulted. See also MacHale's discussion, *The Life and Work of George Boole*, 54–56, which includes lengthy quotes from the correspondence.

40. The evidence is suggestive but circumstantial. For example, in a letter to Bromhead on July 24, 1839, Boole writes that "On a recent journey while spending a night in Cambridge I had the good fortune to meet with a gentleman of the name of Gregory who was kind enough to look over a copy of the paper which I enclosed to you some time ago." Boole to Bromhead, July 24, 1839 (GG 4D 6 IX).

41. Boole to Bromhead, July 24, 1839 (GG 4D 6 IX)

42. Boole to Bromhead, January 6, 1841 (GG 4D 6 XII). The Lacroix is S. F. Lacroix, *Traité du calcul différentiel et du calcul intégral*, 2nd ed., tome second (Paris: Courcier, 1814). Boole is referring to the Public Library of the University of Cambridge, to which he had only limited access because he was never a Cambridge undergraduate.

43. George Boole, "On a General Method of Analysis," *Philosophical Transactions of the Royal Society of London* 134 (1844): 225–282. On Boole's report on the library, see MacHale, *The*

Life and Work of George Boole, 37–38. MacHale also references *A Report on the Library of the Mechanics Institute* (1846) in his bibliography (see MacHale, *The Life and Work of George Boole*, 37–38, 322), which he reports as being held in the Local History Collection of the Lincolnshire Central Library, but the librarians at the Lincolnshire Central Library were unable to locate it for me.

44. Hill, *Victorian Lincoln*, 227. Giving the cost in a relative contemporary value is very difficult, but a good estimate is around one pound or two dollars per week. See the *MeasuringWorth* website at https://www.measuringworth.com/ukcompare/.

45. This transformation of logic from a "moribund subject" to a living science was seen as miraculous in the sense of something like the raising of Lazarus. See, e.g., Sir William Hamilton, "Logic. In Reference to the Recent English Treatises on that Science" [*Edinburgh Review*, 1833]. Reproduced in *Discussions on Philosophy and Literature, Education and University Reform. Chiefly from the Edinburgh Review; Corrected, Vindicated, Enlarged, in Notes and Appendices* (New York: Harper & Brothers, 1861), 128. For a brief overview of Victorian logic including an excellent short bibliographic note, see Ivor Grattan-Guinness, "Victorian Logic from Whately to Russell," in *Mathematics in Victorian Britain*, ed. Raymond Flood, Adrian Rice, and Robin Wilson (Oxford: Oxford University Press, 2011), 359–374. The bibliographic note is on p. 450.

46. On the making of Whately's book, see Pietro Corsi, "The Heritage of Dugald Stewart: Oxford Philosophy and the Method of Political Economy," *Nuncius Annali di Storia della Scienza* (1987): 98–105, on 108–109. Copleston gave his own notes on the topic along with the assignment of writing a logic to Whately. Other contributors included John Henry Newman and Nassau Senior. Richard Whately, *Elements of Logic* (1827; Delmar: Scholars' Facsimiles & Reprints, 1975). Richard Whately, "Logic," in *Encyclopaedia Metropolitana*, Vol. 1 [Pure sciences, vol. 1], ed. Edward Smedley, Hugh James Rose, and Henry John Rose (London, B. Fellowes, F and J Rivington et al., 1845), 193–240.

47. First articulated in De Morgan's paper "On the Structure of the Syllogism," *Transactions of the Cambridge Philosophical Society* 8 (1846): 379–406. In "On the Structure," 2, De Morgan calls it the universe of a proposition, or of a name. Luis M. Laita, "Influences on Boole's Logic: The Controversy between William Hamilton and Augustus De Morgan," *Annals of Science* 36(1) (1979): 45–66, on 52. See also Augustus De Morgan, *First Notions of Logic* (London: James Moyes, 1839). Joan Richards, *Mathematical Visions: The Pursuit of Geometry in Victorian England* (Boston: Academic Press, 1988), 177–178.

48. De Morgan discussed this problem with Boole after they had both published works, in 1847, in logic: Boole's *The Mathematical Analysis of Logic: Being an Essay towards a Calculus of Deductive Reasoning* (Cambridge: Macmillan, Barclay and Macmillan; London: George Bell, 1847), and De Morgan's *Formal Logic: Or, the Calculus of Inference, Necessary and Probable* (London: Taylor and Walton, 1847). In a letter to Boole of June 8, 1850, De Morgan wrote that "It is said, and justly, that two negatives, in ordinary Greek do not make an affirmative—but a more emphatic negative. I remember no instance at this moment but the opening of the Orestes. . . . Nevertheless—in Aristotle—(De Interpr. cap.X) two negatives made an affirmative and *three* negatives a negative— . . . Was Aristotle here talking Greek—or making Greek?" Letter 25 in G. C. Smith, *The Boole–De Morgan Correspondence, 1842–1864* (Oxford: Clarendon Press, 1982), 37.

49. A better example is the quantification "not-man," which also only makes logical sense if tied to a universe of discourse. For example, if our universe is all animals then the class of

things which are "not-man" are all animals except those which are men. The quantification of the predicate follows from the consideration that if the class of men is not empty and the class of not-men is not empty, then it follows that some animals are men.

50. On their meeting in 1847 see Smith, *The Boole–De Morgan Correspondence*, 17. On their wide-ranging conversations, Sophia Elizabeth De Morgan, *Memoir of Augustus De Morgan* (London: Longmans Green & Co. 1882), 168. Luis M. Laita, "Influences on Boole's Logic: The Controversy between William Hamilton and Augustus De Morgan," *Annals of Science* 36(1) (1979): 45–66.

51. As Laita has noticed, the controversy inspired Boole to study De Morgan's paper "On the Structure of the Syllogism," *Transactions of the Cambridge Philosophical Society* 8 (1846): 379–406, where he first introduced the universe of discourse or universe of a proposition or name. See Laita, "Influences on Boole's Logic," 52. He also certainly read Augustus De Morgan, *First Notions of Logic* (London: James Moyes, 1839).

52. George Boole, *The Mathematical Analysis of Logic* (Bristol: Thoemmes Press, 1998), 5.

53. The operation + in Boole's algebra of logic represents the aggregate of classes x and y. In order for $x + y$ to have meaning for Boole, x and y had to be mutually exclusive, which in this case is the class of things that are blue and those that are not blue.

54. The class of $(y + z)$ things is the selection of all round and square things from the universe of objects. $x(y + z)$ is, then, the selection of all blue round things and blue square things, which is exactly equivalent to selecting all blue round things and then all blue square things, or $xy + yz$.

55. Consider the easiest case, a syllogism of the form all animals are mortal, all humans are animals, therefore, all humans are mortal. If we let x = the class of all animals, y = all humans, and z = all mortal beings, then $xz = x$ (the class of mortal animals is all animals, or all animals are mortal) and $yx = y$ (all humans are animals). Multiplying together, we obtain $xy = yx^2z$. But $x^2 = x$ so $y = yz$, or all humans are mortal.

56. Boole, *The Mathematical Analysis of Logic*, 4.

57. Boole, *The Mathematical Analysis of Logic*, 7. George Boole, *An Investigation of the Laws of Thought on Which Are Founded the Mathematical Theories of Logic and Probabilities* (1854; New York: Dover, 1958), esp. 399–424.

58. Including not only De Morgan's *First Notions of Logic*, but also books cited in *The Mathematical Analysis* such as Robert Gordon Latham's *First Outlines of Logic Applied to Grammar and Etymology* (London: Taylor and Walton, 1847).

59. In 1846, Boole had recommended that the institute's library acquire a copy of Whately's *Elements of Logic*, but it does not appear in the catalog published in 1849. MacHale, *Life and Work of George Boole*, 38.

60. Boole's book did not appear in the printed catalog, but he did celebrate it in the 1849 Lincoln and Lincolnshire Mechanics' Institute Report and a copy of "'Johnston's Physical Atlas', contributors to which are members of your [Boole's] institution," was promised to him (*Lincoln and Lincolnshire Mechanics' Institute Report* 1849, 2, Lincoln Central Library, UP. 1035). Perhaps Boole reciprocated with a copy of his book (or the later *Laws of Thought*, which, if acquired after 1849, would account for why it did not appear in the catalog).

61. De Morgan to Boole, May 31, 1847, Smith, *The Boole–De Morgan Correspondence*, 22.

62. "Draft a. De Morgan to Boole, not sent," Smith, *The Boole–De Morgan Correspondence*, 24. There are, of course, also important differences between Boole and De Morgan's book. For a more philosophical appraisal, see Daniel D. Merrill, *Augustus De Morgan and the Logic of Relations* (Dordrecht: Kluwer, 2011).

63. Adrian Rice, "Augustus De Morgan: Historian of Science," *History of Science* 34 (1996): 201–240, on 222.

64. Sophia Elizabeth De Morgan, *Memoir of Augustus De Morgan* (London: Longmans Green & Co. 1882), 364.

65. Sophia De Morgan, *Memoir of Augustus De Morgan*, 58. I especially want to acknowledge the generous way in which Joan Richards has both shared her writings and participated in some rich conversations both physical and virtual about Victorian libraries in general and De Morgan's in particular. An important written source for my thinking about De Morgan's library has been her unpublished chapter, "Searcher after Things Mental," from a draft of some chapters from the book Richards was then writing on the Frend/De Morgan families now published as Joan L. Richards, *Generations of Reason: A Family's Search for Meaning in Post Newtonian England* (New Haven: Yale University Press, 2021).

66. Sophia De Morgan, *Memoir of Augustus De Morgan*, 105. Cataloging library and museum collections was a pressing problem in the nineteenth century, and De Morgan was skeptical that any of the classificatory schemes used to organize catalogues were satisfactory. In his testimony to British museum commissioners, he was skeptical of all schemes except that of an "alphabetical plan." See, De Morgan's testimony, *Report of the Commissioners Appointed to Inquire into the Constitution and Government of the British Museum*, 375-384, on 378. See also Augustus De Morgan, "Libraries and Catalogues," Quarterly Review 72 (1843): 1-25. For a late Victorian librarian's view of De Morgan's dismissal of classificatory schemes, see James Duff Brown, *Manual of Library Classification and Shelf Arrangement* (London: Library Supply Company, 1901), 89-91. De Morgan's recommendation for an "alphabetical plan" was for a major research library not his own private library which was sorted into categories. Karen Attar, "Augustus De Morgan (1806-71), His Reading and His Library," *The Edinburgh History of Reading, vol. 2, Modern Readers*, ed., Mary Hammond (Edinburgh: Edinburgh University Press, 2020), 62-82, on p. 70. For further discussion of these points see, Kevin Lambert, *Symbols and Things: Material Mathematics in the Eighteenth and Nineteenth Centuries* (Pittsburgh: Pittsburgh University Press, 2021), 108–09.

67. Rice, "Augustus De Morgan," 222.

68. Sophia De Morgan, *Memoir of Augustus De Morgan*, 58. The books bought by the London University are now gathered together in the special collections of its Senate House Library.

69. Augustus De Morgan, "On the Earliest Printed Almanacs," *Companion to the Almanac for 1846* (London: Charles Knight, 1846) 1–31, on 1. The most significant problem with reading De Morgan's histories is that they are scattered through articles in a number of different publications, including the SDUK publisher Charles Knight's *Penny Cyclopedia*, the literary journal the *Athenaeum*, and more obscure places such as Knight's *Companion to the Almanac*, which was often bound together with years of the SDUK's *British Almanac*. Although some of his articles are gathered together in the delightful *Budget of Paradoxes*, we still await a more comprehensive collection.

70. De Morgan, "On the Earliest Printed Almanacs," 2.

71. De Morgan, *Arithmetical Books from the Invention of Printing to the Present Time Being Brief Notices of a Large Number of Works Drawn Up from Actual Inspection* (London: Taylor and Walton, 1847), xiii. This was not De Morgan's first work of mathematical bibliography. In the *Companion to the Almanac of 1843* (London: Charles Knight, 1843), De Morgan published a detail account of over 250 items: "References for the History of the Mathematical Sciences," 40–65.

72. Richards, ch. 15 draft, "Searcher after Things Mental," 2–3.
73. Augustus De Morgan to Sir John Herschel, October 16, 1832, reproduced in Sophia De Morgan, *Memoir of Augustus De Morgan*, 75
74. An important center for exactly this view about the relationship between progress and historical research was Cambridge University, where De Morgan was educated. Duncan Forbes, *The Liberal Anglican Idea of History* (London: Cambridge University Press, 1952).
75. De Morgan, *Arithmetical Books*, xiv.
76. This is not the modern, or even the Victorian sense of the word, which also suggests something absurd or self-contradictory. Instead, De Morgan insisted in what he understood to be the fifteenth-century definition of the word, which was what we might simply call an unorthodox view.
77. De Morgan had a historically sophisticated perspective on Copernicus, for which see Rice, "Augustus De Morgan," 207–210.
78. Kevin Lambert, "A Natural History of Mathematics: George Peacock and the Making of English Algebra," *Isis* 104 (2013): 278–302.
79. Augustus De Morgan, *The Differential and Integral Calculus, Containing Differentiation, Integration, Development, Series, Differential Equations, Differences, Summation, Equations of Differences, Calculus of Variations, Definite Integrals,—with applications to Algebra, Plane Geometry, Solid Geometry, and Mechanics Also, Elementary Illustrations of the Differential and Integral Calculus* (London: Library of Useful Knowledge: Baldwin and Cradock, 1842), vii. I would like to thank Joan Richards for bringing this to my attention.
80. De Morgan, *The Differential and Integral Calculus*, 224.
81. Augustin-Louis Cauchy, *Cours d'analyse*, in *Oeuvres completes d'Augustin Cauchy*, publieés sous la direction scientifique de l'Académie (Paris: Gauthier-Vilars), series 2, vol. 3, p. 114. Judith V. Grabiner, *The Origins of Cauchy's Rigorous Calculus* (New York: Dover, 2005), 100.
82. And in fashion, he was right. See Florian Cajori, "Augustus De Morgan on Divergent Series," *Bulletin of the American Mathematical Society* 27 (1920): 77–81. Augustus De Morgan, "On Divergent Series, and Various Points in Analysis Connected with Them," *Transactions of the Cambridge Philosophical Society* 8, part II (1843): 182–203.
83. As Deleuze and Guattari put it: "The fact is that the two kinds of science (State and Nomadic) have different modes of formalization and State science continually imposes its form of sovereignty on the inventions of nomadic science." *A Thousand Plateaus*, 362.

Chapter 7

1. Margaret W. Rossiter, *Women Scientists in America: Struggles and Strategies to 1940* (Baltimore: Johns Hopkins University Press, 1982), 53–57 and especially 335–336 n 7. Fleming's speech appeared in Williamina Paton Fleming, "A Field for 'Woman's Work' in Astronomy," *Astronomy and Astrophysics* 12 (1893): 688–689.
2. David Alan Grier, *When Computers Were Human* (Princeton, NJ: Princeton University Press, 2005).
3. Pamela Mack, "Strategies and Compromises: Women in Astronomy at Harvard College Observatory, 1870–1920," *Journal of the History of Astronomy* 21 (1990): 65–75; and Pamela E. Mack, "Straying from Their Orbits: Women in Astronomy in America," in *Women of*

Science: Righting the Record, ed. G. Kass-Simon and Patricia Farnes (Bloomington: Indiana University Press, 1990), 72–116.

4. W. S. Gilbert and Arthur Sullivan, "H.M.S. Pinafore," in *The Complete Annotated Gilbert & Sullivan*, intro. and ed. Ian Bradley (Oxford: Oxford University Press, 1996), 113–186.

5. Winslow S. Upton, *The Observatory Pinafore (1879)*, transcribed by Jonathan McDowell (Harvard-Smithsonian Center for Astrophysics, 1994), http://hea-www.harvard.edu/~jcm/html/play.html. Also, see the full manuscripts and limited correspondence available in: Harvard College Observatory, "Records Relating to the Observatory Pinafore, 1879–1966," Collections of the Harvard University Archives, UAV 630.495.2.

6. N. R. Johnson, "Information Infrastructure as Rhetoric: Tools for Analysis," *Poroi* 8, no. 1 (2012): 1–3; and Estee Beck, "A Theory of Persuasive Computer Algorithms for Rhetorical Code Studies," *Enculturation* 23 (2016): n.p., http://enculturation.net/a-theory-of-persuasive-computer-algorithms.

7. N. Katherine Hayles, *My Mother Was a Computer: Digital Subjects and Literary Texts* (Chicago: University of Chicago Press, 2005), 1–14.

8. Andrew Fiss and Laura Kasson Fiss, "Laughing out of Math Class: The Vassar *Mathematikado* and Nineteenth-Century Women's Education," *Configurations* 27, no. 3 (Summer 2019): 301–329. For Gilbert and Sullivan scholarship on humor, see Carolyn Williams, *Gilbert and Sullivan: Gender, Genre, Parody* (New York: Columbia University Press, 2011), 97–121; and Laura Kasson Fiss, "'This Particularly Rapid, Unintelligible Patter': Patter Songs and the Word-Music Relationship," in *The Cambridge Companion to Gilbert and Sullivan*, ed. David Eden and Meinhard Saremba (Cambridge: Cambridge University Press, 2009), 98–108.

9. Upton, *The Observatory Pinafore (1879)*.

10. David Brewster, *Memoirs of the Life, Writings, and Discoveries of Sir Isaac Newton, Volume 2* (Boston: Little, Brown, and Co., 1855), 162.

11. Brewster, *Sir Isaac Newton*, esp. 157–186.

12. Caroline Herschel achieved some celebrity for the role. See Richard Holmes, *The Age of Wonder: How the Romantic Generation Discovered the Beauty and Terror of Science* (New York: Random House, 2008), 163–210.

13. Solon I. Bailey, *The History and Work of Harvard Observatory, 1839 to 1927* (New York: McGraw Hill, 1931), 35; Bessie Zaban Jones and Lyle Gifford Boyd, *The Harvard College Observatory: The First Four Directorships, 1939–1919* (Cambridge, MA: Belknap Press of Harvard University Press, 1971), 189–193, 176–210; and Roger Sherman Hoar, "The Pickering Polaris Attachment," *Journal of the United States Artillery* 50 (1919): 230–236.

14. Jones and Boyd, *The Harvard College Observatory*, 386–387; Dorrit Hoffleit, *The Education of American Women Astronomers before 1960* (Cambridge, MA: American Association of Variable Star Observers, 1994), 13; Grier, *When Computers Were Human*, 82; and Dava Sobel, *The Glass Universe: How the Ladies of the Harvard Observatory Took the Measure of the Stars* (New York: Viking, 2016), 9, 120–122.

15. Mack, "Strategies and Compromises," 65–75; and "Anna Winlock," in *Biographical Dictionary of Women in Science, Volume 2*, ed. Marilyn Bailey Oglive and Joy Dorothy Harvey (New York: Routledge, 2000), 1388–1389. Also see Jones and Boyd, *The Harvard College Observatory*, 386–387; Hoffleit, *Education of American Women Astronomers*, 44; Grier, *When Computers Were Human*, 82; and Sobel, *The Glass Universe*, 9, 30, 90.

16. Grier names his fifth chapter about human computers "A Carpet for the Computing Room," referring to their room in the Harvard Observatory. Grier, *When Computers Were Human*, 72–88.

17. See Kevin Lambert, "Material Mathematics: British Algebra as Algorithmic Mathematics," Chapter 6 in this volume.

18. Upton, *Observatory Pinafore (1879)*.

19. Such budget worries appear in many histories of the observatory. See Bailey, *The History and Work of Harvard Observatory*, 35; and Jones and Boyd, *The Harvard College Observatory*. For the director's perspective on such matters, see Edward C. Pickering, *Statement of Work Done at the Harvard College Observatory during the Years 1877–1882* (Cambridge, MA: John Wilson & Son. University Press, 1882).

20. Upton, *The Observatory Pinafore (1879)*, Act II.

21. About the staff's educational levels, see Jones and Boyd, *The Harvard College Observatory*, 189–193.

22. Upton, *The Observatory Pinafore (1879)*, Act I.

23. Williams, *Gilbert and Sullivan*, 114–116.

24. Little is known about Joseph McCormack, as he died soon after. See Jones and Boyd, *The Harvard College Observatory*, 189–193.

25. Upton, *The Observatory Pinafore (1879)*.

26. Observatory Director Harlow Shapley discusses the 1929 changes in a letter to Winslow Upton's daughter: H. Shapley to E. Upton, January 23, 1930, in the Harvard College Observatory, "Records Relating to the Observatory Pinafore, 1879–1966," Collections of the Harvard University Archives, UAV 630.495.2, also hosted here: http://hea-www.harv ard.edu/~jcm/html/shapley.html.

27. Upton, *The Observatory Pinafore (1879)*.

28. Williams, *Gilbert and Sullivan*, 107–109.

29. Mack, "Straying from Their Orbits," 76. Also see Helen Wright, *Sweeper in the Sky: The Life of Maria Mitchell, First Woman Astronomer in America* (New York: Macmillan, 1950), 71; Henry Albers, ed., *Maria Mitchell: A Life in Journals and Letters* (Clinton Corners, NY: College Avenue Press, 2001); and Renée Bergland, *Maria Mitchell and the Sexing of Science: An Astronomer among the American Romantics* (Boston: Beacon, 2008), xi.

30. Jones and Boyd, *The Harvard College Observatory*, 386–387; Grier, *When Computers Were Human*, 82; and Sobel, *The Glass Universe*, 274.

31. Judy Wajcman, *Feminism Confronts Technology* (University Park: Pennsylvania State University Press, 1991), 85, 104, 116; Ruth Schwartz Cowan, *More Work for Mother: The Ironies of Household Technology from the Open Hearth to the Microwave* (New York: Basic Books, 1983); and Adrian Forty, *Objects of Desire: Design and Society 1750–1980* (London: Thames and Hudson, 1986).

32. Sally L. Hacker, *"Doing It the Hard Way": Investigations of Gender and Technology*, ed. Dorothy E. Smith and Susan M. Turner (Boston: Unwin Hyman, 1990), esp. 163–174.

33. Upton, *The Observatory Pinafore (1879)*, Act I.

34. Upton, *The Observatory Pinafore (1879)*, Act I; also reproduced in Grier, *When Computers Were Human*, 87.

35. Upton, *The Observatory Pinafore (1879)*, Act I; also reproduced in Grier, *When Computers Were Human*, 86.

36. Upton, *The Observatory Pinafore (1879)*, Act I.

37. Williams, *Gilbert and Sullivan*, 101–104.

38. Upton, *The Observatory Pinafore (1879)*, Act I.

39. For more theoretical views of humor, see Laura Kasson Fiss, "The Idler's Club: Humor and Sociability in the Age of New Journalism," *Victorian Periodicals Review* 49, no. 3 (2016): 415–430.

40. For annotations about the *Observatory Pinafore* editions, see Upton, *Observatory Pinafore (1879)*.

41. Upton, *The Observatory Pinafore (1879)*.

42. Much of the 1929 materials appears in Harvard College Observatory, "Records Relating to the Observatory Pinafore, 1879–1966," Collections of the Harvard University Archives, UAV 630.495.2. Also see Sobel, *The Glass Universe*, 226–227.

43. Annie Jump Cannon, "Williamina Paton Fleming," *Science* 33 (June 30, 1911): 987–988; Mack, "Straying from Their Orbits," 92–94; Sobel, *The Glass Universe*, 9–10; and the Harvard University Library Open Collections Program, "Williamina Paton Stevens Fleming (1857–1911)," *Women Working, 1800–1930*, last modified 2017, http://ocp.hul.harvard.edu/ww/fleming.html.

44. Jones and Boyd, *The Harvard College Observatory*, 235; Mack, "Straying from Their Orbits," 92; and Sobel, *The Glass Universe*, 12.

45. Mack, "Straying from Their Orbits"; and Sobel, *The Glass Universe*, 105.

46. Rossiter, *Women Scientists in America*, 53–57 and esp. 335–336 n 7. Fleming, "A Field for 'Woman's Work' in Astronomy."

47. Upton, *The Observatory Pinafore (1879)*, Act I.

48. "Resurrected from the Files of Long Ago," *Cambridge Chronicle*, January 17, 1930, hosted at: http://hea-www.harvard.edu/~jcm/html/chron.html.

49. Upton, *The Observatory Pinafore (1879)*, Act I.

50. The texts remain part of the Harvard University Library, in the astronomy, mathematics, and history of science collections.

51. H. Shapley to E. Upton, January 23, 1930.

52. Shapley to E. Upton, January 23, 1930.

53. Upton, *The Observatory Pinafore (1879)*, Act I. Also see Jones and Boyd, *The Harvard College Observatory*, 191; and Sobel, *The Glass Universe*, 226.

54. Jones and Boyd, *The Harvard College Observatory*, 189–193; Grier, *When Computers Were Human*, 85–87; and Sobel, *The Glass Universe*, 226–227.

55. Grier, *When Computers Were Human*, 85.

56. The surprising change appears in H. Shapley to E. Upton, January 23 1930.

57. Robert Hariman, "Political Parody and Public Culture," *Quarterly Journal of Speech* 94, no. 3 (2008): 247–272; Fiss and Fiss, "Laughing out of Math Class."

58. Avianne Tan, "Apollo 11's Source Code Has Tons of Easter Eggs, Including an Ignition File Labeled 'Burn Baby Burn,'" *ABC News*, July 12, 2016.

59. James E. Dobson and Rena J. Mosteirin, *Moonbit* (Goleta, CA: Punctum Books, 2019).

Chapter 8

1. For global histories of seed trade and fungal ecologies and the broader imperial contexts of US agriculture, see Courtney Fullilove, *Profit of the Earth: The Global Seeds of American Agriculture* (Chicago: University of Chicago Press, 2017); and Anna Lowenhaupt Tsing, *Mushroom at the End of the World: On the Possibility of Life in Capitalist Ruins* (Princeton, NJ: Princeton University Press, 2017).

2. See Sydney Mintz, *Sweetness and Power: The Place of Sugar in Modern History* (New York: Viking, 1986); Kyla Wazana Tompkins, "Sweetness, Capacity, and Energy," *American Quarterly* 71, no. 3 (2019): 849–856. For histories of data, control, and US empire, see: Caitlin Rosenthal, *Accounting for Slavery: Masters and Management* (Cambridge: Harvard University Press, 2018); Toni Morrison, "A Humanist View," Portland State University's Oregon Public Speakers Collection: "Black Studies Center public dialogue. Pt. 2," May 30, 1975 (http://bit.ly/1vO2hLP).

3. For example, see Ralf Gente, Stefan F. Busch, Eva-Maria Stübling, Lorenz Maximilian Schneider, Christian B. Hirschmann, Jan C. Balzer, and Martin Koch, "Quality Control of Sugar Beet Seeds with THz Time-Domain Spectroscopy," *IEEE Transactions on Terahertz Science and Technology* 6, no. 5 (2016): 754–756.

4. Control is a critically important concept in the history of computing. For a cornerstone text on the history of control in information society, see James R. Beniger, *The Control Revolution: Technological and Economic Origins of the Information Society* (Cambridge, MA: Harvard University Press, 1989); Michael Adas, *Machines as the Measure of Men: Science, Technology, and Ideologies of Western Dominance* (Ithaca, NY: Cornell University Press, 1989); Carl Mitcham, *Thinking through Technology: The Path between Engineering and Philosophy* (Chicago: University of Chicago Press, 1994). For a comprehensive, long-view study of efficiency, see Jennifer Karns Alexander, *Mantra of Efficiency: From Waterwheel to Social Control* (Baltimore: Johns Hopkins University Press, 2008); see also Samuel P. Hays, *Conservation and the Gospel of Efficiency: The Progressive Conservation Movement, 1890–1920* (Cambridge, MA: Harvard University Press, 1959); Samuel Haber, *Efficiency and Uplift: Scientific Management in the Progressive Era* (Chicago: University of Chicago Press, 1964); Robert Kanigel, *The One Best Way: Frederick Winslow Taylor and the Enigma of Efficiency* (New York: Viking, 1997).

5. My term "computing landscapes" is introduced at length in Theodora Dryer, *Designing Certainty: The Rise of Algorithmic Computing in the Age of Anxiety, 1920–1970* (PhD diss., University of California, San Diego, 2019).

6. Deborah Fitzgerald, *Every Farm a Factory: The Industrial Ideal in American Agriculture* (New Haven, CT: Yale University Press, 2003).

7. For an account of the expansion of factory networks across the United States, see David E. Nye, *Electrifying America: Social Meanings of a New Technology* (Cambridge, MA: MIT Press, 1990).

8. For a comprehensive study of this inspection regime, see Paul J. Miranti, "Corporate Learning and Quality Control at the Bell System, 1877–1929," *Business History Review* 79, no. 1 (2005): 54; for a discussion of Walter Shewhart in history of computing, see David Alan Grier, "Programming and Planning," *IEE Annals of the History of Computing* 33, no. 1 (2011): 86–88.

9. Miranti, "Quality Control at the Bell System," 39–71.

10. Alfred D. Chandler Jr., *The Visible Hand: The Managerial Revolution in American Business* (Cambridge, MA: Harvard University Press, 1977): 79.

11. For comprehensive studies on large-scale control systems in the US context, see James R. Beringer, *The Control Revolution: The Technological and Economic Origins of the Information Society* (Cambridge, MA: Harvard University Press, 1986); David Hounshell, *From American System to Mass Production, 1800–1932* (Baltimore: Johns Hopkins University Press, 1984).

12. For comprehensive studies on the relationship between organizational design and corporate capitalism, see Louis Galambos, "The Emerging Organizational Synthesis in

Modern American History," *Business History Review* 44, no. 3 (Autumn 1970): 279–290; Hunter Heyck, "The Organizational Revolution and the Human Sciences," *Isis* 105, no. 1 (2014): 1–31; David F. Noble, *America by Design: Science, Technology, and the Rise of Corporate Capitalism* (Oxford: Oxford University Press, 1979).

13. W. A. Shewhart, "Quality Control Charts," *Bell System Technical Journal* 5, no. 4 (October 1926): 600.

14. P. C. Mahalanobis, "Walter A. Shewhart and Statistical Quality Control in India," *Sankhyā: The Indian Journal of Statistics (1933–1960)* 9, no. 1 (1948): 52; Karl Pearson, *The Grammar of Science* (Cambridge: Cambridge University Press, 2014), originally published 1895.

15. There is a longer history to integrating mathematics in industry noted by Bell technicians. For example, Bell employee George Camble wrote: "The necessity for mathematics in industry was recognized at least three centuries ago when [Francis] Bacon said: 'For many parts of nature can neither be invented [discovered] with sufficient subtilty nor demonstrated with sufficient perspicuity nor accommodated onto use with sufficient dexterity without the aid and intervening of mathematics.'" See George A. Campbell, "Mathematics in Industrial Research," *Bell System Technical Journal* (3)4 (1924): 550–557.

16. Shewhart, "Quality Control Charts," 601.

17. Shewhart, "Quality Control Charts," 602.

18. Shewhart, "Quality Control Charts," 603.

19. Shewhart, "Some Applications of Statistical Methods," 47.

20. W. A. Shewhart, "Some Applications of Statistical Methods to the Analysis of Physical and Engineering Data," *Bell System Technical Journal* 3, no. 1 (1924): 43–87.

21. Richard Winckel's Western Electric diagram patent of a telephone transmitter captures the many material parts of the transmitter technology, each with its own material constitution and capacity for electricity.

22. See, e.g., John Wishart, "Bibliography of Agricultural Statistics 1931–1933," *Supplement to the Journal of the Royal Statistical Society* 1, no. 1 (1934): 94–106.

23. See H. W. Wiley, "The Sugar Beet: Culture, Seed Development, Manufacture, and Statistics," *Farmers' Bulletin* 52 (1899): 5.

24. Clinton Dewitt Smith and Robert Clark Kedzie, "Sugar Beets in Michigan in 1897," Michigan State Agricultural College Experiment Station, Bulletin 150 (December 1897): 124. Map from "The Theoretical Beet-Sugar Belt of the United States," in H. W. Wiley, "The Sugar Beet: Culture, Seed Development, Manufacture, and Statistics," *Farmers' Bulletin* 52 (1899): 5.

25. See Robert Grimshaw and Lewis Sharpe Ware, "Various Issues," *The Sugar Beet: Scientific Quarterly* 31, no. 1 (1910): 11; "A careful determination made in the Bureau of Soils shows that the so-called sugar beet belt contains a total area of 428,000 sq. miles or 274,000,000 acres. No attempt has been made yet to estimate the available acreage of land lying outside the theoretical belt."

26. George H. Coons, "The Sugar Beet: Product of Science," *Scientific Monthly* 68, no. 3 (1949): 149.

27. Speaking to the industrial-managerial logic of sugar beet production, see F. A. Stilgenbauer, "The Michigan Sugar Beet Industry," *Economic Geography* 3, no. 4 (1927): 486–506; "If the farmer is not amply supplied with working capital, the sugar companies advance cash to the laborers for duties performed in connection with the beet crop as per contract, and deduct these advances from the farmer's credits for beets delivered at the end of the crop

year. Where necessary the companies furnish the farmer with beet seed, fertilizer, and farm implements on the same basis. The hand labor on beets is very great and much labor has to be imported during the summer to care for the industry. Mexican labor is much in evidence in recent years. Child labor is utilized to some extent without any injurious effects." Curtis Marez calls this type of management the "agribusiness gaze"; see Curtis Marez, *Farm Worker Futurism: Speculative Technologies of Resistance* (Minneapolis: Minnesota University Press, 2016).

28. Jim Norris, "Bargaining for Beets: Migrants and Growers in the Red River Valley," *Minnesota History* 58, no. 4 (2002/2003): 199.

29. See Zaragosa Vargas, "Life and Community in the 'Wonderful City of the Magic Motor': Mexican Immigrants in 1920s Detroit," *Michigan Historical Review* 15, no. 1 (1989): 45–68; Dennis Dodin Valdes, *Al Norte: Agricultural Workers in the Great Lakes Region, 1917–1970* (Austin: University of Texas Press, 1991).

30. Sugar beet breeding itself was believed to skirt problems of computing under limited information as an ideal commodity; for example, it avoided problems with sugar cane production tariffs since sugar cane took 18 months to grow and was incommensurable with accounting logics. Sugar beets were also considered to better withstand weather, one of the greatest variables in sugar beet analysis.

31. Joshua Bernhardt, "The Flexible Tariff and the Sugar Industry," *American Economic Review* 16, no. 1 (1926): 182–191.

32. W. G. Cochran, "Graduate Training in Statistics," *American Mathematical Monthly* 53, no. 4 (1946): 193–199.

33. Olynthus B. Clark, "Keeping Them on the Farm: The Story of Henry C. Wallace, the Secretary of Agriculture, Who Has Spent His Life Studying, Teaching, Improving and Practising Farming [*sic*]," *The Independent* 105 (1921): 333.

34. Clark, "Keeping Them on the Farm," 333.

35. Henry C. Wallace, *Our Debt and Duty to the Farmer* (New York: Century Co., 1925).

36. Thorstein Veblen, *The Theory of the Leisure Class: An Economic Study of Institutions* (New York: Dover Thrift Editions, 1994).

37. Unknown, "Plan to Introduce Graduate Studies at the USDA Graduate School in 1921" (1921). Special Collections, USDA National Agricultural Library, accessed March 26, 2018.

38. Alfred Friendly, "Agriculture's School Is on the Upgrade," *Washington Daily News*, August 30, 1938. USDA Graduate School Collection. Special Collections, National Agricultural Library.

39. *Lectures and Promotional Materials, 1921–1976.* USDA Graduate School Collection. Special Collections, National Agricultural Library.

40. Franklin D. Roosevelt, Inaugural Address of 1933 (Washington, DC: National Archives and Records Administration, 1988).

41. Walter A. Shewhart, *Statistical Method from the Viewpoint of Quality Control* (Washington DC: The Graduate School, USDA, 1939).

42. W. Edwards Deming, "Foreword from the Editor," in *Statistical Method from the Viewpoint of Quality Control* (Washington, DC: The Graduate School, USDA, 1939), iv. USDA Graduate School Collection. Special Collections, National Agricultural Library.

43. Deming, "Foreword," iii.

44. Deming, "Foreword," iv.

45. Deming, "Foreword," v.

46. W. Edwards Deming and Raymond T. Birge, *On the Statistical Theory of Errors*, 1934. USDA Graduate School Collection. Special Collections, National Agricultural Library.

47. John R. Mohler, *Address to the Graduate School: Scientific Research*, November 20, 1936, USDA Graduate School Collection. Special Collections, National Agricultural Library.

48. Lectures 1927–1940, USDA Graduate School Collection. Special Collections, National Agricultural Library.

49. Graduate School, U.S. Department of Agriculture, *Special Series of Lectures on Department of Agriculture Objective*. Lectures 1927–1940, USDA Graduate School Collection. Special Collections, National Agricultural Library.

50. John R. Mohler, *Address to the Graduate School: Scientific Research*, November 20, 1936, USDA Graduate School Collection. Special Collections, National Agricultural Library.

51. Mohler, *Address to the Graduate School*.

52. Mohler, *Address to the Graduate School*.

53. For a leading example of this rhetoric, see Thurman W. Arnold, *The Folklore of Capitalism* (New Haven, CT: Yale University Press, 1937).

54. In attendance were various statisticians, administrators, economists, and scientists concerned with the state of democracy, including émigré physicist Albert Einstein.

55. Unknown, *Washington Times*, February 23, 1938. USDA Graduate School Collection. Special Collections, National Agricultural Library.

56. Quoted in George A. Reisch, *How the Cold War Transformed Philosophy of Science: To the Icy Slopes of Logic* (Cambridge: Cambridge University Press, 2005).

57. A. F. Woods, *Letter to Statistically Minded Research Workers in Washington*, February 1938. USDA Graduate School Collection. Special Collections, National Agricultural Library.

58. A. F. Woods, *Letter to Statistically Minded*, 1938.

59. Alfred Friendly, "Agriculture's School Is on the Upgrade," *Washington Daily News*, August 30, 1938. USDA Graduate School Collection. Special Collections, National Agricultural Library.

60. Graduate School Publications. USDA Graduate School Collection. Special Collections, National Agricultural Library.

61. Charles F. Sarle, "The Agricultural Statistics Program," address, 25th session, International Statistical Conferences, Washington, DC, September 13, 1947.

62. On the early Cold War sugar beet industry, see H. S. Owens, "Production and Utilization of Sugar Beets," *Economic Botany* 5, no. 4 (1951): 348–366; H. G. Stigler and R. T. Burdick, "The Economics of Sugar-Beet Mechanization," Bulletin 411-A (Fort Collins, CO: Agricultural Extension Service, April 1950); H. B. Walker, "A Resume of Sixteen Years of Research in Sugar-Beet Mechanization," *Agricultural Engineering* (1948): 425–430; P. B. Smith, "A Survey of Sugar-Beet Mechanization" (Fort Collins, CO, 1950, mimeographed); R. K. Smith, "State Frontiers in Agricultural Statistics: Discussion," *Journal of Farm Economics* 31, no. 1 (1949): 304–308.

63. Boris C. Swerling, "United States Sugar Policy and Western Development," *Proceedings of the Annual Meeting (Western Farm Economics Association)* 24 (1951): 7–11.

64. For an example of an SQC algorithm, see A. Smirnov, B. N. Holben, T. F. Eck, and O. Dubovik, "Cloud-Screening and Quality Control Algorithms for the AERONET Database," *Remote Sensing of Environment* 73, no. 3 (2000): 337–349.

Chapter 9

1. Jacob Cohen, "The Earth Is Round (p<.05)," *American Psychologist* 49, no. 12 (December 1994): 997–1003, on 997.
2. R. A. Fisher, *The Design of Experiments* (Edinburgh: Oliver and Boyd, 1935), esp. 8–9, 13–18.
3. Cohen, "The Earth Is Round," 997.
4. Lancelot T. Hogben, *Statistical Theory: The Relationship of Probability, Credibility, and Error* (London: George Allen & Unwin, 1957).
5. Denton E. Morrison and Ramon E. Henkel, eds, *The Significance Test Controversy—A Reader* (Chicago: Aldine, 1970).
6. William W. Rozeboom, "The Fallacy of the Null-Hypothesis Significance Test," *Psychological Bulletin* 57, no. 5 (1960): 416–428, on 416.
7. David Bakan, "The Test of Significance in Psychological Research," *Psychological Bulletin* 66, no. 6 (December 1966): 423–437, on 423 and 436.
8. Allen L. Edwards, *Statistical Methods for the Behavioral Sciences* (New York: Rinehart, 1954), vii.
9. James E. Wert, Charles O. Neidt, and J. Stanley Ahmann, *Statistical Methods in Educational and Psychological Research* (New York: Appleton-Century-Crofts, 1954), 103, 123–129.
10. Frederick A. Ekeblad, *The Statistical Method in Business: Applications of Probability and Inference to Business and other Problems* (New York: Wiley, 1962), vii.
11. Rozeboom, "The Fallacy of the Null-Hypothesis Significance Test," 424.
12. Among a growing literature on algorithms, see Rob Kitchin, "Thinking Critically about and Researching Algorithms," *Information, Communication and Society* 20, no. 1 (2017): 14–29; and Maarten Bullynck, "Histories of Algorithms: Past, Present and Future," *Historia Mathematica* 43, no. 3 (2016): 332–341.
13. On importing notions of religious ritual into analysis of techno-science, see David F. Noble, *The Religion of Technology: The Divinity of Man and the Spirit of Invention* (New York: Knopf, 1997); and Sandrine Thérèse and Brian Martin, "Shame, Scientist! Degradation Rituals in Science," *Prometheus* 28, no. 2 (2010): 97–110, accessed February 7, 2018.
14. Statisticians were repeating a widely held interpretation of "algorithm": Paul Erickson, Judy L. Klein, Lorraine Daston, Rebecca Lemov, Thomas Sturm, and Michael D. Gordin, *How Reason Almost Lost Its Mind: The Strange Career of Cold War Rationality* (Chicago: University of Chicago Press, 2013), 29–30.
15. Gerd Gigerenzer, Zeno Swijtink, Theodore Porter, Lorraine Daston, John Beatty, and Lorenz Krüger, *The Empire of Chance: How Probability Changed Science and Everyday Life* (Cambridge: Cambridge University Press, 1989), 90–109.
16. Theodor[e] D. Sterling, "Publication Decisions and Their Possible Effects on Inferences Drawn from Tests of Significance—Or Vice Versa," *Journal of the American Statistical Association* 54, no. 285 (March 1959): 30–34; a more recent article has recapitulated Sterling's point, but with a pithier title and updated statistical arguments: John P.A. Ioannidis, "Why Most Published Research Findings are False," *PLOS Medicine* 2, no. 8 (August 2005): 0696-0701, https://doi.org/10.1371/journal.pmed.0020124.
17. George W. Snedecor, *Statistical Methods: Applied to Experiments in Agriculture and Biology* (Ames: Iowa State College Press, 1946 [fourth ed.; first published 1937]); Gerd Gigerenzer and David J. Murray, *Cognition as Intuitive Statistics* (Hillsdale, NJ: Lawrence Erlbaum Associates, 1987), 21.

18. Gigerenzer and Murray, *Cognition as Intuitive Statistics*, 20.
19. For frequencies in the 1950s, see Sterling, "Publication Decisions"; and for the role of analysis of variation (and the teaching of statistics in psychology graduate programs), see Anthony J. Rucci and Ryan D. Tweney, "Analysis of Variance and the 'Second Discipline' of Scientific Psychology: A Historical Account," *Psychological Bulletin* 87, no. 1 (1980): 166–184. For a similar account of the spread of statistical analysis, see Theodore M. Porter, *Trust in Numbers: The Pursuit of Objectivity in Science and Public Life* (Princeton, NJ: Princeton University Press, 1995), 200–213.
20. Kurt Danziger, *Constructing the Subject: Historical Origins of Psychological Research* (Cambridge: Cambridge University Press, 1990), esp. 147.
21. His views are discussed in Gigerenzer and Murray, *Cognition as Intuitive Statistics*, esp. 27.
22. Arthur W. Melton, "Editorial," *Journal of Experimental Psychology* 64, no. 6 (December 1962): 553–557, on 554, cited in Gigerenzer et al., *The Empire of Chance*, 209.
23. Hogben, *Statistical Theory*, 325–326.
24. Gigerenzer and Murray, *Cognition as Intuitive Statistics*, 27.
25. For these claims in particular, see Porter, *Trust in Numbers*, 228–229; Gigerenzer et al., *Empire of Chance*, 210.
26. Louis: J. Rosser Matthews, *Quantification and the Quest for Medical Certainty* (Princeton, NJ: Princeton University Press, 1995), ch. 1–4; Lind: Iain Milne, "Who Was James Lind, and What Exactly Did He Achieve?" *JLL Bulletin* (2012), accessed January 11, 2018, http://www.jameslindlibrary.org/articles/who-was-james-lind-and-what-exactly-did-he-achieve/; Farr: Michael Donnelly, "William Farr and Quantification in Nineteenth-Century English Public Health," in *Body Counts: Medical Quantification in Historical and Sociological Perspective*, ed. Gérard Jorland, Annick Opinel, and George Weisz (Montreal: McGill-Queen's University Press, 2005), 251–265; Graunt: A. M. Endres, "The Functions of Numerical Data in the Writings of Graunt, Petty, and Davenant," *History of Political Economy* 17, no. 2 (1985): 245–264; and Kenneth J. Rothman, "Lessons from John Graunt," *Lancet* 347, no. 8993 (January 6, 1996): 37–39; Legendre, Gauss, Laplace, and Bernoulli are discussed in Gigerenzer et al., *Empire of Chance*.
27. M. A. Quetelet, *A Treatise on Man and the Development of His Faculties*, reprint ed. (New York: Burt Franklin, 1968 [1842]), esp. ix, 33.
28. Émile Durkheim, *On Suicide*, trans. Robin Buss (New York: Penguin, 2006 [1897]), esp. 21–22 and 343.
29. For Pearson's influence, see Theodore M. Porter, *Karl Pearson: The Scientific Life in a Statistical Age* (Princeton, NJ: Princeton University Press, 2004); and Stephen M. Stigler, *The History of Statistics: The Measurement of Uncertainty before 1900* (Cambridge, MA: Harvard University Press, 1986), esp. 300–363.
30. E.g., inherited traits: Francis Galton, *Natural Inheritance* (London: Macmillan, 1889); quality control: Student [William Gosset], "The Probable Error of a Mean," *Biometrika* 6, no. 1 (March 1908): 1–25.
31. G. Udny Yule, *An Introduction to the Theory of Statistics* (London: Griffin; Philadelphia: Lippincott, 1911), 5.
32. Yule, *Introduction to the Theory of Statistics*, 5.
33. Raymond Pearl, *Introduction to Medical Biometry and Statistics* (Philadelphia: Saunders, 1923), 21.
34. Henry E. Garrett, *Statistics in Psychology and Education* (New York: Longmans, Green and Co., 1926), 50–53 and 96–99.

35. Garrett, *Statistics in Psychology and Education*, 163–168, 224–231.

36. Quetelet, *Treatise on Man*, esp. 68–71, 93–100.

37. Lorraine Daston, "The Empire of Observation, 1600–1800," in *Histories of Scientific Observation*, ed. Lorraine Daston and Elizabeth Lunbeck (Chicago: University of Chicago Press, 2011), 81–113, on 82.

38. Theodore M. Porter, "Reforming Vision: The Engineer Le Play Learns to Observe Society Sagely," in *Histories of Scientific Observation*, ed. Daston and Lunbeck, 281–302.

39. Fisher, *Design of Experiments*, 3.

40. An excellent summary of both the theory and the controversy still remains Gigerenzer et al., *Empire of Chance*, 70–122; for a modern interpretation of the mathematical details, see Prakash Gorroochurn, *Classic Topics on the History of Modern Mathematical Statistics* (Hoboken, NJ: Wiley, 2016), ch. 5.

41. Charles E. Rosenberg, "Science, Technology, and Economic Growth: The Case of the Agricultural Experiment Station Scientist, 1975–1914," *Agricultural History* 45, no. 1 (January 1971): 1–20; Giuditta Parolini, "In Pursuit of a Science of Agriculture: The Role of Statistics in Field Experiments," *History and Philosophy of the Life Sciences* 37, no. 3 (2015): 261–281.

42. "Obituary: William Sealy Gosset, 1876–1937," *Journal of the Royal Statistical Society* 101, no. 1 (1938): 248–251.

43. See Theodora Dryer's contribution to this collection (chapter 8).

44. Henry E. Garrett, *Statistics in Psychology and Education*, 6th ed. (New York: McKay, 1966), esp. 213, 218, 247.

45. Nathan Mantel, "A Personal Perspective on Statistical Techniques for Quasi-Experiments," in *On the History of Statistics and Probability*, ed. D. B. Owen (New York: Dekker, 1976), 103–129.

46. For the language of "disproving," see Fisher, *Design of Experiment*, 19; Popper's book was originally published in German in 1935 and translated as Karl Popper, *Logic of Scientific Discovery* (London: Hutchinson, 1959); see also Jordi Cat, "The Unity of Science," *Stanford Encyclopedia of Philosophy* , ed. Edward N. Zalta (Fall 2017 ed.), accessed October 7, 2019, https://plato.stanford.edu/archives/fall2017/entries/scientific-unity/.

47. John L. Rudolph, "Epistemology for the Masses: The Origins of 'The Scientific Method' in American Schools," *History of Education Quarterly* 45, no. 3 (Fall 2005): 341–376.

48. For a definitive rebuttal of any such thing as a scientific method among working scientists in this period, see Steven Shapin, "How to Be Antiscientific," in *The One Culture? A Conversation about Science*, ed. Jay A. Labinger and Harry Collins (Chicago: University of Chicago Press, 2001), 99–115.

49. Paul E. Meehl, "Wanted—A Good Cookbook," in *Psychodiagnosis: Selected Paper* (Minneapolis: University of Minnesota Press, 1972 [1955]), 63–80, quotations on 66 and 79.

50. Meehl, "Wanted—A Good Cookbook," 77–78.

51. See, e.g., Hunter Heyck, *Age of System: Understanding the Development of Modern Social Science* (Baltimore: Johns Hopkins University Press, 2015); even the otherwise excellent Gigerenzer et al., *Empire of Chance* does not situate these developments within Cold War science.

52. Erickson et al., *How Reason Almost Lost Its Mind*, 3, 30.

53. Heyck, *Age of System*, 131.

54. Robert Dorfman, Paul A. Samuelson, and Robert M. Solow, *Linear Programming and Economic Analysis* (New York: McGraw-Hill, 1958), 1–4.

55. John von Neumann and Oskar Morgenstern, *Theory of Games and Economic Behavior*, 2nd ed (Princeton, NJ: Princeton University Press, 1947 [1944]), 8–10.

56. von Neumann and Morgenstern, *Theory of Games*, 79.

57. Paul Erickson, *The World the Game Theorists Made* (Chicago: University of Chicago Press), 97–98.

58. Erickson et al., *How Reason Almost Lost Its Mind*, 176–177.

59. Daniel Kahneman, Paul Slovic, and Amos Tversky, eds, *Judgment under Uncertainty: Heuristics and Biases* (Cambridge: Cambridge University Press, 1982).

60. Erickson, *World the Game Theorists Made*, 81.

61. W. Edwards Deming, *Statistical Adjustment of Data* (New York: John Wiley, 1938), 30.

62. Though first developed under classified Defense Department contracts, the theory was later published as Abraham Wald, *Sequential Analysis* (New York: Wiley, 1947). For initial design and applications to ordnance, see Statistical Research Group, *Sequential Analysis in Inspection and Experimentation*, Report No. 255, AMP Report 30.2R (New York: Columbia University Press, 1945); and more broadly, Patti Wilger Hunter, "Connections, Context, and Community: Abraham Wald and the Sequential Probability Ratio Test," *The Mathematical Intelligencer* 26, no. 1 (December 2004): 25–33.

63. Abraham Wald, "Statistical Decision Functions Which Minimize the Maximum Risk," *Annals of Mathematics*, 2nd ser., 46, no. 2 (April 1945): 265–280, and *Statistical Decision Functions* (New York: Wiley, 1950).

64. L. J. Savage, "Theory of Statistical Decision," *Journal of the American Statistical Association* 46, no. 253 (March 1951): 55–67, on 55.

65. Leonard J. Savage, *The Foundations of Statistics*, 2nd rev ed. (New York: Dover, 1972 [1954]), 6.

66. Joan Fisher Box, "R. A. Fisher and the Design of Experiments, 1922–1926," *American Statistician* 34, no. 1 (February 1980): 1–7, on 3.

67. Erickson, *World the Game Theorists Made*, 19.

68. Erickson, *World the Game Theorists Made*, 98–99.

69. Erickson, *World the Game Theorists Made*, 65.

70. Harold Hotelling, "Abraham Wald," *American Statistician* 5, no. 1 (February 1951): 18–19.

71. Thomas S. Ferguson, "The Development of the Decision Model," in *On the History of Statistics and Probability*, ed. D. B. Owen (New York: Dekker, 1976), 333–346, on 336.

72. Heyck, *Age of System*, 138.

73. Hallam Stevens has tracked how the situation changed in the 1980s, as biology (among other fields) gradually became data- and statistics-dependent: Hallam Stevens, *Life out of Sequence: A Data-Driven History of Bioinformatics* (Chicago: University of Chicago Press, 2013); more broadly, he argues it was only *after* 1970 that algorithms fundamentally changed the way the practice of the natural sciences (as opposed to the earlier transformation in the social sciences): Hallam Stevens, "A Feeling for the Algorithm: Working Knowledge and Big Data in Biology," *Osiris* 32 (2017): 151–174, esp. 152.

74. Gigerenzer and Murray, *Cognition as Intuitive Statistics*, 19–28.

75. Michael Coy Acree, "Theories of Statistical Inference in Psychological Research: A Historico-Critical Study" (PhD diss., Clark University, 1978), on 1.

76. Kurt Danziger, *Constructing the Subject: Historical Origins of Psychological Research* (Cambridge: Cambridge University Press), 68–100, quotation on p. 85.

77. Kurt Danziger, "Statistical Method and the Historical Development of Research Practice in American Psychology," in *The Probabilistic Revolution: Volume 2: Ideas in the Sciences*, ed.

Lorenz Krüger, Gerd Gigerenzer, and Mary S. Morgan (Cambridge, MA: MIT Press, 1987) 35–47, on 46. See also Gigerenzer and Murray, *Cognition as Intuitive Statistics*, 27; and James H. Capshew, *Psychologists on the March: Science, Practice, and Professional Identity in America, 1929–1969* (Cambridge: Cambridge University Press, 1999), 212–217.

78. Gigerenzer and Murray, *Cognition as Intuitive Statistics*, 186–187. Emphasis added.

79. Danziger, *Constructing the Subject*, 147–154.

80. Rozeboom, "The Fallacy of the Null-Hypothesis Significance Test," 424.

81. This section draws on Christopher J. Phillips, "The Taste Machine: Sense, Subjectivity, and Statistics in the California Wine World," *Social Studies of Science* 46, no. 3 (2016): 461–481.

82. Steven Shapin, "A Taste of Science: Making the Subjective Objective in the California Wine World," *Social Studies of Science* 46, no. 3 (2016): 436–460.

83. Maynard A. Amerine and Edward B. Roessler, *Wines: Their Sensory Evaluation* (San Francisco: Freeman, 1976), 60–62.

84. See Morrison and Henkel, eds, *Significance Test Controversy*, 59–64.

85. Henry K. Beecher, *Measurement of Subjective Responses: Quantitative Effects of Drugs* (New York: Oxford, 1959), and "Pain: One Mystery Solved," *Science* 151, no. 3712 (1966): 840–841; on Beecher, see Noémi Tousignant, *Pain and the Pursuit of Objectivity: Pain-Measurement Technologies in the United States, c. 1890–1975* (PhD thesis, McGill University, 2006), and "The Rise and Fall of the Dolorimeter: Pain, Analgesics, and the Management of Subjectivity in Mid-Twentieth-Century United States," *Journal of the History of Medicine and Allied Sciences* 66, no. 2 (2011): 145–179.

86. Cf. Porter, *Trust in Numbers*, 229.

87. Harry M. Marks, *The Progress of Experiment: Science and Therapeutic Reform in the United States, 1900–1930* (Cambridge: Cambridge University Press, 1997); William G. Rothstein, *Public Health and the Risk Factor: A History of an Uneven Medical Revolution* (Rochester, NY: University of Rochester Press, 2003); Gérard Jorland, Annick Opinel, and George Weisz, eds., *Body Counts: Medical Quantification in History and Sociological Perspectives* (Montreal: McGill–Queen's University Press, 2005).

88. Jeremy A. Greene, *Prescribing by Number: Drugs and the Definition of Disease* (Baltimore: Johns Hopkins University Press, 2007); Laura Bothwell, "The Emergence of the Randomized Controlled Trial: Origins to 1980" (PhD diss., Columbia University, 2014); Daniel Carpenter, *Reputation and Power: Organizational Image and Pharmaceutical Regulation at the FDA* (Princeton, NJ: Princeton University Press, 2010), esp. 269–280.

89. Danziger, *Constructing the Subject*, 77–78.

90. Paul E. Meehl, "Theoretical Risks and Tabular Asterisks: Sir Karl, Sir Ronald, and the Slow Progress of Soft Psychology," *Journal of Consulting and Clinical Psychology* 46 (1978): 806–834, on 817.

91. The literature on objectivity/subjectivity has been one realm in which "epistemic virtue" is widely discussed, e.g., Lorraine Daston and Peter Galison, *Objectivity* (New York: Zone Books, 2010), 39–42. My account is more heavily influenced by Steven Shapin, "The Sciences of Subjectivity," *Social Studies of Science* 42, no. 2 (2011): 170–184.

92. As some advocates claim, the "messiness" of big data is its virtue: Viktor Mayer-Schönberger and Kenneth Cukier, *Big Data: A Revolution That Will Transform How We Live, Work, and Think* (Boston: Houghton Mifflin, 2013), 32–49.

93. The replacement of explicit causal mechanisms with "secret" calculations continues to fuel criticism of algorithms, e.g., Cathy O'Neil, *Weapons of Math Destruction: How Big Data Increases Inequality and Threatens Democracy* (New York: Crown, 2016); and Frank

Pasquale, *The Black Box Society: The Secret Algorithms That Control Money and Information* (Cambridge, MA: Harvard University Press, 2015).

94. For an early summary of the possibilities of available software, see D. F. Andrews, "Developing Examples for Learning Statistics; Data and Computing," *International Statistical Review* 41, no. 2 (1973): 225–228; and Norman H. Nie, Dale H. Bent, and C. Hadlai Hull, *SPSS: Statistical Package for the Social Sciences* (New York: McGraw-Hill, 1970).

95. Jon Agar, "What Difference Did Computers Make?," *Social Studies of Science* 36, no. 6 (December 2006): 869–907. His claim mainly applies to the earliest forms of electronic computing; in other respects, electronic computers have had a qualitatively different effect: Stevens, "Feeling for the Algorithm," 154–157; Stephanie Dick, "AfterMath: The Work of Proof in the Age of Human-Machine Collaboration," *Isis* 102 (2011): 494–505.

Chapter 10

1. Xindong Wu et al., "Top 10 Algorithms in Data Mining," *Knowledge and Information Systems* 14, no. 1 (December 4, 2007): 1–37, https://doi.org/10.1007/s10115-007-0114-2. See now Adrian Mackenzie, *Machine Learners: Archaeology of a Data Practice* (Cambridge, MA: MIT Press, 2018), 127–137.

2. Xinran He et al., "Practical Lessons from Predicting Clicks on Ads at Facebook" (New York: ACM Press, 2014), 1–9, https://doi.org/10.1145/2648584.2648589.

3. Leo Breiman et al., *Classification and Regression Trees* (Boca Raton, FL: Chapman & Hall, 1984), viii. For skepticism about such claims, see Jon Agar, "What Difference Did Computers Make?," *Social Studies of Science* 36, no. 6 (December 1, 2006): 869–907, https://doi.org/10.1177/03063 12706073450; and the measured pushback in Hallam Stevens, *Life out of Sequence: A Data-Driven History of Bioinformatics* (Chicago: University of Chicago Press, 2013).

4. See, e.g., Virginia Eubanks, *Automating Inequality: How High-Tech Tools Profile, Police, and Punish the Poor* (New York: St. Martin's Press, 2017); Frank Pasquale, *The Black Box Society: The Secret Algorithms That Control Money and Information* (Cambridge, MA: Harvard University Press, 2015), among many others. For a breakdown of different humanist accounts of opacity, see Jenna Burrell, "How the Machine 'Thinks': Understanding Opacity in Machine Learning Algorithms," *Big Data & Society* 3, no. 1 (January 5, 2016): 4–5, https://doi.org/10.1177/2053951715622512. For a machine learning viewpoint with a taxonomy of different forms of "interpretable," see Zachary Chase Lipton, "The Mythos of Model Interpretability," *CoRR* abs/1606.03490 (2016), http://arxiv.org/abs/1606.03490. And from the philosophy of science, Kathleen A. Creel, "Transparency in Complex Computational Systems," *Philosophy of Science* 87 (2020): 4.

5. See the call for genealogy in Massimo Mazzotti, "Algorithmic Life," *Los Angeles Review of Books*, accessed April 17, 2017, https://lareviewofbooks.org/article/algorithmic-life/.

6. For implementation and the materiality of the machine, see Paul Edwards, *A Vast Machine: Computer Models, Climate Data, and the Politics of Global Warming* (Cambridge, MA: MIT Press, 2010), esp. ch. 5. on computational friction; and Stephanie Dick, "Of Models and Machines: Implementing Bounded Rationality," *Isis* 106, no. 3 (2015): 623–634.

7. For an initial survey of this in the United States and beyond, see Matthew L. Jones, "How We Became Instrumentalists (Again): Data Positivism since World War II," *Historical Studies in the Natural Sciences* 48, no. 5 (November 1, 2018): 673–684, https://doi.org/10.1525/

hsns.2018.48.5.673. See also Adrian Mackenzie, "The Production of Prediction: What Does Machine Learning Want?," *European Journal of Cultural Studies* 18, no. 4–5 (2015): 429–445. Ann Johnson, "Rational and Empirical Cultures of Prediction," in *Mathematics as a Tool*, ed. Johannes Lenhard and Martin Carrier, vol. 327 (Cham: Springer International Publishing, 2017), 23–35, https://doi.org/10.1007/978-3-319-54469-4_2.

8. On the question of scaling algorithms, see Matthew L. Jones, "Querying the Archive: Database Mining from Apriori to Page-Rank," in *Science in the Archives: Pasts, Presents, Futures*, ed. Lorraine Daston (Chicago: Chicago University Press, 2016), 311–328.

9. On writing the history of machine learning, see Aaron Plasek, "On the Cruelty of Really Writing a History of Machine Learning," *IEEE Annals of the History of Computing* 38, no. 4 (December 2016): 6–8, https://doi.org/10.1109/MAHC.2016.43.

10. For a survey of more code-based approaches to studying algorithms, see Rob Kitchin, "Thinking Critically about and Researching Algorithms," *Programmable City Working Paper 5*, 2014, http://dx.doi.org/10.2139/ssrn.2515786.

11. Nick Seaver, "Knowing Algorithms," in *A Field Guide for Science & Technology Studies*, ed. Janet Vertesi and David Ribes (Princeton, NJ: Princeton University Press, 2019), 419. See also his "Algorithms as Culture: Some Tactics for the Ethnography of Algorithmic Systems," *Big Data & Society* 4, no. 2 (December 2017), https://doi.org/10.1177/2053951717738104.

12. "Scikit-Learn/Tree.Py at 14031f65d144e3966113d3daec836e443c6d7a5b · Scikit-Learn/Scikit-Learn," accessed April 15, 2017, https://github.com/scikit-learn/scikit-learn/blob/14031f6/sklearn/tree/tree.py#L508.

13. For this synthesis, see Phillips (chapter 9, this volume); and Gerd Gigerenzer, ed., *The Empire of Chance: How Probability Changed Science and Everyday Life*, Ideas in Context (Cambridge: Cambridge University Press, 1989), ch. 3.

14. John A. Sonquist and James N. Morgan, *The Detection of Interaction Effects: A Report on a Computer Program for the Selection of Optional Combinations of Explanatory Variables* (Ann Arbor: Institute for Social Research, University of Michigan, 1964), iii. See also John A. Sonquist, Elizabeth Lauh Baker, and James N. Morgan, *Searching for Structure; an Approach to Analysis of Substantial Bodies of Micro-Data and Documentation for a Computer Program* (Ann Arbor: Survey Research Center, University of Michigan, 1973).

15. Sonquist and Morgan, *The Detection of Interaction Effects*, 2.

16. Hillel J. Einhorn, "Alchemy in the Behavioral Sciences," *Public Opinion Quarterly* 36, no. 3 (1972): 368–369. For these critiques, see Mackenzie, *Machine Learners*, 129–130.

17. Einhorn, "Alchemy in the Behavioral Sciences," 369.

18. James N. Morgan and Frank M. Andrews, "A Comment on Einhorn's 'Alchemy in the Behavioral Sciences,'" *Public Opinion Quarterly* 37, no. 1 (1973): 127.

19. Quoted in Eric Siegel, *Predictive Analytics: The Power to Predict Who Will Click, Buy, Lie, or Die* (Hoboken, NJ: Wiley, 2016), 175–176.

20. Breiman et al., *Classification and Regression Trees*, 7. For the practitioners' retrospective, see the videos at https://www.salford-systems.com/videos/conferences/cart-founding-fathers/.

21. Breiman et al., *Classification and Regression Trees*, 7.

22. Breiman et al., *Classification and Regression Trees*, 8.

23. Breiman et al., *Classification and Regression Trees*, viii.

24. Breiman et al., *Classification and Regression Trees*, 7.

25. For reasons of space, I'm not discussing two additional distinguished authors of the 1984 book, Richard Olshen and Charles Stone, both of whom profoundly deepened the mathematical statistics of the volume.

26. Richard Olshen and Leo Breiman, "A Conversation with Leo Breiman," *Statistical Science* 16, no. 2 (2001): 196.

27. Olshen and Breiman, "A Conversation with Leo Breiman," 188.

28. Leo Breiman, "[A Report on the Future of Statistics]: Comment," *Statistical Science* 19, no. 3 (2004): 411–411.

29. John W. Tukey, "The Future of Data Analysis," *Annals of Mathematical Statistics* 33, no. 1 (March 1962): 1–67, https://doi.org/10.1214/aoms/1177704711, at 6. See Colin Mallows, "Tukey's Paper after 40 Years," *Technometrics* 48, no. 3 (2006): 319–325.

30. Luisa T. Fernholz et al., "A Conversation with John W. Tukey and Elizabeth Tukey," *Statistical Science* 15, no. 1 (2000): 85.

31. Leo Breiman and William S. Meisel, "Empirical Techniques for Analyzing Air Quality and Metereological Data. Part III. Short-Term Changes to Ground Lever Ozone Concentration: An Empirical Analysis," Environmental Monitoring Series (Environmental Sciences Research Laboratory, Environmental Protection Agency, June 1976), 5.

32. Breiman and Meisel, "Empirical Techniques," 13.

33. Breiman et al., *Classification and Regression Trees*, 5.

34. Breiman et al., *Classification and Regression Trees*, 6.

35. William S. Meisel and Leo Breiman, "Topics in the Analysis and Optimization of Complex Systems. Appendix B. Tree Structured Classification Methods," Final report to AFOSR (Technology Service Corporation, February 28, 1977), 4, http://oai.dtic.mil/oai/oai?verb= getRecord&metadataPrefix=html&identifier=ADA038209.

36. Meisel and Breiman, "Topics in the Analysis and Optimization of Complex Systems," 4.

37. Meisel and Breiman, "Topics in the Analysis and Optimization of Complex Systems," 4.

38. Breiman et al., *Classification and Regression Trees*, 23.

39. Breiman et al., *Classification and Regression Trees*, 37.

40. N. I. Fisher, "A Conversation with Jerry Friedman," *Statistical Science* 30, no. 2 (May 2015): 271, https://doi.org/10.1214/14-STS509.

41. Fisher, "A Conversation with Jerry Friedman," 276.

42. Fisher, "A Conversation with Jerry Friedman," 276.

43. J. H. Friedman, "A Recursive Partitioning Decision Rule for Nonparametric Classification," *IEEE Transactions on Computers* C–26, no. 4 (April 1977): 4, https://doi.org/10.1109/ TC.1977.1674849.

44. Fisher, "A Conversation with Jerry Friedman," 276.

45. Leo Breiman, "Statistical Modeling: The Two Cultures," *Statistical Science* 16, no. 3 (2001): 201.

46. Joseph Adam November, *Biomedical Computing: Digitizing Life in the United States* (Baltimore: Johns Hopkins University Press, 2012), 259–268.

47. Edward Feigenbaum, Oral History of Edward Feigenbaum, interview by Nils Nilsson, 20, 27 2007, 62–63, http://archive.computerhistory.org/resources/access/text/2013/05/ 102702002-05-01-acc.pdf; D. E. Forsythe, "Engineering Knowledge: The Construction of Knowledge in Artificial Intelligence," *Social Studies of Science* 23, no. 3 (August 1, 1993): 445–477, https://doi.org/10.1177/0306312793023003002.

48. J. R. Quinlan, "Discovering Rules by Induction from Large Collections of Examples," in *Expert Systems in the Micro-Electronic Age*, ed. Donald Michie (Edinburgh: Edinburgh University Press, 1979), 168.

49. Donald Michie, "Expert Systems Interview," *Expert Systems* 2, no. 1 (1985): 22.

50. Keki B. Irani et al., "Applying Machine Learning to Semiconductor Manufacturing," *IEEE Expert* 8, no. 1 (1993): 41.

51. J. Ross Quinlan, "Induction over Large Data Bases," Stanford Heuristic Programming Project Memo HPP 79-14 (Stanford University, May 1979), 1.

52. Quinlan, "Induction over Large Data Bases," 4.

53. Nathan Ensmenger, "Is Chess the Drosophila of Artificial Intelligence? A Social History of an Algorithm," *Social Studies of Science* 42, no. 1 (February 1, 2012): 5–30, https://doi.org/10.1177/0306312711424596.

54. Quinlan, "Induction over Large Data Bases," 3.

55. Quinlan, "Induction over Large Data Bases," 9.

56. Earl B. Hunt, Janet Martin, and Philip J. Stone, *Experiments in Induction* (New York: Academic Press, 1966), 10.

57. In a letter, Quinlan explained: "I sat in on a course given by Donald Michie [also visiting Stanford at that time] and became intrigued with a task he proposed, namely, learning a rule for deciding the result of a simple chess endgame. ID3 started out as a recoding of Buz's [that is, Earl B. Hunt's] CLS, but I changed some of the innards (such as the criterion for splitting a set of cases) and incorporated the iterative approach that allowed ID3 to handle the then-enormous set of 29,000 training cases." Nils J. Nilsson, *The Quest for Artificial Intelligence: A History of Ideas and Achievements* (Cambridge; New York: Cambridge University Press, 2010), 504. The algorithm works by selecting a "window," a subset of the training data, and then producing a tree that correctly classifies everything in the subset correctly. Then that tree is used to classify the rest of the training data. It then collects instances it has misclassified, and attempts to construct a superior tree capable of correctly classifying that additional labeled data.

58. J. R. Quinlan, "Induction of Decision Trees," *Machine Learning* 1, no. 1 (1986): 89–91.

59. Quinlan, "Induction of Decision Trees," 92.

60. Cao Feng and D. Michie, "Machine Learning of Rules and Trees," in *Machine Learning, Neural and Statistical Classification*, ed. Donald Michie et al. (Upper Saddle River, NJ: Ellis Horwood, 1994), 51, http://dl.acm.org/citation.cfm?id=212782.212787.

61. Feng and Michie, "Machine Learning of Rules and Trees," 51, my italics.

62. Ron Kohavi and Ross Quinlan, "Decision Tree Discovery," in *Handbook of Data Mining and Knowledge Discovery* (updated October 1999), 267–276, http://ai.stanford.edu/~ron nyk/treesHB.pdf. The software in C is available at rulequest.com/Personal/c4.5r8.tar.gz.

63. These claims rest on a survey of the major data mining journal *KDD*. For the challenge of the "literature" in contemporary science, see Christopher M. Kelty and Hannah Landecker, "Ten Thousand Journal Articles Later: Ethnography of 'The Literature' in Science," *Empiria*, no. 18 (2009): 173–192.

64. For his biography, see, e.g., Usama Fayyad, "Personal Observations of a Data Mining Disciple A Data Miner's Story—Getting to Know the Grand Challenges of Research," http://slideplayer.com/slide/6409052/.

65. Usama Mohammad Fayyad, "On the Induction of Decision Trees for Multiple Concept Learning" (Ann Arbor: University of Michigan, 1992), 3.

66. Fayyad, "On the Induction of Decision Trees," 6.

67. Usama M. Fayyad, S. George Djorgovski, and Nicholas Weir, "From Digitized Images to Online Catalogs Data Mining: A Sky Survey," *AI Magazine* 17, no. 2 (1996): 54.

68. Usama Fayyad, "Taming the Giants and the Monsters: Mining Large Databases for Nuggets of Knowledge," *Database Programming and Design Magazine* 11, no. 3 (1998). https://fayyad.com/taming-the-giants-and-the-monsters-mining-large-databases-for-nuggets-of-knowledge/

69. John Shafer, Rakeeh Agrawal, and Manish Mehta, "SPRINT: A Scalable Parallel Classifier for Data Mining," in *Proceedings of the 22nd International Conference on Very Large Data Bases* (San Mateo, CA: Morgan Kaufmann, 1996), 554.

70. Wray Buntine, "IND: Creation and Manipulation of Decision Trees from Data," accessed April 22, 2017, https://ti.arc.nasa.gov/opensource/projects/ind/.

71. Ronny Kohavi and Dan Sommerfield, "MLC++: Machine Learning Library in C++. SGIMLC++ Utilities 2.0," October 7, 1996.

72. README file, S. Murthy, *The OC1 Decision Tree System*, n.d., http://ccb.jhu.edu/software/oc1/oc1.tar.gz.

73. See Ronny Kohavi and Chia-Hsin LI, "TDDTInducer.c," March 1, 1995.

74. See Kohavi and LI, "TDDTInducer.c."

75. Peter Huber, "From Large to Huge: A Statistician's Reactions to KDD & DM," *KDD* 97 (1997): 307.

76. See the special "Algorithms and Culture" special issue of Big Data & Society, 2017, http://journals.sagepub.com/page/bds/collections/algorithms-in-culture.

77. U. M. Fayyad et al., "Machine Learning of Expert System Rules: Applications to Semiconductor Manufacturing," in *Collected Notes on the Workshop for Pattern Discovery in Large Databases (NASA Ames, January 14–15, 1991)* (Moffett Field, CA: NASA Ames Research Center, 1991), 26.

78. For an introduction and taxonomy of forms of ensemble learning, see Martin Sewell, "Ensemble Learning," August 2008, http://machine-learning.martinsewell.com/ensembles/ensemble-learning.pdf.

79. Leo Breiman and Nong Shang, "Born Again Trees," n.d.

80. Breiman, "Statistical Modeling," 208.

81. [redacted] OPC-MCR-GCHQ, "HIMR Data Mining Research Problem Book," September 20, 2011.

82. For the labor history of machine learning, see, e.g., Lilly Irani, "Justice for 'Data Janitors,'" *Public Books* (blog), January 15, 2015, http://www.publicbooks.org/justice-for-data-janitors/.

Bibliography

Introduction

Ames, Morgan G. 2018. "Deconstructing the Algorithmic Sublime." *Big Data & Society* 5(1): 1–4. Special issue on Algorithms in Culture.

Barthes, Roland. 1972. *Mythologies*. London: Paladin.

Beer, David. 2016. "Introduction: The Social Power of Algorithms." *Information, Communication & Society* 20(1): 1–13. Special issue on the Social Power of Algorithms.

Benjamin, Ruha. 2019. *Race after Technology: Abolitionist Tools for the New Jim Code*. N.p.: Wiley.

Bertoloni Meli, Dominico. 2006. *Thinking with Objects: The Transformation of Mechanics in the Seventeenth Century*. Baltimore: Johns Hopkins University Press.

Bowker, Geoffrey C., and Susan Leigh Star. 1999. *Sorting Things Out: Classification and Its Consequences*. Cambridge, MA: MIT Press.

Braverman, Harry. 1974. *Labor and Monopoly Capital*. New York and London: Monthly Review Press.

Broussard, Meredith. 2018. *Artificial Unintelligence: How Computers Misunderstand the World*. Cambridge, MA: MIT Press.

Burrell, Jenna. 2016. "How the Machine 'Thinks': Understanding Opacity in Machine Learning Algorithms." *Big Data & Society* 3(1): 1–12.

Chabert, Jean-Luc. 1999. *A History of Algorithms: From the Pebble to the Microchip*. Berlin: Springer.

Daston, Lorraine. 2011. *"Rules Rule: From Enlightenment Reason to Cold War Rationality."* Una's Lecture at the Townsend Center for the Humanities, University of California, Berkeley, April 2011.

Dick, Stephanie. 2011. "AfterMath: The Work of Proof in the Age of Human-Machine Collaboration." *Isis* 102(3): 494–505.

Dick, Stephanie. 2015. *After Math: (Re)configuring Minds, Proof, and Computing in the Postwar United States*. PhD diss., Harvard University.

Dourish, Paul. 2017. *The Stuff of Bits: An Essay on the Materialities of Information*. Cambridge, MA: MIT Press.

Dryer, Theodora. 2018. "Algorithms under the Reign of Probability." *IEEE Annals of the History of Computing* 40(1): 93–96.

Dryer, Theodora. 2019. *Designing Certainty: The Rise of Algorithmic Computing in an Age of Anxiety, 1920–1970*. PhD diss., University of California San Diego.

Dupuy, Jean-Pierre. 2009. *On the Origins of Cognitive Science: The Mechanization of the Mind*. Cambridge, MA: MIT Press.

Edwards, Paul. 2010. *A Vast Machine: Computer Models, Climate Data, and the Politics of Global Warming*. Cambridge, MA: MIT Press.

Erickson, Paul, Judy L. Klein, Lorraine Daston, Rebecca Lemov, Thomas Sturm, and Michael D. Gordin. 2013. *How Reason Almost Lost Its Mind: The Strange Career of Cold War Rationality*. Chicago: Chicago University Press.

Eubanks, Virginia. 2018. *Automating Inequality: How High-Tech Tools Profile, Police, and Punish the Poor*. New York: St. Martin's Press.

Gangadharan, Seeta Peña, and Jedrzej Niklas. 2019. "Decentering Technology in Discourse on Discrimination." *Information, Communication & Society* 22(7): 882–899.

Gillespie, Tarleton. 2014. "The Relevance of Algorithms." In *Media Technologies*, edited by Tarleton Gillespie, Pablo Boczkowski, and Kirsten Foot, 167–193. Cambridge, MA: MIT Press.

Jasanoff, Sheila. 2015. "Future Imperfect: Science, Technology, and the Imaginations of Modernity." In *Dreamscapes of Modernity: Sociotechnical Imaginaries and the Fabrication of Power*, edited by Sheila Jasanoff and Sang-Hyun Kim, 1–33. Chicago: University of Chicago Press.

Langlois, Ganaele. 2014. *Meaning in the Age of Social Media*. New York: Palgrave Macmillan.

Law, John. 1986. "On the Methods of Long Distance Control: Vessels, Navigation, and the Portuguese Route to India." In *Power, Action and Belief: A New Sociology of Knowledge?*, edited by John Law, 234–263. *Sociological Review Monograph 32*. London: Routledge.

Law, John. 1994. *Organizing Modernity: Social Ordering and Social Theory*. Oxford: Blackwell.

Light, Jennifer. 1999. "When Computers Were Women." *Technology and Culture* 40(3): 455–483.

Lubar, Steven. 1992. "'Do Not Fold, Spindle or Mutilate': A Cultural History of the Punch Card." *Journal of American Culture* 15(4): 43–55.

MacKenzie, Donald. 2006. *An Engine, Not a Camera: How Financial Models Shape Markets*. Cambridge, MA: MIT Press.

MacKenzie, Donald, and Judy Wajcman. 1985. *The Social Shaping of Technology*. Maidenhead: Open University Press.

Mager, Astrid. 2014. "Defining Algorithmic Ideology: Using Ideology Critique to Scrutinize Corporate Search Engines." *tripleC: Communication, Capitalism & Critique. Open Access Journal for a Global Sustainable Information Society* 12(1): 28–39.

Marx, Karl. 1867. *Capital. Volume I: The Process of Production of Capital*. Available online at https://marxists.org/archive/marx/works/1867-c1/ch15.htm#S10.

Mosco, Vincent. 2004. *The Digital Sublime: Myth, Power, and Cyberspace*. Cambridge, MA: MIT Press.

Nye, David E. 1996. *American Technological Sublime*. Cambridge, MA: MIT Press.

Noble, Safiya. 2018. *Algorithms of Oppression*. New York: New York University Press.

O'Neil, Cathy. 2016. *Weapons of Math Destruction*. New York: Crown Random House.

Pasquale, Frank. 2015. *The Black Box Society: The Secret Algorithms That Control Money and Information*. Cambridge, MA: Harvard University Press.

Roberts, Sarah. 2019. *Behind the Screen: Content Moderation in the Shadows of Social Media*. New Haven, CT: Yale University Press.

Scott, James C. 1999. *Seeing Like a State: How Certain Schemes to Improve the Human Condition Have Failed*. New Haven, CT: Yale University Press.

Stark, Luke. 2018. "Algorithmic Psychometrics and the Scalable Subject." *Social Studies of Science* 48(2): 204–231. https://doi.org/10.1177/0306312718772094.

Steiner, Christopher. 2012. *Automate This: How Algorithms Came to Rule Our World*. New York: Penguin.

Van Couvering, Elizabeth. 2008. "The History of the Internet Search Engine: Navigational Media and the Traffic Commodity." In *Web Search: Information Science and Knowledge Management*, edited by A. Spink and M. Zimmer, vol 14, 177–206. Berlin and Heidelberg: Springer.

Widman, Jeff. n.d. "fig. 2.1." Archived version available at http://web.archive.org/web/2022040 9144546/http://edgerank.net/.

Winner, Langdon. 1978. *Autonomous Technology: Technics-out-of-Control as a Theme in Political Thought*. Cambridge, MA: MIT Press.

Wittgenstein, Ludwig. 1953. *Philosophical Investigations*. Oxford: Blackwell.

Ziewitz, Malte. 2016. "Governing Algorithms: Myth, Mess, and Methods." *Science, Technology, & Human Values* 41(1): 3–16. https://doi.org/10.1177/0162243915608948.

Chapter 1

Arnaud, Sophie. *La voix de la nature dans l'oeuvre de Jacques Peletier du Mans (1517–1582)*. Paris: Honoré Champion, 2005.

Axworthy, Angela. *Le Mathématicien renaissant et son savoir. Le statut des mathématiques selon Oronce Fine*. Paris: Classiques Garnier, 2016.

Cardano, Girolamo. *Artis magnae, sive, De regulis algebraicis, liber unus*. Nuremberg: Johannes Petreius, 1545.

Cardano, Girolamo. *Artis magnae, sive, De regulis algebraicis, liber unus*. Edited by Massimo Tamborini. Milan: FrancoAngeli, 2011.

Cardano, Girolamo. *The Rules of Algebra*. Translated by T. Richard Witmer. Mineola, NY: Dover Publications, Inc., 1992. [reissue of 1968 edition]

Cifoletti, Giovanna. "The Creation of the History of Algebra in the Sixteenth Century." In *L'Europe mathématique: Histoires, Mythes, Identités*, edited by Catherine Goldstein, Jeremy Gray, and Jim Ritter, 123–142. Paris: Éditions de la Maison des sciences de l'homme, 1996.

Cifoletti, Giovanna. "From Valla to Viète: The Rhetorical Reform of Logic and Its Use in Early Modern Algebra." *Early Science and Medicine* 11, no. 4 (2006): 390–423.

Cifoletti, Giovanna. "Mathematics and Rhetoric: Jacques Peletier, Guillaume Gosselin, and the Making of the French Algebraic Tradition." PhD diss., Princeton University, 1992.

Cifoletti, Giovanna. "La question de l'algèbre: Mathématiques et rhétorique des hommes de droit dans la France du 16e siècle." *Annales: Histoire, Sciences Sociales* 50, no. 6 (November–December 1995): 1385–1416.

Cifoletti, Giovanna. "L'utile de l'entendement et l'utile de l'action: discussion sur l'utilité des mathématiques au xvie siècle." *Revue de synthèse* 4, no. 2–4 (April–December 2001): 503–520.

Coquard, Jean-Marie. "Mathématiques et dialectique dans l'oeuvre de Simon Stevin: l'intérêt des series de problèmes." *SHS Web of Conferences* 22, no. 12 (2015).

Davis, Natalie Zemon. "Mathematicians in Sixteenth-Century French Academies: Some Further Evidence." *Renaissance News* 11 (1958): 3–10.

Davis, Natalie Zemon. "Peletier and Beza Part Company." *Studies in the Renaissance* 11 (1964): 188–222.

Davis, Natalie Zemon "Sixteenth-Century French Arithmetics on the Business Life." *Journal of the History of Ideas* 21, no. 1 (1960): 18–48.

De Risi, Vincenzo, ed. *Mathematizing Space: The Objects of Geometry from Antiquity to the Early Modern Age*. Cham et al.: Springer/Birkhäuser, 2015.

Dijksterhuis, E. J. *Simon Stevin: Science in the Netherlands around 1600*. Translated by C. Dikshoorn. The Hague: Martinus Nijhoff, 1970.

Diophantus of Alexandria. *Opera omnia* .... Edited by Paul Tannery. 2 vols. Leipzig: B. G. Teubner, 1893–1895.

Diophantus of Alexandria. *Rerum arithmeticarum libri sex*. Edited by Wilhelm Xylander. Basel: Eusebius and Nicolas Episcopius, 1575.

Forcadel, Pierre. *L'Arithmeticque*. Paris: Guillaume Cavellat, 1557.

Freguglia, Paolo. "Viète Reader of Diophantus: An Analysis of Zeteticorum Libri Quinque." *Bollettino di storia delle scienze matematiche* 28, no. 1 (2008): 51–95.

Gosselin, Guillaume. *De arte magna libri IV Traité d'algèbre suivi de Praelectio/Leçon sur la mathématique*. Edited and translated by Odile Le Guillou-Kouteynikoff. Paris: Les Belles Lettres, 2016.

Goulding, Robert. *Defending Hypatia: Ramus, Savile, and the Renaissance Rediscovery of Mathematical History*. Dordrecht: Springer, 2010.

Grafton, Anthony. *Cardano's Cosmos: The Worlds and Works of a Renaissance Astrologer*. Cambridge, MA: Harvard University Press, 1999.

Grafton, Anthony, and Lisa Jardine. *From Humanism to the Humanities: Education and the Liberal Arts in Fifteenth- and Sixteenth-Century Europe.* Cambridge, MA: Harvard University Press, 1986.

Guicciardini, Niccolò. *Isaac Newton on Mathematical Certainty and Method.* Cambridge, MA: MIT Press, 2009.

Hérigone, Pierre. *Cursus Mathematicus, nova, brevi et clara methodo demonstratus … Tomus Primus.* Paris: Henry Le Gras, 1634.

Høyrup, Jens. "The Formation of a Myth: Greek Mathematics—Our Mathematics." In *L'Europe mathématique: Histoires, mythes, identités,* edited by Catherine Goldstein, Jeremy Gray, and Jim Ritter, 103–119. Paris: Éditions de la Maison des sciences de l'homme, 1996.

Jardine, Nicholas. "The Epistemology of the Sciences." In *The Cambridge History of Renaissance Philosophy,* edited by Charles Schmitt, Quentin Skinner, Eckhard Kessler, and Jill Kraye, 685–711. Cambridge: Cambridge University Press, 1988.

Kaplan, Abram. "Analysis and Demonstration: Wallis and Newton on Mathematical Presentation." *Notes and Records of the Royal Society* 74, no. 2 (2018): 447–468.

Klein, Jacob. *Greek Mathematical Thought and the Origin of Algebra.* Translated by Eva Brann. Mineola, NY: Dover Publications, Inc., 1992. [reissue of 1968 edition]

Loget, François. "L'algèbre en France au XVIe siècle." In *Pluralité de l'algèbre à la Renaissance,* edited by Sabine Rommevaux, Maryvonne Spiesser, and Maria Rosa Massa Esteve, 69–101. Paris: Honoré Champion, 2012.

Loget, François. "De l'algèbre comme art à l'algèbre pour l'enseignement: Les manuels de Pierre de La Ramée, Bernard Salignac et Lazare Schöner." *Revue de synthèse,* ser. 6, 132, no. 4 (2011): 495–527.

Mahoney, Michael. "The Beginnings of Algebraic Thought in the Seventeenth Century." In *Descartes: Philosophy, Mathematics and Physics,* edited by Stephen Gaukroger, 141–168. Sussex: Harvester Press, 1980.

Malcolm, Noel, and Jacqueline Stedall. *John Pell (1611–1685) and His Correspondence with Sir Charles Cavendish: The Mental World of an Early Modern Mathematician.* Oxford: Oxford University Press, 2004.

Mancosu, Paolo. "Aristotelian Logic and Euclidean Mathematics: Seventeenth-Century Developments of the Quaestio de Certitudine Mathematicarum." *Studies in the History and Philosophy of Science* 23, no. 2 (1992): 241–265.

Margolin, Jean-Claude. "L'Enseignement des mathématiques en France (1540–70): Charles de Bovelles, Fine, Peletier, Ramus." In *French Renaissance Studies 1540–70: Humanism and the Encyclopedia,* edited by Peter Sharratt, 109–155. Edinburgh: Edinburgh University Press, 1976.

Marr, Alexander, ed. *The Worlds of Oronce Fine: Mathematics, Instruments, and Print in Renaissance France.* Donington: Shaun Tyas, 2009.

Massa Esteve, Maria Rosa. "The Role of Symbolic Language in the Transformation of Mathematics." *Philosophica* 82 (2012): 153–193.

Meskens, Ad. *Travelling Mathematics—The Fate of Diophantos' Arithmetic.* Basel: Springer, 2010.

Morse, Joann Stephanie. "The Reception of Diophantus' 'Arithmetic' in the Renaissance." PhD diss., Princeton University, 1981.

Moyer, Ann. "Reading Boethius on Proportion: Renaissance Editions, Epitomes, and Versions of the Arithmetic and Music." In *Proportions: Science, Musique, Peinture & Architecture,* edited by Sabine Rommevaux, Philippe Vendrix, and Vasco Zara, 51–68. Paris: Honoré Champion, 2012.

Neal, Katherine. *From Discrete to Continuous: The Broadening of Number Concepts in Early Modern England.* Dordrecht: Springer, 2002.

Netz, Reviel. "Proclus' Division of the Mathematical Proposition into Parts: How and Why Was It Formulated?" *Classical Quarterly* 49, no. 1 (1999): 282–303.

Netz, Reviel. "Reasoning and Symbolism in Diophantus: Preliminary Observations." In *The History of Mathematical Proof in Ancient Traditions*, edited by Karine Chemla, 327–361. Cambridge: Cambridge University Press, 2012.

Netz, Reviel. *The Shaping of Deduction in Greek Mathematics*. Cambridge: Cambridge University Press, 1999.

Oosterhoff, Richard J. *Making Mathematical Culture: University and Print in the Circle of Lefèvre d'Étaples*. Oxford: Oxford University Press, 2018.

Oughtred, William. *Key to the Mathematics*. Translated by Robert Wood. London: Thomas Harper for Richard Whitaker, 1647.

Panza, Marco. "What Is New and What Is Old in Viète's Analysis Restituta and Algebra Nova, and Where Do They Come From?" *Revue d'histoire des mathématiques* 13 (2007): 85–153.

Parshall, Karen Hunger. "A Plurality of Algebras, 1200–1600: Algebraic Europe from Fibonacci to Clavius." *BSHM Bulletin: Journal of the British Society for the History of Mathematics* 32, no. 1 (2017): 2–16.

Peletier, Jacques. *L'algebre*. Lyon: Jean de Tournes, 1554.

Peletier, Jacques. *L'Aritmetique . . . departie in quatre Livres*. Poitiers: [Jean de Marnef], 1549.

Proclus. *In Primum Euclidis Elementorum Librum Commentarii*, 203 ff.; *A Commentary on the First Book of Euclid's Elements*. Translated by Glenn Morrow, 159–162. Princeton, NJ: Princeton University Press, 1970.

Pycior, Helena. *Symbols, Impossible Numbers, and Geometric Entanglements*. Cambridge: Cambridge University Press, 1997.

Rabouin, David. *Mathesis Universalis: L'idée de mathématique universelle d'Aristote à Descartes*. Paris: Presses universitaires de France, 2009.

Rashed, Roshdi. *The Development of Arabic Mathematics: Between Arithmetic and Algebra*. Translated by A. F. W. Armstrong. Dordrecht, Boston, and London: Kluwer Academic Publishers, 1994.

Rommevaux, Sabine, Maryvonne Spiesser, and Maria Rosa Massa Esteve, eds. *Pluralité de l'algèbre à la Renaissance*. Paris: Honoré Champion, 2012.

Rommevaux, Sabine, Philippe Vendrix, and Vasco Zara, eds. *Proportions. Science, Musique, Peinture & Architecture*. Turnhout: Brepols Publishers, 2012.

Stedall, Jacqueline. *From Cardano's "Great Art" to Lagrange's "Reflections": Filling a Gap in the History of Algebra*. Zürich: European Mathematics Society, 2011.

Stevin, Simon. *L'Arithmetique de Simon Stevin de Bruges: Contenant les computations des nombres Arithmetiques ou vulgaires: Aussi l'Algebre, avec les equations de cinc quantitez. Ensemble les quatre premiers livres d'Algebre de Diophante d'Alexandrie, maintenant premierement traduicts en François*. Leiden: Christophe Plantin, 1585. Van Egmond, Warren. "How Algebra Came to France." In *Mathematics from Manuscript to Print 1300–1600*, edited by Cynthia Hay, 127–144. Oxford: Clarendon Press, 1988.

Viète, François. *The Analytic Art*. Translated by T. Richard Witmer. Kent, OH: Kent State University Press, 1983.

Viète, François. *Opera Mathematica*. Edited by Franz van Schooten. Leiden: Elsevier, 1646.

Wallis, John. *Correspondence*. Edited by Philip Beeley and Christoph Scriba. 4 vols. Oxford: Oxford University Press, 2003–.

Wallis, John. *Opera Mathematica*. 3 vols. Oxford: Sheldon Theater, 1693–1699.

Chapter 2

Alexander, Amir. "The Imperialist Space of Elizabethan Mathematics." *Studies in the History and Philosophy of Science* 26, no. 4 (1995): 559–591.

Allen, Valerie. *On Farting: Language and Laughter in the Middle Ages*. New York: Palgrave Macmillan, 2007.

Barany, Michael J. "God, King, and Geometry: Revisiting the Introduction to Cauchy's Cours d'analyse." *Historia Mathematica* 38 (2011): 368–388.

Barany, Michael J. "Integration by Parts: Wordplay, Abuses of Language, and Modern Mathematical Theory on the Move." *Historical Studies in the Natural Sciences* 48, no. 3 (2018): 259–299.

Barany, Michael J. "Savage Numbers and the Evolution of Civilization in Victorian Prehistory." *British Journal for the History of Science* 47, no. 2 (2014): 239–255.

Barany, Michael J. "Translating Euclid's Diagrams into English, 1551–1571." In *Philosophical Aspects of Symbolic Reasoning in Early Modern Mathematics*, edited by Albrecht Heeffer and Martin Van Dyck, 125–163. London: College Publications, 2010.

Bourbaki, Nicolas. *Elements of Mathematics: Theory of Sets*. Paris: Hermann, 1968.

Boyarin, Daniel. *Socrates and the Fat Rabbis*. Chicago: University of Chicago Press, 2009.

Cajori, Florian. *A History of Mathematical Notations*, Vol. I: *Notations in Elementary Mathematics*. London: Open Court, 1928.

Chaucer, Geoffrey. *The Canterbury Tales*. Edited by F. N. Robinson (1957), University of Michigan Corpus of Early English Prose and Verse, https://quod.lib.umich.edu/c/cme/CT/1:3.6?rgn=div2;view=fulltext, accessed 2019.

Cowles, Henry. *A Method Only: The Evolving Meaning of Science in the United States, 1830–1910*. PhD diss., Princeton University, 2015.

Denniss, John. "Learning Arithmetic: Textbooks and Their Users in England 1500–1900." In *Oxford Handbook of the History of Mathematics*, edited by Eleanor Robson and Jacqueline Stedall, 448–467. Oxford: Oxford University Press, 2009.

Denniss, John, and Fenny Smith. "Robert Recorde and His Remarkable Arithmetic." In *Robert Recorde: The Life and Times of a Tudor Mathematician*, edited by Gareth Roberts and Fanny Smith, 25–38. Cardiff: University of Wales Press, 2012.

Easton, Joy B. "The Early Editions of Robert Recorde's Ground of Artes." *Isis* 58, no. 4 (1967): 515–532.

"Equals Sign." https://en.wikipedia.org/w/index.php?title=Equals_sign, accessed 2017.

Heninger, S. K., Jr. "Tudor Literature of the Physical Sciences." *Huntington Library Quarterly* 32, no. 2 (1969): 101–133.

Horkheimer, Max, and Theodor W. Adorno. *Dialectic of Enlightenment: Philosophical Fragments*. Translated by Edmund Jephcott. Edited by Gunzelin Schmid Noerr. Stanford, CA: Stanford University Press, 2002.

Høyrup, Jens. "Hesitating Progress—The Slow Development toward Algebraic Symbolization in Abbacus- and Related Manuscripts, c. 1300 to c. 1550." In *Philosophical Aspects of Symbolic Reasoning*, edited by Albrecht Heeffer and Maarten Van Dyck, 3–56. London: College Publications, 2010.

Johnson, Francis R. "Latin versus English: The Sixteenth-Century Debate over Scientific Terminology." *Studies in Philology* 41, no. 2 (1944): 109–135.

Johnson, Francis R., and Sanford V. Larkey. "Robert Recorde's Mathematical Teaching and the Anti-Aristotelian Movement." *Huntington Library Quarterly* 7 (1935): 59–87.

Johnston, Stephen. "Recorde, Robert (c. 1512–1558)." In *Oxford Dictionary of National Biography*. Oxford: Oxford University Press, 2004. doi:10.1093/ref:odnb/23241. https://www-oxforddnb-com.libproxy.berkeley.edu/view/10.1093/ref:odnb/9780198614128.001.0001/odnb-9780198614128-e-23241.

Jones, Matthew. *Reckoning with Matter: Calculating Machines, Innovation, and Thinking about Thinking from Pascal to Babbage*. Chicago: University of Chicago Press, 2016.

Jones, Matthew L. "How We Became Instrumentalists (Again): Data Positivism since World War II." *Historical Studies in the Natural Sciences* 48, no. 5 (2018): 673–684.

Kaye, Joel. "Money and Administrative Calculation as Reflected in Scholastic Natural Philosophy." In *Arts of Calculation: Quantifying Thought in Early Modern Europe*, edited by David Glimp and Michelle R. Warren, 1–18. New York: Palgrave Macmillan, 2004.

Klein, David, and R. James Milgram. "The Role of Long Division in the K–12 Curriculum." 2000. http://www.csun.edu/~vcmth00m/longdivision.pdf (accessed 2017).

Lynch, Michael. "Extending Wittgenstein: The Pivotal Move from Epistemology to the Sociology of Science." In *Science as Practice and Culture*, edited by A. Pickering, 215–265. Chicago: University of Chicago Press.

MacKenzie, Donald. *Mechanizing Proof: Computing, Risk, and Trust.* Cambridge, MA: MIT Press, 2001.

Mancosu, Paolo. *Philosophy of Mathematics and Mathematical Practice in the Seventeenth Century.* Oxford: Oxford University Press, 1996.

Mathews, Jay. "10 Myths (Maybe) about Learning Math." *Washington Post*, May 31, 2005.

Middle English Dictionary, The. http://quod.lib.umich.edu/cgi/m/mec/med-idx, accessed 2017.

O'Brien, Timothy D. "'Ars-Metrik': Science, Satire and Chaucer's Summoner." *Mosaic: An Interdisciplinary Critical Journal* 23, no. 4 (1990): 1–22.

O'Connor, J. J., and E. F. Robertson. "Recorde Summary." *MacTutor History of Mathematics.* http://www-history.mcs.st-and.ac.uk/Mathematicians/Recorde.html, accessed 2017.

OED Online. Oxford University Press, 2016.

Olson, Glending. "Measuring the Immeasurable: Farting, Geometry, and Theology in the Summoner's Tale." *Chaucer Review* 43, no. 4 (2009): 414–427.

Powell, Nia M. W. "The Welsh Context of Robert Recorde." In *Robert Recorde: The Life and Times of a Tudor Mathematician*, edited by Gareth Roberts and Fanny Smith, 123–144. Cardiff: University of Wales Press, 2012.

Recorde, Robert. *The grou[n]d of artes teachyng the worke and practise of arithmetike, moch necessary for all states of men.* London: R. Wolfe, 1543. STC (2nd ed.) 20797.5, accessed on Early English Books Online.

Recorde, Robert. *The ground of artes teaching the woorke and practise of arithmetike, both in whole numbres and fractions, after a more easyer and exacter sorte than any lyke hath hytherto beene set forth: with divers new additions.* London: R. Wolfe, 1558. STC (2nd ed.) 20799.5, accessed on Early English Books Online (2017).

Roberts, Gordon. *Robert Recorde: Tudor Scholar and Mathematician.* Cardiff: University of Wales Press, 2016.

Sellers, Patricia A. "The Trouble with Long Division." *Teaching Children Mathematics* 16, no. 9 (2010): 516–520.

Stockton, Will. "'I Am Made an Ass': Falstaff and the Scatology of Windsor's Polity." *Texas Studies in Literature and Language* 49, no. 4 (2007): 340–360.

Taylor, E. G. R. *The Mathematical Practitioners of Tudor & Stuart England.* Cambridge: Cambridge University Press, 1954.

Thomas, Keith. "Numeracy in Early Modern England." *Transactions of the Royal Historical Society* 37 (1987): 103–132.

Wilde, Lisa. "'Whiche elles shuld farre excelle mans mynde': Numerical Reason in Robert Recorde's Ground of Artes (1543)." *Journal of the Northern Renaissance* 6 (2014). http://www.northernrenaissance.org/whiche-elles-shuld-farre-excelle-mans-mynde-numerical-reason-in-robert-recordes-ground-of-artes-1543/.

Williams, Jack. *Robert Recorde: Tudor Polymath, Expositor and Practitioner of Computation, History of Computing.* London: Springer Verlag, 2011.

Wittgenstein, Ludwig. *Remarks on the Foundations of Mathematics.* Translated by G. E. M. Anscombe. Edited by G. H. von Wright, R. Rhees, and G. E. M. Anscombe. Oxford: Basil Blackwell, 1956.

Chapter 3

Alexander, Amir. *Infinitesimal: How a Dangerous Mathematical Theory Shaped the Modern World*. (New York: Farrar, Straus & Giroux/Scientific American, 2014.

Berkeley, George. *The Analyst, or A Discourse Addressed to an Infidel Mathematician*. Dublin: S. Fuller, 1734.

Boylan, Michael. "Henry More's Space and the Spirit of Nature." *Journal of the History of Philosophy* 18, no. 4 (October 1980): 395–405.

Cavalieri, Bonaventura. *Exercitationes Geometricae Sex*. Book 1. Bologna: Iacob Monti, 1647.

Cavalieri, Bonaventura. *Geometria Indivisibilibus Libri VI*. Bologna: Clementis Ferroni, 1635.

Galileo Galilei. *The Discoveries and Opinions of Galileo*. Edited and translated by Stillman Drake. New York: Anchor Books, 1957.

Gascoigne, John. "The Universities and the Scientific Revolution: The Case of Newton and Restoration Cambridge." *History of Science* 23, no. 4 (December 1985): 391–434.

Guicciardini, Niccolò. *Isaac Newton on Mathematical Certainty and Method*. Cambridge, MA: MIT Press, 2009.

Jacob, Margaret C. *The Newtonians and the English Revolution, 1689–1720*. New York: Gordon and Breach, 1990.

Loria, Gino, and Giuseppe Vassura, eds. *Opere di Evangelista Torricelli*. Vol. 1. Faenza: G. Montanari, 1919–1944.

Mandelbrote, Scott. "The Uses of Natural Theology in Seventeenth-Century England." *Science in Context* 20, no. 3 (2007): 451–480.

McGuire, J. E. "Force, Active Principles, and Newton's Invisible Realm." *Ambix* 15, no. 3 (1968): 154–208.

McGuire, J. E., and Martin Tamny. *Certain Philosophical Questions: Newton's Trinity Notebook*. Cambridge: Cambridge University Press, 1983.

McGuire, J. E., and P. M. Rattansi. "Newton and the 'Pipes of Pan.'" *Notes and Records of the Royal Society of London* 21, no. 2 (December 1966): 108–143.

Newton, Isaac. *The Mathematical Papers of Isaac Newton*. Edited by D. T. Whiteside. Cambridge: Cambridge University Press, 1967.

Newton, Isaac. *The Principia: Mathematical Principles of Natural Philosophy*. Edited and translated by I. Bernard Cohen and Anne Whitman. Berkeley: University of California Press, 1999.

Panza, Marco. "From Velocities to Fluxions." In *Interpreting Newton: Critical Essays*, edited by Andrew Janiak and Eric Schliesser, 219–254. Cambridge: Cambridge University Press, 2012.

Rattansi, P. M. "Reason in Sixteenth and Seventeenth Century Natural Philosophy." In *Changing Perspectives in the History of Science*, edited by M. Teich and R. M. Young, 148–166. London: Heinemann Educational, 1973.

Sailor, Danton B. "Cudworth and Descartes." *Journal of the History of Ideas* 23, no. 1 (January–March 1962): 133–140.

Shapin, Steven. "Of Gods and Kings: Natural Philosophy and Politics in the Leibniz-Clarke Disputes." *Isis* 72, no. 2 (June 1981): 187–215.

Shapin, Steven. *A Social History of Truth: Civility and Science in Seventeenth Century England*. Chicago: University of Chicago Press, 1995.

Shapin, Steven, and Simon Schaffer. *Leviathan and the Air-Pump: Hobbes, Boyle, and the Experimental Life*. Princeton, NJ: Princeton University Press, 1985.

Shapiro, Barbara J. *Probability and Certainty in Seventeenth Century England*. Princeton, NJ: Princeton University Press, 1983.

Sprat, Thomas. *History of the Royal Society of London*. London, 1667.

Chapter 4

Adorno, Theodor, and Max Horkheimer. *The Dialectic of Enlightenment*. 2nd ed. London: Verso 1997.

Baker, Keith Michael. *Condorcet. From Natural Philosophy to Social Mathematics*. Chicago: University of Chicago Press, 1974.

Baker, Keith Michael. "Enlightenment and the Institution of Society: Notes for a Conceptual History." In *Civil Society. History and Possibilities*, ed. Sudip Kaviraj and Sunil Khalnani, 84–104. Cambridge: Cambridge University Press, 2001.

Bailly, Jean Sylvain. *Histoire de l'astronomie moderne depuis la fondation de l'école d'Alexandrie, jusqu'à l'époque de M.D.CC.XXX [M.D.CC.XXXII]*. 3 vols. Paris, 1779-1782; new edition, 1785.

Boutier, Jean. *Les Plans de Paris*. 2nd ed. Paris: Bibliothèque nationale de France, 2007.

Caparrini, Sandro, and Craig Fraser. "Mechanics in the Eighteenth Century." In *The Oxford Handbook of the History of Physics*, ed. Jed Buchwald and Robert Fox, 358–405. Oxford: Oxford University Press, 2013.

Charles, Loic. "The Visual History of the *Tableau Économique*." *European Journal of the History of Economic Thought* 10, no. 4 (December 2003): 527–550.

Commercium epistolicum D. Johannis Collins, *et aliorum de analysi promota: jussu Societatis Regiæ in lucem editum*. London: Pearson, 1712.

Daston, Lorraine. *Classical Probability and the Enlightenment*. Princeton, NJ: Princeton University Press, 1995.

Dictionnaire de l'Académie Française. 4th ed. Paris: Chez la Veuve B. Brunet, 1762.

Dictionnaire de l'Académie Française. 6th ed. Paris: Firmin Didot Frères, 1835.

Foucault, Michel. *The Birth of Biopolitics. Lectures at the Collège de France 1978-1979*. New York: Picador, 2010.

Foucault, Michel. *Discipline and Punish: The Birth of the Prison*. Translated by Alan Sheridan. New York: Vintage Books, 1995.

Foucault, Michel. *On the Government of the Living. Lectures at the Collège de France 1979-1980*. New York: Picador, 2016.

Foucault, Michel. *Security, Territory, Population. Lectures at the Collège de France 1977-78*. New York: Picador, 2009.

Foucault, Michel. *Society Must Be Defended. Lectures at the Collège de France 1975-76*. New York: Picador, 2003.

Fraser, Craig. *Calculus and Analytical Mechanics in the Age of Enlightenment*. Aldershot, Hampshire: Ashgate, 1997.

Genovese, Elizabeth Fox. *The Origins of Physiocracy: Economic Revolution and Social Order in Eighteenth-Century France*. Ithaca, NY: Cornell University Press, 1976.

Le Grand Robert de la langue française: dictionnaire alphabétique et analogique de la langue française. Edited by Alain Rey. 2nd ed. Paris: Le Robert, 2001.

Guerlac, Henry. *Lavoisier. The Crucial Year. The Background and Origin of His First Experiments on Combustion in 1772*. Ithaca, NY: Cornell University Press, 1961.

Guicciardini, Niccolò. *The Development of Newtonian Calculus in Britain, 1700-1800*. Cambridge: Cambridge University Press, 1990.

Guicciardini, Niccolò. *Newton on Mathematical Certainty and Method*. Cambridge, MA: MIT Press, 2009.

Hacking, Ian. *The Taming of Chance*. Cambridge: Cambridge University Press, 1990.

Hall, A. R. *Philosophers at War: The Quarrel between Newton and Leibniz*. Cambridge: Cambridge University Press, 1980.

Hankins, Thomas. *Science and the Enlightenment*. Cambridge: Cambridge University Press, 1985.

Heilbron, John L. *Electricity in the 17th & 18th Centuries: A Study in Early Modern Physics.* Berkeley: University of California Press, 1979.

Huguet, Edmond. *Dictionnaire de la langue française du seizième siècle.* Paris: Honoré Champion, 1925–1956.

Jesseph, Douglas *Berkeley's Philosophy of Mathematics* (Chicago: University of Chicago Press, 1993).

Kaplan, Steven L. *Bread, Politics, and Political Economy in the Reign of Louis XV.* 2 vols. Dordrecht: Springer, 1976.

Kaplan, Steven L. "The Famine Plot Persuasion in Eighteenth-Century France." *Transactions of the American Philosophical Society.* Philadelphia: American Philosophical Society, 1982.

Kim, Mi Gyung. *Affinity, That Elusive Dream: A Genealogy of the Chemical Revolution.* Cambridge, MA: MIT Press, 2003.

Kline, Morris. *Mathematics: The Loss of Certainty.* Oxford: Oxford University Press, 1980.

Lagrange, Joseph-Louis. *Mécanique analytique.* Paris: La Veuve Desaint, 1788.

Ledoux, Claude Nicolas. *L'architecture considérée sous le rapport de l'art, des moeurs et de la legislation.* Paris: H. L. Perronneau, 1804.

List, Christian. "Social Choice Theory." In *The Stanford Encyclopedia of Philosophy* (Winter 2013 Edition).

Mancosu, Paolo. *The Philosophy of Mathematics & Mathematical Practice in the Seventeenth Century.* Oxford: Oxford University Press, 1999.

Mauro, Tullio de, ed. *Grande Dizionario Italiano dell'Uso.* Torino: UTET, 1999–2007.

Meek, Ronald. *The Economics of Physiocracy.* Cambridge, MA: Harvard University Press, 1962.

Mémoire sur la réformation de la police de France. Soumis au roi en 1749 par m, Guillauté, officier de la Maréchaussée de l'Ile-de-France. Illustré de 28 dessins de Gabriel de Saint-Aubin. Introduction et notes par Jean Seznec. Paris: Editions Hermann, 1974.

Miller, Judith. *Mastering the Market: The State and the Grain Trade in Northern France, 1700–1860.* Cambridge: Cambridge University Press, 1998.

Oxford English Dictionary Online. Oxford University Press, December 2019.

Pelletier, Monique. *Les cartes des Cassini. La science au service de l'Etat et des regions.* Paris: Comité des travaux historiques et scientifiques, 2002.

Philosophical Transactions of the Royal Society of London 29, no. 342 (December 1714).

Piaget, Jean. *Epistémologie des Sciences de l'Homme.* Paris: Gallimard, 1971.

Schabas, Margaret. *The Natural Origins of Economics.* Chicago: University of Chicago Press, 2007.

Schliesser, Eric. *Adam Smith: Systematic Philosopher and Public Thinker.* Oxford: Oxford University Press, 2017.

Shank, J. B. *Before Voltaire: The French Origins of "Newtonian" Mechanics, 1680–1715.* Chicago: University of Chicago Press, 2018.

Shank, J. B. "A French Jesuit in the Royal Society of London: Father Louis-Bertrand de Castel, S.J. and Enlightenment Mathematics, 1720–1735." *Journal of Early Modern Studies* 1 (November 2012): 151–184.

Smith, Adam. *An Inquiry Concerning the Nature and Causes of the Wealth of Nations.* London: Methuen & Co. Ltd., 1776.

Tommaseo, Niccolò. *Dizionario della lingua italiana.* Torino: Soc. L'Unione Tipografico-Editrice, 1865.

Vidler, Anthony. *Claude-Nicolas Ledoux: Architecture and Social Reform at the End of the Ancien Régime.* Cambridge, MA: MIT Press, 1990.

Weintraub, E. Roy. *How Economics Became a Mathematical Science.* Durham, NC: Duke University Press, 2002.

Weisstein, Eric W. "Binet's Fibonacci Number Formula." *MathWorld.* http://mathworld.wolfram.com/BinetsFibonacciNumberFormula.html.

Weulersse, Georges. *La Physiocratie à la Fin du Règne de Louis XV*. Paris: Presses Universitaires de France, 1959.

Weulersse, Georges. *La Physiocratie sous les Ministres Turgot et Necker*. Paris: Presses Universitaires de France, 1950.

Winch, Donald. *Riches and Poverty: An Intellectual History of Political Economy in Britain, 1750–1834*. Cambridge: Cambridge University Press, 1996.

Chapter 5

Abridgment of the Public Permanent Laws of Virginia. 1796. Augustine Davis.

Adams, Daniel. 1828. *The Scholar's Arithmetic; or, Federal Accountant* . . . Printed by John Prentiss.

Allen, Nathaniel. 1805. *Cyphering Book*. Worcester, MA: Penmanship Collection, American Antiquarian Society.

Arnold, Alfred. 1842. *Arnold's Ready Reckoner*. Providence: Knowles & Vosk.

Barlow, Peter, George Peacock, Dionysius Lardner, Sir George Biddell Airy, H. P. Hamilton, A. Levy, Augustus De Morgan, and Henry Mosley. 1847. *Encyclopaedia of Pure Mathematics*. London: R. Griffin and Co.

Baucom, Ian. 2005. *Specters of the Atlantic: Finance Capital, Slavery, and the Philosophy of History*. Durham, NC: Duke University Press.

Benjamin, Ruha. 2019. *Race after Technology: Abolitionist Tools for the New Jim Code*. Cambridge; Medford, MA: Polity.

Blunt, Joseph. 1822. *The Merchant's and Shipmaster's Assistant*. London: E. M. Blunt.

Blunt, Joseph. 1832. *The Merchant's and Shipmaster's Assistant*. New York: Blunt.

Bragg, George F. 1914. *Men of Maryland*, Baltimore: Church Advocate Press. Documenting the American South, North American Slave Narratives, University of North Carolina, docsouth. unc.edu/neh/bragg/.

Calvert, John. 1875. *Calvert's Pocket Wages Table, 56½ Hours, and Compendium of the New Factory Act, 1874*. Manchester: John Heywood.

Carruthers, Bruce G., and Wendy Nelson Espeland. 1991. "Accounting for Rationality: Double-Entry Bookkeeping and the Rhetoric of Economic Rationality." *American Journal of Sociology* 97(1): 31–69.

Christian, Brian, and Tom Griffiths. 2016. *Algorithms to Live By: The Computer Science of Human Decisions*. New York: Henry Holt.

Cocker, Edward, and J. Hawkins. 1702. *Cocker's Arithmetick*. London: W. Richardson.

Cocker, Edward. 1769. *Cocker's Arithmetic. 51st ed*. Dublin: Isaac Jackson.

Cohen, Patricia Cline. 1982. *A Calculating People: The Spread of Numeracy in Early America*. Chicago: University of Chicago Press.

Cook, Eli. 2017. *The Pricing of Progress: Economic Indicators and the Capitalization of American Life*. Cambridge, MA: Harvard University Press.

Cronon, William. 1992. *Nature's Metropolis: Chicago and the Great West*. New York: W. W. Norton.

Dryer, Theodora. 2019. "The New Critical History of Surveillance and Human Data." *Historical Studies in the Natural Sciences* 49(5): 556–565. https://doi.org/10.1525/hsns.2019.49.5.556.

Espeland, Wendy Nelson, and Mitchell L. Stevens. 1992. "Commensuration as a Social Process." *Annual Review of Sociology* 24: 313–343.

Equiano, Olaudah. 1789. *The Interesting Narrative of the Life of Olaudah Equiano; or, Gustavus Vassa, the African. Vol. 1*. London: The Author. Documenting the American South, North

American Slave Narratives, University of North Carolina, http://docsouth.unc.edu/neh/equiano1/.

Eubanks, Virginia. 2018. *Automating Inequality: How High-Tech Tools Profile, Police, and Punish the Poor.* New York: St. Martin's Press.

Fenning, Daniel. 1783. *The British youth's instructor.* 10th ed., revised by W. Burbidge, Schoolmaster, in Salisbury. London, M.DCC.LXXXIII., 3–4 and frontmatter. Eighteenth Century Collections Online.

The "Fifty-One Hour" Wages Reckoner, by a Retired Banker. 1872. Edinburgh: Johnston, Hunter, & Co.

Finn, Ed. 2017. *What Algorithms Want: Imagination in the Age of Computing.* Cambridge, MA: MIT Press.

Glasse, Hannah, 1760. *The Servant's Directory or House-Keeper's Companion.* London, Eighteenth Century Collections Online.

Hacking, Ian. 1999. *The Taming of Chance.* Cambridge: Cambridge University Press.

Huestis, C. P. 1847. *The Mechanics' and Laborers' Ready Reckoner.* New York: C. P. Huestis.

Jefferson, Thomas. 1790. *Report of the Secretary of State on the Subject of Establishing a Uniformity in the Weights, Measures and Coins of the United States.* New York: Francis Childs and John Swaine.

Kimber, Emmor. 1805. *Arithmetic Made Easy to Children: Being a Collection of Useful and Familiar Examples Methodically Arranged under Their Respective Heads.* Philadelphia: Kimber, Conrad, & Company.

Lincoln, Abraham. 1826. "The Single Rule of Three." Fragment from Lincoln's copy book. Providence. John Hay Library, https://repository.library.brown.edu/studio/item/bdr:72542/.

Lincoln, Abraham. 1895. *Words of Lincoln: Including Several Hundred Opinions of His Life and Character by Eminent Persons of This and Other Lands.* Edited by Osborn H. Oldroyd. Washington, DC: O. H. Oldroyd.

Littell, William. 1809. *The Statute Law of Kentucky . . .* Vol. I. Frankfort, KY: William Hunter.

MacKenzie, Donald A. 2006. *An Engine, Not a Camera: How Financial Models Shape Markets.* Cambridge, MA: MIT Press.

Marx, Karl. 1887. *Capital: A Critical Analysis of Capitalist Production.* London: George Allen & Unwin.

McCusker, John J. 2005. "The Demise of Distance: The Business Press and the Origins of the Information Revolution in the Early Modern Atlantic World." *American Historical Review* 110(2): 295–321.

Member of the Philadelphia Bar. 1838. *The Form Book.* Philadelphia: Haswell, Barrington & Haswell.

Montefiore, Joshua. 1811. *The American Trader's Compendium.* Philadelphia: S. R. Fisher.

Noble, Safiya Umoja. 2018. *Algorithms of Oppression: How Search Engines Reinforce Racism.* New York: New York University Press.

Offley, G. W. 1859. *A Narrative of the Life and Labors of the Rev. G. W. Offley.* Hartford: n.p. Documenting the American South, North American Slave Narratives, University of North Carolina, http://docsouth.unc.edu/neh/offley/.

O'Neil, Cathy. 2017. *Weapons of Math Destruction: How Big Data Increases Inequality and Threatens Democracy.* New York: Broadway Books.

Page, David P. 1841. "Advancement in the Means and Methods of Public Instruction." In *Annual Meeting: Proceedings, Constitution, List of Active Members, and Addresses,* 107–142. American Institute of Instruction.

Pasquale, Frank. 2015. *The Black Box Society: The Secret Algorithms That Control Money and Information.* Cambridge, MA: Harvard University Press.

Pennsylvania German Society. 1907. "Record of Indentures of Individuals Bound out as Apprentices Servants Etc and of German and Other Redemptioners in the Office of the Mayor of the City of Philadelphia October 3 1771 to October 5 1773." *Proceedings and Addresses*. Philadelphia: American Philosophical Society.

Pennington, James W. C. 1849. *The Fugitive Blacksmith*. London: Charles Gilpin. Documenting the American South, North American Slave Narratives, University of North Carolina, http://docsouth.unc.edu/neh/penning49/.

Pennsylvania State Education Association. 1854. *Pennsylvania School Journal* 3: 107–108.

Place, Francis. 1972. *The Autobiography of Francis Place: 1771–1854*. Edited by Mary Thale. New York: Cambridge University Press.

Poovey, Mary. 2008. *Genres of the Credit Economy: Mediating Value in Eighteenth- and Nineteenth-Century Britain*. Chicago: University of Chicago Press.

Porter, Theodore M. 1995. *Trust in Numbers: The Pursuit of Objectivity in Science and Public Life*. Princeton, NJ: Princeton University Press.

Public Statutes at Large of the State of Ohio: From the Close of Chase's Statutes, February 1833 to the Present Time. And a Supplement Containing All Laws Passed prior to February 1833 Which Are Now in Force. 1854. Edited by Maskell E. Curwen. Vol. III. Cincinnati: E. Morgan and Co.

Rothenberg, Winifred. 1992. *From Market-Places to a Market Economy: The Transformation of Rural Massachusetts, 1750–1850*. Chicago: University of Chicago Press.

Schumpeter, Joseph. 1950. *Capitalism, Socialism, and Democracy*. 3rd ed. New York: Harper.

Smith, David Eugene, and Frank Swetz. 1987. *Capitalism and Arithmetic: The New Math of the 15th Century, Including the Full Text of the Treviso Arithmetic of 1478, Translated by David Eugene Smith*. La Salle, IL: Open Court.

Smyth, William. 1810. *The New American Clerk's Instructor*. Philadelphia: James Sharan.

Sombart, Werner. 1967. *The Quintessence of Capitalism: A Study of the History and Psychology of the Modern Business Man*. New York: H. Fertig.

Southern Literary Messenger. 1839. Vol. 5. Richmond, VA: T. W. White.

Steiner, Christopher. 2012. *Automate This: How Algorithms Came to Rule Our World*. New York: Portfolio/Penguin.

Stokes, William. 1879. *Stokes's Rapid Arithmetic*. London: Houlston & Sons.

Sunstein, Cass R. 1994. "Incommensurability and Valuation in Law." *Michigan Law Review* 92(4): 779–861.

Thornton, Tamara Plakins. 2016. *Nathaniel Bowditch and the Power of Numbers: How a Nineteenth-Century Man of Business, Science, and the Sea Changed American Life*. Chapel Hill: University of North Carolina Press.

Wardhaugh, Benjamin. 2012. *Poor Robin's Prophecies: A Curious Almanac, and the Everyday Mathematics of Georgian Britain*. Oxford: Oxford University Press.

Weber, Max. 1950. *General Economic History*. Glencoe, IL: Free Press.

Webster, Noah, and Edmund Henry Barker. 1830. *A Dictionary of the English Language*. 3rd ed. New York: Black and Young.

Weston, William. 1754. *The Complete Merchant's Clerk; or, British and American Compting-House*. London: Charles Rivington.

Wickman, Tom. 2011. "Arithmetic and Afro-Atlantic Pastoral Protest: The Place of (in)numeracy in Gronniosaw and Equiano." *Atlantic Studies* 8(2): 189–212.

Zakim, Michael. 2006. "Bookkeeping as Ideology." *Common-Place* 6(3). http://commonplace.online/article/bookkeeping-as-ideology/.

Zakim, Michael. 2006. "The Business Clerk as Social Revolutionary; or, A Labor History of the Nonproducing Classes." *Journal of the Early Republic* 26(4): 563–603.

Zelizer, Viviana A. Rotman. 1994. *The Social Meaning of Money*. New York: Basic Books.

Chapter 6

Allaire, Patricia, and Robert E. Bradley. "Symbolical Algebra as a Foundation for Calculus: D. F. Gregory's Contribution." *Historia Mathematica* 29 (2002): 395–426.

Black, Alistair. "The Victorian Information Society: Surveillance, Bureaucracy, and Public Librarianship in 19th-Century Britain." *The Information Society* 17 (2001): 63–80.

Boole, George. *An Address on the Genius and Discoveries of Sir Isaac Newton*. Lincoln: Gazette Office, 1835.

Boole, George. *An Investigation of the Laws of Thought on Which Are Founded the Mathematical Theories of Logic and Probabilities*. 1854; New York: Dover, 1958.

Boole, George. "On a General Method of Analysis." *Philosophical Transactions of the Royal Society of London* 134 (1844): 225–282.

Boole, George. "On Certain Theorems in the Calculus of Variations." *Cambridge Mathematical Journal II* Cambridge: E. Johnson (1841): 97–102.

Boole, George. *Studies in Logic and Probability*. Edited by R. Rhees. London: Watts and Co., 1952.

Boole, George. *The Mathematical Analysis of Logic: Being an Essay towards a Calculus of Deductive Reasoning*. Cambridge: Macmillan, Barclay and Macmillan; London: George Bell, 1847.

Boole, George. *The Mathematical Analysis of Logic*. Bristol: Thoemmes Press, 1998.

Boole, George. *Treatise on Differential Equations*. Cambridge: Macmillan & Co., 1859.

Boole, George, to John Lubbock, February 16, 1849. Lubbock Papers in the Royal Society Archive, LUB. B. 362.

Boole, Mary Everest. *Collected Works*. Edited by E. M. Cobham. 4 vols. London: The C. W. Daniel Co., 1931.

Cajori, Florian. "Augustus De Morgan on Divergent Series." *Bulletin of the American Mathematical Society* 27 (1920): 77–81.

Cauchy, Augustin-Louis. *Oeuvres completes d'Augustin Cauchy, publieés sous la direction scientifique de l'Académie*. Series 2, vol. 3. Paris: Gauthier-Vilars.Corsi, Pietro. "The Heritage of Dugald Stewart: Oxford Philosophy and the Method of Political Economy." *Nuncius Annali di Storia della Scienza* (1987): 98–105.

Crilly, Tony. "The Cambridge Mathematical Journal and Its Descendants: The Linchpin of a Research Community in the Early and Mid-Victorian Age." *Historia Mathematica* 31 (2004): 455–497.

Crilly, Tony, and A. J. Crilly. *Arthur Cayley: Mathematician Laureate of the Victorian Age*. Baltimore: Johns Hopkins University Press, 2006.

De Landa, Manuel. *A New Philosophy of Society: Assemblage Theory and Social Complexity*. London: Bloomsbury, 2006.

De Morgan, Augustus. *Arithmetical Books from the Invention of Printing to the Present Time Being Brief Notices of a Large Number of Works Drawn Up from Actual Inspection*. London: Taylor and Walton, 1847.

De Morgan, Augustus. *The Differential and Integral Calculus, Containing Differentiation, Integration, Development, Series, Differential Equations, Differences, Summation, Equations of Differences, Calculus of Variations, Definite Integrals,—with applications to Algebra, Plane Geometry, Solid Geometry, and Mechanics Also, Elementary Illustrations of the Differential and Integral Calculus*. London: Library of Useful Knowledge: Baldwin and Cradock, 1842.

De Morgan, Augustus. *First Notions of Logic*. London: James Moyes, 1839.

De Morgan, Augustus. *Formal Logic: Or, the Calculus of Inference, Necessary and Probable*. London: Taylor and Walton, 1847.

De Morgan, Augustus. *In the Companion to the Almanac of 1843*. London: Charles Knight, 1843.

De Morgan, Augustus. "On Divergent Series, and Various Points in Analysis Connected with Them." *Transactions of the Cambridge Philosophical Society* 8, part II (1843): 182–203.

De Morgan, Augustus. "On the Earliest Printed Almanacs." In *Companion to the Almanac for 1846*, 1–31. London: Charles Knight, 1846.

De Morgan, Augustus. "On the Structure of the Syllogism." *Transactions of the Cambridge Philosophical Society* 8 (1846): 379–406.

De Morgan, Sophia Elizabeth. *Memoir of Augustus De Morgan*. London: Longmans Green & Co. 1882.

Deleuze, Gilles, and Félix Guattari. *A Thousand Plateaus: Capitalism and Schizophrenia*. Translated by Brian Massumi. Minneapolis: University of Minneapolis Press, 1987.

Dubbey, J. M. "Babbage, Peacock and Modern Algebra." *Historia Mathematica* 4 (1977): 295–302.

Dubbey, J. M. "The Introduction of the Differential Notation to Great Britain." *Annals of Science* 19 (1963): 37–48.

Dubbey, J. M. *The Mathematical Work of Charles Babbage*. New York: Cambridge University Press, 1978.

Durand-Richard, M. J. "Genèse de l'algèbre symbolique en Angleterre: Une influence possible de John Locke." *Revue d'histoire des sciences* 43 (1990): 129–180.

Enros, C. J. "The Analytical Society: Mathematics at Cambridge University in the Early Nineteenth-Century." PhD diss., University of Toronto, 1979.

Enros, C. J. "The Analytical Society (1812–1813): Precursor to the Renewal of Cambridge Mathematics." *Historia Mathematica* 10 (1983): 24–47.

Enros, C. J. "Cambridge University and the Adoption of Analytics in Early Nineteenth-Century England." In *Social History of Nineteenth-Century Mathematics*, edited by H. Mehrtens, H. Bos, and I Schneider, 135–148. Boston/Basel/Stuttgart: Birkhauser, 1981.Epstein, David, and Silvio Levy. "Experimentation and Proof in Mathematics." *Notices of the AMS* 42 (1995): 670–674.

Epstein, David, Silvio Levy, and Rafael de la Llave. "About This Journal," *Experimental Mathematics* 1 (1992): 1–3

Forbes, Duncan. *The Liberal Anglican Idea of History*. London: Cambridge University Press, 1952.

Foucault, Michel. *Discipline & Punish: The Birth of the Prison*. 2nd ed. New York: Random House, 1995.

Franklin, Kathryn J., James A. Johnson, and Emily Miller Bonney. "Towards Incomplete Archeologies?" In *Incomplete Archaeologies: Assembling Knowledge in the Past and the Present*, edited by Emily Miller Bonney, Kathryn J. Franklin, and James A. Johnson, ix–xvii. Oxford: Oxbow Books, 2016.

Fyfe, Aileen. *Steam Powered Knowledge: William Chambers and the Business of Publishing*. Chicago: University of Chicago Press, 2012.

Giebelhausen, Michaela, and Tim Berringer, eds. *Writing the Pre-Raphaelites: Text, Context, Subtext*. New York: Routledge, 2016.

Grabiner, Judith. *The Origins of Cauchy's Rigorous Calculus*. New York: Dover, 1981.

Grabiner, Judith V. *The Origins of Cauchy's Rigorous Calculus*. New York: Dover, 2005.

Grattan-Guinness, Ivor. "Victorian Logic from Whately to Russell." In *Mathematics in Victorian Britain*, edited by Raymond Flood, Adrian Rice, and Robin Wilson, 359–374. Oxford: Oxford University Press, 2011.

Gray, Jeremy. "Overstating Their Case? Reflections on British Pure Mathematics in the 19th Century." In *Mathematics in Victorian Britain*, edited by Raymond Flood, Adrian Rice, and Robin Wilson, 397–414. Oxford: Oxford University Press, 2011.

Gregory, D. F. *The Mathematical Writings of Duncan Farquharson Gregory*. Edited by William Walton. Cambridge: Deighton, Bell & Co.; London: Bell and Daldy, 1865.

Gregory, D. F. "On the Real Nature of Symbolic Algebra." *Transactions of the Royal Society of Edinburgh* 14 (1840): 208–216,

Hacking, Ian. *Why Is There Philosophy of Mathematics at All?* Cambridge: Cambridge University Press, 2014.

Hamilton, Sir William. "Logic. In Reference to the Recent English Treatises on that Science." [*Edinburgh Review*, 1833].

Harel, David. *Algorithmics: The Spirit of Computing.* 2nd ed. New York: Addison Wesley, 1992.

Hill, Sir Francis. *Victorian Lincoln.* Cambridge: Cambridge University Press, 1974.

Hilton, Boyd. *A Mad Bad and Dangerous People? England 1783–1846.* Oxford: Clarendon Press, 2013.

Inkster, Ian, ed. *The Steam Intellect Societies—Essays on Culture, Education and Industry circa 1820–1914.* Nottingham: Department of Adult Education, University of Nottingham, 1985.

Laita, Luis. "The Influence of Boole's Search for a Universal Method of Analysis on the Creation of His Logic." *Annals of Science* 34 (1977): 163–176.

Laita, Luis M. "Influences on Boole's Logic: The Controversy between William Hamilton and Augustus De Morgan." *Annals of Science* 36, no. 1 (1979): 45–66.

Lambert, Kevin. "A Natural History of Mathematics: George Peacock and the Making of English Algebra." *Isis* 104 (2013): 278–302.

Lambert, Kevin. "Victorian Stained Glass as Memorial: An Image of George Boole." In *Visions of the Industrial Age, 1830–1914: Modernity and the Anxiety of Representation in Europe*, edited by Minsoo Kang and Amy Woodson Boulton, 205–226. Aldershot/Burlington: Ashgate, 2008.

Latham, Gordon. *First Outlines of Logic Applied to Grammar and Etymology.* London: Taylor and Walton, 1847.

Latour, Bruno. *Pandora's Hope: Essays on the Reality of Science Studies.* Cambridge, MA: Harvard University Press, 1999.

Latour, Bruno. *Reassembling the Social: An Introduction to Actor-Network-Theory.* Oxford: Oxford University Press, 2007.

MacHale, Desmond. *The Life and Work of George Boole: A Prelude to the Digital Age.* Cork: Cork University Press, 2014.

MacKenzie, Donald. *Mechanized Proof: Computing, Risk, and Trust.* Cambridge, MA: MIT Press, 2003.

Merrill, Daniel D. *Augustus De Morgan and the Logic of Relations.* Dordrecht: Kluwer, 2011.

Merzbach, Uta C., and Carl B. Boyer. *A History of Mathematics.* 3rd ed. Hoboken, NJ: John Wiley & Sons, 2011.

Nahin, Paul J. *The Logician and the Engineer: How George Boole and Claude Shannon Created the Information Age.* Princeton, NJ, and Oxford: Princeton University Press, 2013.

Prettejohn, Elizabeth. *The Art of the Pre-Raphaelites.* Princeton, NJ: Princeton University Press, 2000.

Rice, Adrian. "Augustus De Morgan: Historian of Science." *History of Science* 34 (1996): 201–240.

Richards, Joan L. "The Art and Science of British Algebra." *Historia Mathematica* 7 (1980): 343–365

Richards, Joan L. "Augustus De Morgan, the History of Mathematics, and the Foundations of Algebra." *Isis* 78 (1987): 7–30.

Richards, Joan L. *Generations of Reason: A Family's Search for Meaning in Post Newtonian England.* New Haven, CT: Yale University Press, 2021.

Richards, Joan L. "God, Truth, and Mathematics in Nineteenth-Century England." In *The Invention of Physical Science*, edited by Mary Jo Nye, Joan Richards, and Richard H. Stuewer, 51–78. Dordrecht: Kluwer, 1992.

Richards, Joan L. *Mathematical Visions: The Pursuit of Geometry in Victorian England.* Boston: Academic Press, 1988.

Richards, Joan L. "Rigor and Clarity: Foundations of Mathematics in France and England, 1800–1840." *Science in Context* 4 (1991): 297–319.

Rotman, Brian. "Counting on Non-Euclidean Fingers." In *Mathematics as Sign: Writing, Imagining, Counting*, 125–153. Stanford, CA: Stanford University Press, 2000.

Royle, John Edward. "Mechanics' Institutes and the Working Classes, 1840–1860." *The Historical Journal* 14, no. 2 (1971): 305–321.

Schaffer, Simon. "Babbage's Dancer and the Impresarios of Mechanism." In *Cultural Babbage: Technology, Time and Invention*, edited by Francis Spufford and Jennifer S. Uglow, 52–80. London: Faber & Faber, 1996.

Schorske, Carl E. *Thinking with History: Explorations in the Passage to Modernism*. Princeton, NJ: Princeton University Press, 1998.

Secord, James A. *Victorian Sensation: The Extraordinary Publication, Reception, and Secret Authorship of Vestiges of the Natural History of Creation*. Chicago: University of Chicago Press, 2000.

Shapin, Steven, and Barry Barnes. "Science, Nature and Control: Interpreting Mechanics Institutes." *Social Studies of Science* 7 (1977): 31–74.

Smith, G. C. *The Boole–De Morgan Correspondence, 1842–1864*. Oxford: Clarendon Press, 1982.

Topham, Jonathan. "A Textbook Revolution." In *Books and the Sciences in History*, edited by Marina Frasca Spada and Nicholas Jardine, 317–337. Cambridge/New York: Cambridge University Press, 2000.

Topham, J. R. "Two Centuries of Cambridge Publishing and Bookselling: A Brief History of Deighton, Bell, and Co. 1778–1998, with a Checklist of the Archive." *Transactions of the Cambridge Bibliographical Society* 11, no. 3 (1998): 350–403.

Turing, Alan. "On Computable Numbers, with an Application to the Entscheidungsproblem." *Proceedings of the London Mathematical Society* 2, no. 1 (1937): 230–265.

Tylecote, Mabel. *The Mechanics' Institutes of Lancashire and Yorkshire before 1851*. Manchester: Manchester University Press, 1957.

Warwick, Andrew. *Masters of Theory: Cambridge and the Rise of Mathematical Physics*. Chicago: University of Chicago Press, 2003.

Werner, Marcia. *Pre-Raphaelite Painting and Nineteenth-Century Realism*. Cambridge: Cambridge University Press, 2005.

Whately, Richard. *Elements of Logic* 1827; Delmar: Scholars' Facsimiles & Reprints, 1975.

Whately, Richard. "Logic." In *Encyclopaedia Metropolitana*, edited by Edward Smedley, Hugh James Rose, and Henry John Rose, Vol. 1 [Pure sciences, vol. 1], 193–240. London, 1845.

Chapter 7

Albers, Henry, ed. *Maria Mitchell: A Life in Journals and Letters*. Clinton Corners, NY: College Avenue Press, 2001.

Bailey, Solon I. *The History and Work of Harvard Observatory, 1839 to 1927*. New York: McGraw Hill, 1931.

Beck, Estee. "A Theory of Persuasive Computer Algorithms for Rhetorical Code Studies." *Enculturation* 23 (2016). http://enculturation.net/a-theory-of-persuasive-computer-algorithms.

Bergland, Renée. *Maria Mitchell and the Sexing of Science: An Astronomer among the American Romantics*. Boston: Beacon, 2008.

Brewster, David. *Memoirs of the Life, Writings, and Discoveries of Sir Isaac Newton*. Vol. 2. Boston: Little, Brown, and Co., 1855.

Cambridge Chronicle. "Resurrected from the Files of Long Ago," January 17, 1930. http://hea-www.harvard.edu/~jcm/html/chron.html.

Cowan, Ruth Schwartz. *More Work for Mother: The Ironies of Household Technology from the Open Hearth to the Microwave*. New York: Basic Books, 1983.

Dobson, James E., and Rena J. Mosteirin. *Moonbit*. Goleta, CA: Punctum Books, 2019.

Fiss, Andrew, and Laura Kasson Fiss. "Laughing out of Math Class: The Vassar Mathematikado and Nineteenth-Century Women's Education." *Configurations* 27, no. 3 (Summer 2019): 301–329.

Fiss, Laura Kasson. "The Idler's Club: Humor and Sociability in the Age of New Journalism." *Victorian Periodicals Review* 49, no. 3 (2016): 415–430.

Fiss, Laura Kasson. "'This Particularly Rapid, Unintelligible Patter': Patter Songs and the Word-Music Relationship." In *The Cambridge Companion to Gilbert and Sullivan*, edited by David Eden and Meinhard Saremba, 98–108. Cambridge: Cambridge University Press, 2009.Forty, Adrian. *Objects of Desire: Design and Society 1750-1980*. London: Thames and Hudson, 1986.

Gilbert, W. S., and Arthur Sullivan. "H.M.S. Pinafore." In *The Complete Annotated Gilbert & Sullivan*. Introduced and edited by Ian Bradley, 113–186. Oxford: Oxford University Press, 1996.

Grier, David Alan. *When Computers Were Human*. Princeton, NJ: Princeton University Press, 2005.

Hacker, Sally L. *"Doing It the Hard Way": Investigations of Gender and Technology*. Edited by Dorothy E. Smith and Susan M. Turner. Boston: Unwin Hyman, 1990.

Hariman, Robert. "Political Parody and Public Culture." *Quarterly Journal of Speech* 94, no. 3 (2008): 247–272.

Harvard College Observatory. "Records Relating to the Observatory Pinafore, 1879–1966." Collections of the Harvard University Archives, UAV 630.495.2.

Harvard University Library Open Collections Program. "Williamina Paton Stevens Fleming (1857–1911)." *Women Working, 1800-1930*, last modified 2017, http://ocp.hul.harvard.edu/ww/fleming.html.

Hayles, N. Katherine. *My Mother Was a Computer: Digital Subjects and Literary Texts*. Chicago: University of Chicago Press, 2005.

Hoar, Roger Sherman. "The Pickering Polaris Attachment." *Journal of the United States Artillery* 50 (1919): 230–236.

Hoffleit, Dorrit. *The Education of American Women Astronomers before 1960*. Cambridge, MA: American Association of Variable Star Observers, 1994.

Holmes, Richard. *The Age of Wonder: How the Romantic Generation Discovered the Beauty and Terror of Science*. New York: Random House, 2008.

Johnson, N. R. "Information Infrastructure as Rhetoric: Tools for Analysis." *Poroi* 8, no. 1 (2012): 1–3.

Jones, Bessie Zaban, and Lyle Gifford Boyd. *The Harvard College Observatory: The First Four Directorships, 1939-1919*. Cambridge, MA: Belknap Press of Harvard University Press, 1971.

Mack, Pamela. "Strategies and Compromises: Women in Astronomy at Harvard College Observatory, 1870–1920." *Journal of the History of Astronomy* 21 (1990): 65–75.

Mack, Pamela E. "Straying from Their Orbits: Women in Astronomy in America." In *Women of Science: Righting the Record*, edited by G. Kass-Simon and Patricia Farnes, 72–116. Bloomington: Indiana University Press, 1990.

Oglivie, Marilyn Bailey, and Joy Dorothy Harvey, eds. *Biographical Dictionary of Women in Science*. Volume 2. New York: Routledge, 2000.

Pickering, Edward C. *Statement of Work Done at the Harvard College Observatory during the Years 1877-1882*. Cambridge, MA: John Wilson & Son University Press, 1882.

Rossiter, Margaret W. *Women Scientists in America: Struggles and Strategies to 1940*. Baltimore: Johns Hopkins University Press, 1982.

Shapley, H., to E. Upton, January 23, 1930, in the Harvard College Observatory. "Records Relating to the Observatory Pinafore, 1879–1966." Collections of the Harvard University Archives, UAV 630.495.2, also hosted at http://hea-www.harvard.edu/~jcm/html/shapley.html.

Sobel, Dava. *The Glass Universe: How the Ladies of the Harvard Observatory Took the Measure of the Stars*. New York: Viking, 2016.

Tan, Avianne. "Apollo 11's Source Code Has Tons of Easter Eggs, Including an Ignition File Labeled 'Burn Baby Burn.'" ABC News, July 12, 2016.

Upton, Winslow S. *The Observatory Pinafore (1879)*. Transcribed by Jonathan McDowell. Harvard-Smithsonian Center for Astrophysics, 1994. http://hea-www.harvard.edu/~jcm/html/play.html.

Wajcman, Judy. *Feminism Confronts Technology*. University Park: Pennsylvania State University Press, 1991.

Williams, Carolyn. *Gilbert and Sullivan: Gender, Genre, Parody*. New York: Columbia University Press, 2011.

Wright, Helen. *Sweeper in the Sky: The Life of Maria Mitchell, First Woman Astronomer in America*. New York: Macmillan, 1950.

Chapter 8

Adas, Michael. *Machines as the Measure of Men: Science, Technology, and Ideologies of Western Dominance*. Ithaca, NY: Cornell University Press, 1989.

Annual Report of the Secretary of the State Board of Agriculture of the State of Michigan. United States: Robert Smith Print. Company, state printers and binders, 1899.

Arnold, Thurman W. *The Folklore of Capitalism*. New Haven, CT: Yale University Press, 1937.

Beniger, James R. *The Control Revolution: Technological and Economic Origins of the Information Society*. Cambridge, MA: Harvard University Press, 1986.

Bernhardt, Joshua. "The Flexible Tariff and the Sugar Industry." *American Economic Review* 16, no. 1 (1926): 182–191.

Campbell, George A. "Mathematics in Industrial Research." *Bell System Technical Journal* 3, no. 4 (1924): 550–557.

Chandler, Alfred D., Jr. *The Visible Hand: The Managerial Revolution in American Business*. Cambridge, MA: Harvard University Press, 1977.

Clark, Olynthus B. "Keeping Them on the Farm: The Story of Henry C. Wallace, the Secretary of Agriculture, Who Has Spent His Life Studying, Teaching, Improving and Practising Farming [*sic*]." *The Independent* 105 (1921): 333ff.

Cochran, W.G. "Graduate Training in Statistics." *American Mathematical Monthly* 53, no. 4 (1946): 193–199.

Coons, George H. "The Sugar Beet: Product of Science." *Scientific Monthly* 68, no. 3 (1949): 149–164.

Deming, W. Edwards. "Foreword from the Editor." In *Statistical Method from the Viewpoint of Quality Control*. Washington, DC: The Graduate School, USDA, 1939.

Deming, W. Edwards, and Raymond T. Birge. *On the Statistical Theory of Errors, 1934*. USDA Graduate School Collection. Special Collections, National Agricultural Library.

Dryer, Theodora. *Designing Certainty: The Rise of Algorithmic Computing in the Age of Anxiety, 1920–1970*. PhD diss., University of California, San Diego, 2019.

Fitzgerald, Deborah. *Every Farm a Factory: The Industrial Ideal in American Agriculture*. New Haven, CT: Yale University Press, 2003.

Friendly, Alfred. "Agriculture's School Is on the Upgrade." *Washington Daily News*, August 30, 1938.

Fullilove, Courtney. *Profit of the Earth: The Global Seeds of American Agriculture.* Chicago: University of Chicago Press, 2017.

Galambos, Louis. "The Emerging Organizational Synthesis in Modern American History." *Business History Review* 44, no. 3 (Autumn 1970): 279–290.

Gente, Ralf, Stefan F. Busch, Eva-Maria Stübling, Lorenz Maximilian Schneider, Christian B. Hirschmann, Jan C. Balzer, and Martin Koch. "Quality Control of Sugar Beet Seeds with THz Time-Domain Spectroscopy." *IEEE Transactions on Terahertz Science and Technology* 6, no. 5 (2016): 754–756.

Grier, David Alan. "Programming and Planning." *IEEE Annals of the History of Computing* 33, no. 1 (2011): 86–88.

Grimshaw, Robert, and Lewis Sharpe Ware. "Various Issues." *The Sugar Beet: Scientific Quarterly* 31, no. 1 (1910): 7ff.

Haber, Samuel. *Efficiency and Uplift: Scientific Management in the Progressive Era.* Chicago: University of Chicago Press, 1964.

Hays, Samuel P. *Conservation and the Gospel of Efficiency: The Progressive Conservation Movement, 1890–1920.* Cambridge, MA: Harvard University Press, 1959.

Heyck, Hunter. "The Organizational Revolution and the Human Sciences." *Isis* 105, no. 1 (2014): 1–31.

Hounshell, David. *From American System to Mass Production, 1800–1932.* Baltimore: Johns Hopkins University Press, 1984.

Kanigel, Robert. *The One Best Way: Frederick Winslow Taylor and the Enigma of Efficiency.* New York: Viking, 1997.

Karns Alexander, Jennifer. *Mantra of Efficiency: From Waterwheel to Social Control.* Baltimore: Johns Hopkins University Press, 2008.

Lottes, P., M. Hoeferlin, S. Sander, M. Müter, P. Schulze, and L. C. Stachniss. "An Effective Classification System for Separating Sugar Beets and Weeds for Precision Farming Applications." In *2016 IEEE International Conference on Robotics and Automation* (2016): 5157–5163. doi: 10.1109/ICRA.2016.7487720.

Mahalanobis, P. C. "Walter A. Shewhart and Statistical Quality Control in India." *Sankhyā: The Indian Journal of Statistics (1933–1960)* 9, no. 1 (1948): 51–60.

Marez, Curtis. *Farm Worker Futurism: Speculative Technologies of Resistance.* Minneapolis: Minnesota University Press, 2016.

Mintz, Sydney. *Sweetness and Power: The Place of Sugar in Modern History.* New York: Viking, 1986.

Miranti, Paul J. "Corporate Learning and Quality Control at the Bell System, 1877–1929." *Business History Review* 79, no. 1 (2005): 39–72.

Mitcham, Carl. *Thinking through Technology: The Path between Engineering and Philosophy.* Chicago: University of Chicago Press, 1994.

Mohler, John R. Address to the Graduate School: Scientific Research, November 20, 1936, USDA Graduate School Collection. Special Collections, National Agricultural Library.

Noble, David F. *America by Design: Science, Technology, and the Rise of Corporate Capitalism.* Oxford: Oxford University Press, 1979.

Norris, Jim. "Bargaining for Beets: Migrants and Growers in the Red River Valley." *Minnesota History* 58, no. 4 (2002/2003): 196–209.

Nye, David E. *Electrifying America: Social Meanings of a New Technology.* Cambridge, MA: MIT Press, 1990.

Owens, H. S. "Production and Utilization of Sugar Beets." *Economic Botany* 5, no. 4 (1951): 348–366.

Pearson, Karl. *The Grammar of Science.* Cambridge: Cambridge University Press, 2014.

"Plan to Introduce Graduate Studies at the USDA Graduate School in 1921" (1921). Special Collections, USDA National Agricultural Library, accessed March 26, 2018. https://www.nal.usda.gov/exhibits/speccoll/items/show/8765.

Reisch, George A. *How the Cold War Transformed Philosophy of Science: To the Icy Slopes of Logic*. Cambridge: Cambridge University Press, 2005.

Roosevelt, Franklin D. "Inaugural Address of 1933." Washington, DC: National Archives and Records Administration, 1988.

Sarle, Charles F. "The Agricultural Statistics Program." Address, 25th session, International Statistical Conferences, Washington, DC, September 13, 1947. https://www.nass.usda.gov/Education_and_Outreach/Reports,_Presentations_and_Conferences/Yield_Reports/The%20Agricultural%20Statistics%20Program.pdf.

Shewhart, W. A. "Quality Control Charts." *Bell System Technical Journal* 5, no. 4 (October 1926): 593–603.

Shewhart, W. A. "Some Applications of Statistical Methods to the Analysis of Physical and Engineering Data." *Bell System Technical Journal* 3, no. 1 (1924): 43–87.

Shewhart, Walter A. *Statistical Method from the Viewpoint of Quality Control*. Washington DC: The Graduate School, USDA, 1939.

Smirnov, A., B. N. Holben, T. F. Eck, and O. Dubovik. "Cloud-Screening and Quality Control Algorithms for the AERONET Database." *Remote Sensing of Environment* 73, no. 3 (2000): 337–349.

Smith, Clinton DeWitt, and Robert Clark Kedzie. "Sugar Beets in Michigan in 1897." *Michigan State Agricultural College Experiment Station*, Bulletin 150 (December 1897).

Smith, P. B. "A Survey of Sugar-Beet Mechanization." Fort Collins, CO, 1950, mimeographed.

Smith, R. K. "State Frontiers in Agricultural Statistics: Discussion." *Journal of Farm Economics* 31, no. 1 (1949): 304–308.

Stigler, H. G., and R. T. Burdick. "The Economics of Sugar-Beet Mechanization." Bulletin 411-A. Fort Collins, CO: Agricultural Extension Service, April 1950.

Stilgenbauer, F. A. "The Michigan Sugar Beet Industry." *Economic Geography* 3, no. 4 (1927): 486–506.

Swerling, Boris C. "United States Sugar Policy and Western Development." *Proceedings of the Annual Meeting* (Western Farm Economics Association) 24 (1951): 7–11.

Tompkins, Kyla Wazana. "Sweetness, Capacity, and Energy." *American Quarterly* 71, no. 3 (2019): 849–856.

Tsing, Anna Lowenhaupt. *Mushroom at the End of the World: On the Possibility of Life in Capitalist Ruins*. Princeton, NJ: Princeton University Press, 2017.

Valdes, Dennis Dodin. *Al Norte: Agricultural Workers in the Great Lakes Region, 1917–1970*. Austin: University of Texas Press, 1991.

Vargas, Zaragosa. "Life and Community in the 'Wonderful City of the Magic Motor': Mexican Immigrants in 1920s Detroit." *Michigan Historical Review* 15, no. 1 (1989): 45–68.

Veblen, Thorstein. *The Theory of the Leisure Class: An Economic Study of Institutions*. New York: Dover Thrift Editions, 1994.

Walker, H. B. "A Resume of Sixteen Years of Research in Sugar-Beet Mechanization." *Agricultural Engineering* (1948): 425–430

Wallace, Henry C. *Our Debt and Duty to the Farmer*. New York: Century Co., 1925.

Wiley, H. W. "The Sugar Beet: Culture, Seed Development, Manufacture, and Statistics." *Farmers' Bulletin* 52 (1899): 5. Accessed from UNT Libraries Government Documents Department. https://digital.library.unt.edu/ark:/67531/metadc85519/.

Winckel, Richard C. "Telephone Transmitter." US Patent US1603300A (1924).

Wishart, John. "Bibliography of Agricultural Statistics 1931–1933." *Supplement to the Journal of the Royal Statistical Society* 1, no. 1 (1934): 94–106.

Woods, A. F. Letter to Statistically Minded Research Workers in Washington, February 1938. USDA Graduate School Collection. Special Collections, National Agricultural Library.

Chapter 9

Acree, Michael Coy. "Theories of Statistical Inference in Psychological Research: A Historico-Critical Study." PhD diss., Clark University, 1978.

Agar, Jon. "What Difference Did Computers Make?" *Social Studies of Science* 36, no. 6 (December 2006): 869–907.

Amerine, Maynard A., and Edward B. Roessler. *Wines: Their Sensory Evaluation*. San Francisco: Freeman, 1976.

Andrews, D. F. "Developing Examples for Learning Statistics; Data and Computing." *International Statistical Review* 41, no. 2 (1973): 225–228.

Bakan, David. "The Test of Significance in Psychological Research." *Psychological Bulletin* 66, no. 6 (December 1966): 423–437.

Beecher, Henry K. *Measurement of Subjective Responses: Quantitative Effects of Drugs*. New York: Oxford, 1959.

Beecher, Henry K. "Pain: One Mystery Solved." *Science* 151, no. 3712 (1966): 840–841.

Bothwell, Laura. "The Emergence of the Randomized Controlled Trial: Origins to 1980." PhD diss., Columbia University, 2014.

Box, Joan Fisher. "R. A. Fisher and the Design of Experiments, 1922–1926." *American Statistician* 34, no. 1 (February 1980): 1–7.

Bullynck, Maarten. "Histories of Algorithms: Past, Present and Future." *Historia Mathematica* 43, no. 3 (2016): 332–341.

Capshew, James H. *Psychologists on the March: Science, Practice, and Professional Identity in America, 1929–1969*. Cambridge: Cambridge University Press, 1999.

Carpenter, Daniel. *Reputation and Power; Organizational Image and Pharmaceutical Regulation at the FDA*. Princeton, NJ: Princeton University Press, 2010.

Cat, Jordi. "The Unity of Science." In *Stanford Encyclopedia of Philosophy*, edited by Edward N. Zalta (Fall 2017 ed.), accessed October 7, 2019. https://plato.stanford.edu/archives/fall2017/entries/scientific-unity/.

Cohen, Jacob. "The Earth Is Round (p<.05)." *American Psychologist* 49, no. 12 (December 1994): 997–1003.

Danziger, Kurt. *Constructing the Subject: Historical Origins of Psychological Research*. Cambridge: Cambridge University Press, 1990.

Danziger, Kurt. "Statistical Method and the Historical Development of Research Practice in American Psychology." In *The Probabilistic Revolution*, Vol. 2: *Ideas in the Sciences*, edited by Lorenz Krüger, Gerd Gigerenzer, and Mary S. Morgan, 35–47. Cambridge, MA: MIT Press, 1987.Daston, Lorraine. "The Empire of Observation, 1600–1800." In *Histories of Scientific Observation*, edited by Lorraine Daston and Elizabeth Lunbeck, 81–113. Chicago: University of Chicago Press, 2011.

Daston, Lorraine, and Peter Galison. *Objectivity*. New York: Zone Books, 2010.

Deming, W. Edwards. *Statistical Adjustment of Data*. New York: John Wiley, 1938.

Dick, Stephanie. "AfterMath: The Work of Proof in the Age of Human-Machine Collaboration." *Isis* 102 (2011): 494–505.

Donnelly, Michael. "William Farr and Quantification in Nineteenth-Century English Public Health." In *Body Counts: Medical Quantification in Historical and Sociological Perspective*, edited by Gérard Jorland, Annick Opinel, and George Weisz, 251–265. Montreal: McGill-Queen's University Press, 2005.

Dorfman, Robert, Paul A. Samuelson, and Robert M. Solow. *Linear Programming and Economic Analysis*. New York: McGraw-Hill, 1958.

Durkheim, Émile. *On Suicide*. Translated by Robin Buss. New York: Penguin, 2006 [1897].

Edwards, Allen L. *Statistical Methods for the Behavioral Sciences*. New York: Rinehart, 1954.

Ekeblad, Frederick A. *The Statistical Method in Business: Applications of Probability and Inference to Business and other Problems*. New York: Wiley, 1962.

Endres, A. M. "The Functions of Numerical Data in the Writings of Graunt, Petty, and Davenant." *History of Political Economy* 17, no. 2 (1985): 245–264.

Erickson, Paul. *The World the Game Theorists Made*. Chicago: University of Chicago Press.

Erickson, Paul, Judy L. Klein, Lorraine Daston, Rebecca Lemov, Thomas Sturm, and Michael D. Gordin. *How Reason Almost Lost Its Mind: The Strange Career of Cold War Rationality*. Chicago: University of Chicago Press, 2013.

Ferguson, Thomas S. "The Development of the Decision Model." In *On the History of Statistics and Probability*, edited by D. B. Owen, 333–346. New York: Dekker, 1976.

Fisher, R. A. *The Design of Experiments*. Edinburgh: Oliver and Boyd, 1935.

Galton, Francis. *Natural Inheritance*. London: Macmillan, 1889.

Garrett, Henry E. *Statistics in Psychology and Education*. New York: Longmans, Green and Co., 1926.

Garrett, Henry E. *Statistics in Psychology and Education*. 6th ed. New York: McKay, 1966.

Gigerenzer, Gerd, and David J. Murray. *Cognition as Intuitive Statistics*. Hillsdale, NJ: Lawrence Erlbaum Associates, 1987.

Gigerenzer, Gerd, Zeno Swijtink, Theodore Porter, Lorraine Daston, John Beatty, and Lorenz Krüger. *The Empire of Chance: How Probability Changed Science and Everyday Life*. Cambridge: Cambridge University Press, 1989.

Gorroochurn, Prakash. *Classic Topics on the History of Modern Mathematical Statistics*. Hoboken, NJ: Wiley, 2016.

Gosset, William. "The Probable Error of a Mean." *Biometrika* 6, no. 1 (March 1908): 1–25.

Greene, Jeremy A. *Prescribing by Number: Drugs and the Definition of Disease*. Baltimore: Johns Hopkins University Press, 2007.

Heyck, Hunter. *Age of System: Understanding the Development of Modern Social Science*. Baltimore: Johns Hopkins University Press, 2015.

Hogben, Lancelot T. *Statistical Theory: The Relationship of Probability, Credibility, and Error*. London: George Allen & Unwin, 1957.

Hotelling, Harold. "Abraham Wald." *American Statistician* 5, no. 1 (February 1951): 18–19.

Hunter, Patti Wilger. "Connections, Context, and Community: Abraham Wald and the Sequential Probability Ratio Test." *The Mathematical Intelligencer* 26, no. 1 (December 2004): 25–33.

Jorland, Gérard, Annick Opinel, and George Weisz, eds. *Body Counts: Medical Quantification in History and Sociological Perspectives*. Montreal: McGill–Queen's University Press, 2005.Kahneman, Daniel, Paul Slovic, and Amos Tversky, eds. *Judgment under Uncertainty: Heuristics and Biases*. Cambridge: Cambridge University Press, 1982.

Kitchin, Rob. "Thinking Critically about and Researching Algorithms." *Information, Communication and Society* 20, no. 1 (2017): 14–29.

Mantel, Nathan. "A Personal Perspective on Statistical Techniques for Quasi-Experiments." In *On the History of Statistics and Probability*, ed. D. B. Owen , 103–129. New York: Dekker, 1976.

Marks, Harry M. *The Progress of Experiment: Science and Therapeutic Reform in the United States, 1900–1930*. Cambridge: Cambridge University Press, 1997.

Mayer-Schönberger, Viktor, and Kenneth Cukier. *Big Data: A Revolution That Will Transform How We Live, Work, and Think*. Boston: Houghton Mifflin, 2013.

Meehl, Paul E. "Theoretical Risks and Tabular Asterisks: Sir Karl, Sir Ronald, and the Slow Progress of Soft Psychology." *Journal of Consulting and Clinical Psychology* 46 (1978): 806–834.

Meehl, Paul E. "Wanted—A Good Cookbook." In his *Psychodiagnosis: Selected Papers*, 63–80. Minneapolis: University of Minnesota Press, 1972 [1955].

Melton, Arthur W. "Editorial." *Journal of Experimental Psychology* 64, no. 6 (December 1962): 553–557.

Milne, Iain. "Who Was James Lind, and What Exactly Did He Achieve?" *JLL Bulletin* (2012), accessed January 11, 2018. http://www.jameslindlibrary.org/articles/who-was-james-lind-and-what-exactly-did-he-achieve/.

Morrison, Denton E., and Ramon E. Henkel, eds. *The Significance Test Controversy—A Reader*. Chicago: Aldine, 1970.

Nie, Norman H., Dale H. Bent, and C. Hadlai Hull. *SPSS: Statistical Package for the Social Sciences*. New York: McGraw-Hill, 1970.

Noble, David F. *The Religion of Technology: The Divinity of Man and the Spirit of Invention*. New York: Knopf, 1997.

"Obituary: William Sealy Gosset, 1876–1937." *Journal of the Royal Statistical Society* 101, no. 1 (1938): 248–251.

O'Neil, Cathy. *Weapons of Math Destruction: How Big Data Increases Inequality and Threatens Democracy*. New York: Crown, 2016.

Parolini, Giuditta. "In Pursuit of a Science of Agriculture: The Role of Statistics in Field Experiments." *History and Philosophy of the Life Sciences* 37, no. 3 (2015): 261–281.

Pasquale, Frank. *The Black Box Society: The Secret Algorithms That Control Money and Information*. Cambridge, MA: Harvard University Press, 2015.

Pearl, Raymond. *Introduction to Medical Biometry and Statistics*. Philadelphia: Saunders, 1923.

Phillips, Christopher J. "The Taste Machine: Sense, Subjectivity, and Statistics in the California Wine World." *Social Studies of Science* 46, no. 3 (2016): 461–481.

Popper, Karl. *Logic of Scientific Discovery*. London: Hutchinson, 1959.

Porter, Theodore M. *Karl Pearson: The Scientific Life in a Statistical Age*. Princeton, NJ: Princeton University Press, 2004.

Porter, Theodore M. "Reforming Vision: The Engineer Le Play Learns to Observe Society Sagely." In *Histories of Scientific Observation*, edited by Lorraine Daston and Elizabeth Lunbeck, 281–302. Chicago: University of Chicago Press, 2011.

Porter, Theodore M. *Trust in Numbers: The Pursuit of Objectivity in Science and Public Life*. Princeton, NJ: Princeton University Press, 1995.

Quetelet, M. A. *A Treatise on Man and the Development of His Faculties*. Repr. ed. New York: Burt Franklin, 1968 [1842].

Rosenberg, Charles E. "Science, Technology, and Economic Growth: The Case of the Agricultural Experiment Station Scientist, 1875–1914." *Agricultural History* 45, no. 1 (January 1971): 1–20.

Rosser Matthews, Louis J. *Quantification and the Quest for Medical Certainty*. Princeton, NJ: Princeton University Press, 1995.

Rothman, Kenneth J. "Lessons from John Graunt." *Lancet* 347, no. 8993 (January 6, 1996): 37–39.

Rothstein, William G. *Public Health and the Risk Factor: A History of an Uneven Medical Revolution*. Rochester, NY: University of Rochester Press, 2003.

Rozeboom, William W. "The Fallacy of the Null-Hypothesis Significance Test." *Psychological Bulletin* 57, no. 5 (1960): 416–428.

Rucci, Anthony J., and Ryan D. Tweney. "Analysis of Variance and the 'Second Discipline' of Scientific Psychology: A Historical Account." *Psychological Bulletin* 87, no. 1 (1980): 166–184.

Rudolph, John L. "Epistemology for the Masses: The Origins of 'The Scientific Method' in American Schools." *History of Education Quarterly* 45, no. 3 (Fall 2005): 341–376.

Savage, L. J. "Theory of Statistical Decision." *Journal of the American Statistical Association* 46, no. 253 (March 1951): 55–67.

Savage, Leonard J. *The Foundations of Statistics*. 2nd rev ed. New York: Dover, 1972 [1954].

Shapin, Steven. "How to Be Antiscientific." In *The One Culture? A Conversation about Science*, edited by Jay A. Labinger and Harry Collins, 99–115. Chicago: University of Chicago Press, 2001.

Shapin, Steven. "The Sciences of Subjectivity." *Social Studies of Science* 42, no. 2 (2011): 170–184.

Shapin, Steven. "A Taste of Science: Making the Subjective Objective in the California Wine World." *Social Studies of Science* 46, no. 3 (2016): 436–460.

Snedecor, George W. *Statistical Methods: Applied to Experiments in Agriculture and Biology*. 4th ed. Ames: Iowa State College Press, 1946 [1937].

Statistical Research Group. *Sequential Analysis in Inspection and Experimentation*. Report No. 255, AMP Report 30.2R. New York: Columbia University Press, 1945.

Sterling, Theodor[e] D. "Publication Decisions and Their Possible Effects on Inferences Drawn from Tests of Significance—Or Vice Versa." *Journal of the American Statistical Association* 54, no. 285 (March 1959): 30–34.

Stevens, Hallam. "A Feeling for the Algorithm: Working Knowledge and Big Data in Biology." *Osiris* 32 (2017): 151–174.

Stevens, Hallam. *Life out of Sequence: A Data-Driven History of Bioinformatics*. Chicago: University of Chicago Press, 2013.

Stigler, Stephen M. *The History of Statistics: The Measurement of Uncertainty before 1900*. Cambridge, MA: Harvard University Press, 1986.

Thérèse, Sandrine, and Brian Martin. "Shame, Scientist! Degradation Rituals in Science." *Prometheus* 28, no. 2 (2010): 97–110.

Tousignant, Noémi. "Pain and the Pursuit of Objectivity: Pain-Measurement Technologies in the United States, c. 1890–1975." PhD thesis, McGill University, 2006.

Tousignant, Noémi. "The Rise and Fall of the Dolorimeter: Pain, Analgesics, and the Management of Subjectivity in Mid-Twentieth-Century United States." *Journal of the History of Medicine and Allied Sciences* 66, no. 2 (2011): 145–179.

von Neumann, John, and Oskar Morgenstern. *Theory of Games and Economic Behavior*. 2nd ed. Princeton, NJ: Princeton University Press, 1947 [1944].

Wald, Abraham. *Sequential Analysis*. New York: Wiley, 1947.

Wald, Abraham. *Statistical Decision Functions*. New York: Wiley, 1950.

Wald, Abraham. "Statistical Decision Functions Which Minimize the Maximum Risk." <u>Annals of Mathematics</u>, 2nd ser., 46, no. 2 (April 1945): 265–280.

Wert, James E., Charles O. Neidt, and J. Stanley Ahmann. *Statistical Methods in Educational and Psychological Research*. New York: Appleton-Century-Crofts, 1954.

Yule, G. Udny. *An Introduction to the Theory of Statistics*. London: Griffin; Philadelphia: Lippincott, 1911.

Chapter 10

Agar, Jon. "What Difference Did Computers Make?" *Social Studies of Science* 36, no. 6 (December 1, 2006): 869–907. https://doi.org/10.1177/0306312706073450.

Breiman, Leo. "[A Report on the Future of Statistics]: Comment." *Statistical Science* 19, no. 3 (2004): 411.

Breiman, Leo. "Statistical Modeling: The Two Cultures." *Statistical Science* 16, no. 3 (2001): 201.

Breiman, Leo, et al. *Classification and Regression Trees*. Boca Raton, FL: Chapman & Hall, 1984.

Breiman, Leo, and William S. Meisel. "Empirical Techniques for Analyzing Air Quality and Meteorological Data. Part III. Short-Term Changes to Ground Lever Ozone Concentration: An Empirical Analysis." Environmental Monitoring Series. Environmental Sciences Research Laboratory, Environmental Protection Agency, June 1976.Buntine, Wray.

"IND: Creation and Manipulation of Decision Trees from Data," accessed April 22, 2017. https://ti.arc.nasa.gov/opensource/projects/ind/.

Burrell, Jenna. "How the Machine 'Thinks': Understanding Opacity in Machine Learning Algorithms." *Big Data & Society* 3, no. 1 (January 5, 2016): 4–5. https://doi.org/10.1177/2053951715622512.

Creel, Kathleen A. "Transparency in Complex Computational Systems." *Philosophy of Science* 87 (2020): 568–589.

Dick, Stephanie. "Of Models and Machines: Implementing Bounded Rationality." *Isis* 106, no. 3 (2015): 623–634.

Edwards, Paul. *A Vast Machine: Computer Models, Climate Data, and the Politics of Global Warming.* Cambridge, MA: MIT Press, 2010.

Einhorn, Hillel J. "Alchemy in the Behavioral Sciences." *Public Opinion Quarterly* 36, no. 3 (1972): 368–369.

Ensmenger, Nathan. "Is Chess the Drosophila of Artificial Intelligence? A Social History of an Algorithm." *Social Studies of Science* 42, no. 1 (February 1, 2012): 5–30. https://doi.org/10.1177/0306312711424596.

Eubanks, Virginia. *Automating Inequality: How High-Tech Tools Profile, Police, and Punish the Poor.* New York: St. Martin's Press, 2017.Fayyad, Usama. "Personal Observations of a Data Mining Disciple a Data Miner's Story—Getting to Know the Grand Challenges of Research." http://slideplayer.com/slide/6409052/.

Fayyad, Usama. "Taming the Giants and the Monsters: Mining Large Databases for Nuggets of Knowledge." *Database Programming and Design Magazine* 11, no. 3 (1998). https://fayyad.com/taming-the-giants-and-the-monsters-mining-large-databases-for-nuggets-of-knowledge/.

Fayyad, Usama Mohammad. "On the Induction of Decision Trees for Multiple Concept Learning." Ann Arbor: University of Michigan, 1992.

Fayyad, U. M., et al. "Machine Learning of Expert System Rules: Applications to Semiconductor Manufacturing." In *Collected Notes on the Workshop for Pattern Discovery in Large Databases (NASA Ames, January 14–15, 1991).* Moffett Field, CA: NASA Ames Research Center, 1991.

Fayyad, Usama M., S. George Djorgovski, and Nicholas Weir. "From Digitized Images to Online Catalogs Data Mining: A Sky Survey." *AI Magazine* 17, no. 2 (1996): 51–66.

Feigenbaum, Edward. Oral History of Edward Feigenbaum, interview by Nils Nilsson, June 20 and 27, 2007, 62–63. http://archive.computerhistory.org/resources/access/text/2013/05/102702002-05-01-acc.pdf.

Feng, Cao, and D. Michie. "Machine Learning of Rules and Trees." In *Machine Learning, Neural and Statistical Classification*, edited by Donald Michie et al., 51. Upper Saddle River, NJ: Ellis Horwood, 1994. http://dl.acm.org/citation.cfm?id=212782.212787.

Fernholz, Luisa T., et al. "A Conversation with John W. Tukey and Elizabeth Tukey." *Statistical Science* 15, no. 1 (2000): 79–94.

Fisher, N. I. "A Conversation with Jerry Friedman." *Statistical Science* 30, no. 2 (May 2015): 271. https://doi.org/10.1214/14-STS509.

Forsythe, D. E. "Engineering Knowledge: The Construction of Knowledge in Artificial Intelligence." *Social Studies of Science* 23, no. 3 (August 1, 1993): 445–477. https://doi.org/10.1177/0306312793023003002.

Friedman, J. H. "A Recursive Partitioning Decision Rule for Nonparametric Classification." *IEEE Transactions on Computers* C-26, no. 4 (April 1977): 4. https://doi.org/10.1109/TC.1977.1674849.

Gigerenzer, Gerd, ed. *The Empire of Chance: How Probability Changed Science and Everyday Life, Ideas in Context.* Cambridge: Cambridge University Press, 1989.

He, Xinran, et al. "Practical Lessons from Predicting Clicks on Ads at Facebook." New York: ACM Press, 2014. https://doi.org/10.1145/2648 584.2648 589.

Huber, Peter. "From Large to Huge: A Statistician's Reactions to KDD & DM." *KDD* 97 (1997).

Hunt, Earl B., Janet Martin, and Philip J. Stone. *Experiments in Induction*. New York: Academic Press, 1966.

Irani, Keki B., et al. "Applying Machine Learning to Semiconductor Manufacturing." *IEEE Expert* 8, no. 1 (1993): 41–47.

Irani, Lilly. "Justice for 'Data Janitors.'" *Public Books* (blog), January 15, 2015, http://www.publ icbooks.org/justice-for-data-janitors/.

Johnson, Ann. "Rational and Empirical Cultures of Prediction." In *Mathematics as a Tool*, edited by Johannes Lenhard and Martin Carrier, 23–35. Cham: Springer International Publishing, 2017. https://doi.org/10.1007/978-3-319-54469-4_2.

Jones, Matthew L. "How We Became Instrumentalists (Again): Data Positivism since World War II." *Historical Studies in the Natural Sciences* 48, no. 5 (November 1, 2018): 673–684. https://doi.org/10.1525/hsns.2018.48.5.673.

Jones, Matthew L. "Querying the Archive: Database Mining from Apriori to Page-Rank." In *Science in the Archives: Pasts, Presents, Futures*, edited by Lorraine Daston, 311–328. Chicago: Chicago University Press, 2016.

Kelty, Christopher M., and Hannah Landecker. "Ten Thousand Journal Articles Later: Ethnography of 'The Literature' in Science." *Empiria*, no. 18 (2009): 173–192.

Kitchin, Rob. "Thinking Critically about and Researching Algorithms." Programmable City Working Paper 5, 2014. http://dx.doi.org/10.2139/ssrn.2515786.

Kohavi, Ron, and Ross Quinlan. "Decision Tree Discovery." In *Handbook of Data Mining and Knowledge Discovery* (updated October 1999), 267–276. http://ai.stanford.edu/~ronnyk/treesHB.pdf.

Kohavi, Ronny, and Dan Sommerfield. "MLC++: Machine Learning Library in C++. SGIMLC++ Utilities 2.0." October 7, 1996. https://web.archive.org/web/20010603153442/http://www.sgi.com/tech/mlc/.

Lipton, Zachary Chase. "The Mythos of Model Interpretability." CoRR abs/1606.03490 (2016). http://arxiv.org/abs/1606.03490.

Mackenzie, Adrian. *Machine Learners: Archaeology of a Data Practice*. Cambridge, MA: MIT Press, 2018.

Mackenzie, Adrian. "The Production of Prediction: What Does Machine Learning Want?" *European Journal of Cultural Studies* 18, no. 4–5 (2015): 429–445.

Mallows, Colin. "Tukey's Paper after 40 Years." *Technometrics* 48, no. 3 (2006): 319–325.

Mazzotti, Massimo. "Algorithmic Life." *Los Angeles Review of Books*, accessed April 17, 2017. https://lareviewofbooks.org/article/algorithmic-life/.

Meisel, William S., and Leo Breiman. "Topics in the Analysis and Optimization of Complex Systems. Appendix B. Tree Structured Classification Methods." Final report to AFOSR (Technology Service Corporation, February 28, 1977), 4. http://oai.dtic.mil/oai/oai?verb=getRecord&metadataPrefix=html&identifier=ADA038209.

Michie, Donald. "Expert Systems Interview." *Expert Systems* 2, no. 1 (1985): 20–23.

Morgan, James N., and Frank M. Andrews. "A Comment on Einhorn's 'Alchemy in the Behavioral Sciences.'" *Public Opinion Quarterly* 37, no. 1 (1973): 127–131.

Murthy, S. README file, The OC1 Decision Tree System, n.d. http://ccb.jhu.edu/software/oc1/oc1.tar.gz.

November, Joseph Adam. *Biomedical Computing: Digitizing Life in the United States*. Baltimore: Johns Hopkins University Press, 2012.

Olshen, Richard, and Leo Breiman. "A Conversation with Leo Breiman." *Statistical Science* 16, no. 2 (2001): 196.

Pasquale, Frank. *The Black Box Society: The Secret Algorithms That Control Money and Information*. Cambridge, MA: Harvard University Press, 2015.

Plasek, Aaron. "On the Cruelty of Really Writing a History of Machine Learning." *IEEE Annals of the History of Computing* 38, no. 4 (December 2016): 6–8. https://doi.org/10.1109/MAHC.2016.43.

Pollard, A. W., and G. R. Redgrave, (eds.). *A Short-Title Catalogue of Books Printed in England, Scotland and Ireland, and of English Books Printed Abroad 1475–1640*. 2nd edn. London: The Bibliographical Society. Vol. I (A–H), 620. Vol. II (I–Z), 1976, 504. Vol. III (Indexes, addenda, corrigenda), 1991.

Quinlan, J. R. "Discovering Rules by Induction from Large Collections of Examples." In *Expert Systems in the Micro-Electronic Age*, edited by Donald Michie, 168–201. Edinburgh: Edinburgh University Press, 1979.

Quinlan, J. R. "Induction of Decision Trees." *Machine Learning* 1, no. 1 (1986): 89–91.

Quinlan, J. Ross. "Induction over Large Data Bases." Stanford Heuristic Programming Project Memo HPP 79-14 (Stanford University, May 1979).

"Scikit-Learn/Tree.Py at 14031f65d144e3966113d3daec836e443c6d7a5b · Scikit-Learn/Scikit-Learn," accessed April 15, 2017. https://github.com/scikit-learn/scikit-learn/blob/14031f6/sklearn/tree/tree.py#L508.

Seaver, Nick. "Algorithms as Culture: Some Tactics for the Ethnography of Algorithmic Systems." *Big Data & Society* 4, no. 2 (December 2017). https://doi.org/10.1177/205395171 7738104.

Seaver, Nick. "Knowing Algorithms." In *A Field Guide for Science & Technology Studies*, edited by Janet Vertesi and David Ribes, 419. Princeton, NJ: Princeton University Press, 2019.Sewell, Martin. "Ensemble Learning," August 2008. http://machine-learning.marti nsewell.com/ensembles/ensemble-learning.pdf.

Shafer, John, Rakeeh Agrawal, and Manish Mehta. "SPRINT: A Scalable Parallel Classifier for Data Mining." In *Proceedings of the 22nd International Conference on Very Large Data Bases*. San Mateo, CA: Morgan Kaufmann, 1996.

Siegel, Eric. *Predictive Analytics: The Power to Predict Who Will Click, Buy, Lie, or Die*. Hoboken, NJ: Wiley, 2016.

Sonquist, John A., Elizabeth Lauh Baker, and James N. Morgan. *Searching for Structure; an Approach to Analysis of Substantial Bodies of Micro-Data and Documentation for a Computer Program*. Ann Arbor: Survey Research Center, University of Michigan, 1973.

Sonquist, John A., and James N. Morgan. *The Detection of Interaction Effects: A Report on a Computer Program for the Selection of Optional Combinations of Explanatory Variables*. Ann Arbor: Institute for Social Research, University of Michigan, 1964.Stevens, Hallam. *Life out of Sequence: A Data-Driven History of Bioinformatics*. Chicago: University of Chicago Press, 2013.

Tukey, John W. "The Future of Data Analysis." *Annals of Mathematical Statistics* 33, no. 1 (March 1962): 1–67. https://doi.org/10.1214/aoms/1177704711.

Wu, Xindong, et al. "Top 10 Algorithms in Data Mining." *Knowledge and Information Systems* 14, no. 1 (December 4, 2007): 1–37. https://doi.org/10.1007/s10 115-007-0114-2.

Index